Lecture Notes of the Institute for Computer Sciences, Social Informatics and Telecommunications Engineering 81

Javier Del Ser   Eduard Axel Jorswieck
Joaquin Miguez   Marja Matinmikko
Daniel P. Palomar   Sancho Salcedo-Sanz
Sergio Gil-Lopez (Eds.)

Mobile Lightweight Wireless Systems

Third International ICST Conference,
MOBILIGHT 2011
Bilbao, Spain, May 9-10, 2011
Revised Selected Papers

 Springer

Volume Editors

Javier Del Ser
E-mail: javier.delser@tecnalia.com

Eduard Axel Jorswieck
E-mail: eduard.jorswieck@tu-dresden.de

Joaquin Miguez
E-mail: joaquin.miguez@uc3m.es

Marja Matinmikko
E-mail: marja.matinmikko@vtt.fi

Daniel P. Palomar
E-mail: palomar@ust.hk

Sancho Salcedo-Sanz
E-mail: sancho.salcedo@uah.es

Sergio Gil-Lopez
E-mail: sergio.gil@tecnalia.com

ISSN 1867-8211
ISBN 978-3-642-29478-5
DOI 10.1007/978-3-642-29479-2

e-ISSN 1867-822X
e-ISBN 978-3-642-29479-2

Springer Heidelberg Dordrecht London New York

Library of Congress Control Number: 2012935130

CR Subject Classification (1998): C.2, D.2, C.4, K.6.5, C.2.4, D.4.4

Typesetting: Camera-ready by author, data conversion by Scientific Publishing Services, Chennai, India

Printed on acid-free paper

Springer is part of Springer Science+Business Media (www.springer.com)

Preface

The third edition of the International ICST Conference on Mobile Lightweight Wireless Systems (MOBILIGHT 2011) was held during May 9–10, 2011 at the Barcelo Nervion Hotel in Bilbao, Spain, a beautiful venue internationally recognized for business, trading and its devotion to first-level R&D, as buttressed by more than 200 technology-based companies and R&D centers established in this region. This edition continued a series of scientific events previously held in Athens, Greece (2009) and Barcelona, Spain (2010), all of which were dedicated to the latest advances on avant-garde wireless communication techniques, service provisioning and application deployment for resource-constrained devices and networks. In particular, MOBILIGHT 2011 featured first-rate theoretical and practical research on physical-layer information, data and signal processing for wireless systems (e.g., MIMO schemes), transmission of multimedia content over wireless links, and localization and tracking algorithms and applications especially tailored for devices with stringently limited computational resources.

MOBILIGHT 2011 was organized as a 2-day single-track event with 18 technical presentations, and 3 specialized workshops focused on opportunistic sensing and processing in mobile wireless sensor and cellular networks (MOBISENSE), multimode wireless access networks (MOWAN) and strategic network planning applied to market regulation (NETSTRAT), totalling 34 papers presented during the conference and included in the proceedings. A social dinner was also scheduled on the first day of the conference.

The conference had close to 40 participants from industry and academia from a wide range of countries such as Portugal, France, Romania, Tunisia, Denmark, Sweden, France, Greece, Spain, UK and Italy. A number of submissions from world-wide renowned universities, research centers and institutions were received by our Technical Program Committee (Joaquin, Marja and Daniel), to whom we are truly grateful for their support in the review process. Their expertise, knowledge and experience have been the key to rendering an outstanding MOBILIGHT 2011 technical program. We also thank the invited speakers at our event, Pedro M. Crespo (CEIT and TECNUN, University of Navarra, Spain), Robert Piche (Tampere University of Technology, Finland), Marc Brogle (SAP Research, Zürich, Switzerland) and Paul J.M. Havinga (University of Twente, The Netherlands), for accepting our invitation to give keynotes at the conference. Our gratitude also extends to the representatives of ICST, EAI, CREATE-NET and TECNALIA for sponsoring this event and for their help in arranging all the logistics. A special acknowledgment is due to Elena J. Fezzardi from EAI for running the event smoothly. Last but not least, we the General Chairs would like to thank all the reviewers, whose effort reflects their commitment to the success of this yearly event.

We eagerly look forward to another successful MOBILIGHT edition in 2012 and forthcoming years.

January 2012   Javier Del Ser
                                                              Eduard Jorswieck

Organization

General Co-chairs

Javier Del Ser   Tecnalia Research & Innovation, Spain
Eduard Jorswieck   TU Dresden, Germany

Steering Committee

Imrich Chlamtac   CREATE-NET, Italy
Fabrizio Granelli   University of Trento, Italy
Charalabos Skianis   University of the Aegean, Greece

Technical Program Committee Co-chairs

Joaquin Miguez   University Carlos III Madrid, Spain
Marja Matinmikko   VTT, Finland
Daniel Palomar   Hong Kong University of Science and
       Technology, Hong Kong

Workshop / Special Sessions Co-chairs

Sancho Salcedo-Sanz   University of Alcalá, Spain
Thomas Haustein   FhG-HHI, Germany

Tutorial Co-chairs

Alejandro Ribeiro   University of Pennsylvania, USA
Carlos Escudero   University of A Coruña, Spain

Publications Chair

Sergio Gil-Lopez   Tecnalia Research & Innovation, Spain

Publicity Co-chairs

Christos Verikoukis   CTTC, Spain
Manuel Vazquez   University Carlos III, Spain
Martin Haardt   TU Ilmenau, Germany

Panel Co-chairs

Josep Vidal   Technical University of Catalonia, Spain
Periklis Chatzimisios   Alexander TEI of Thessaloniki, Greece

Local Arrangements Chair

Jose M. Cabero   Tecnalia Research & Innovation, Spain

Conference Coordinator

Elena J. Fezzardi   EAI, Italy
Richard Heffernan   EAI, Italy

Web Chair

Eirini Karapistoli   Aristotle University of Thessaloniki, Greece

Second International Workshop on Multimode Wireless Access Networks (MOWAN 2011)

Eugen Borcoci   Polytechnic University of Bucharest, Romania
Djamal-Eddine Meddour   France Telecom R&D, France

First International Workshop on Network Design, Algorithms and Methodology Applied to Market Regulation and Strategic Studies (NETSTRAT 2011)

José Antonio Portilla-Figueras University of Alcalá, Spain
Juan Eulogio Sánchez   University of Alcalá, Spain

First International Workshop on Opportunistic Sensing and Processing in Mobile Wireless Sensor and Cellular Networks (MOBISENSE 2011)

Hans van den Berg   TNO, The Netherlands
Desislava Dimitrova   University of Bern, Switzerland
Paul Havinga   University of Twente, The Netherlands
Nirvana Meratnia   University of Twente, The Netherlands

Table of Contents

Track 3

Track 4

Capacity Region of the Two-Way Multi-antenna Relay Channel with Analog Tx-Rx Beamforming

Cristian Lameiro, Alfredo Nazábal, Fouad Gholam,
Javier Vía, and Ignacio Santamaría

Communications Engineering Department (DICOM), University of Cantabria,
Santander, 39005, Spain
{lameiro,alfredo,fouad,jvia,nacho}@gtas.dicom.unican.es

Abstract. In this paper we study the multiple-input multiple-output two-way relay channel (MIMO-TWRC) when the nodes use analog beamforming. Following the amplify-and-forward (AF) strategy, the problem consists of finding the transmit and receive beamformers of the nodes and the relay, and the power allocated to each one, that achieve the boundary of the capacity region. We express the optimal node beamformers in terms of the relay beamformers, and show that the capacity region can be efficiently characterized using convex optimization techniques. Numerical examples are provided to illustrate the results of this paper, and to compare the capacity region achieved by analog beamforming against the conventional MIMO schemes that operate at the baseband.

Keywords: two-way relay channel, analog beamforming, convex optimization, capacity region.

1 Introduction

Analog beamforming has crescent interest due to the reduced cost, power consumption and system size in comparison to conventional multiple-input multiple-output (MIMO) schemes that apply beamforming in the digital domain [1]. Conventional MIMO systems require a radio-frequency (RF) chain for each antenna in order to process each data stream at baseband. On the other hand, applying beamforming in the RF domain (what we call hereafter RF-MIMO) entails acquiring and processing a single data stream, and thus the cost and power consumption are significantly reduced [2].

The RF-MIMO architecture is shown in Fig. 1. The transmitter (the receiver operates analogously) applies a set of complex weights, $w[n]$, which represent the gain factors and phase shifts at each antenna, and focus the energy beam in the proper direction. For point-to-point links, the design of the optimal Tx-Rx beamformers for multicarrier transmissions has been thoroughly considered in [2]-[4]. In [5], we proposed an optimal transmission strategy for the two-user RF-MIMO broadcast channel, and extended it to the K-user case and the multiple access channel.

On another front, cooperative and multihop communications have been a research topic of interest in recent years, due to the coverage extension that

J. Del Ser et al. (Eds.): MOBILIGHT 2011, LNICST 81, pp. 1–17, 2012.
© Institute for Computer Sciences, Social Informatics and Telecommunications Engineering 2012

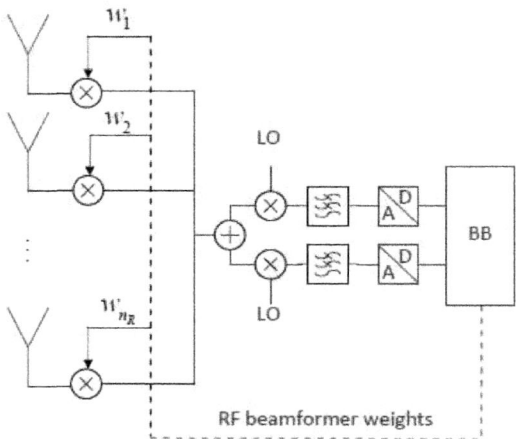

Fig. 1. RF-MIMO transceiver. A single data stream is processed, and thus the cost and power consumption are significantly reduced.

they provide. The two-way relay channel (TWRC) is one of the most basic multihop communication systems. The simplest TWRC consists of two source nodes that exchange information through an assisting relay node. We follow the coding strategy called amplify-and-forward (AF), also adopted in [6], [7]. Hence, only two phases are needed for the exchange of a whole data frame: a multiple access (MAC) phase and a broadcast (BC) phase.

Currently, the TWRC is receiving a great interest and there are many works on optimal transmission strategies and optimal beamforming for the multi-antenna case [6]-[9]. Authors in [6] consider the TWRC with amplify-and-forward (TWRC-AF) strategy when the source nodes are single antenna terminals and the relay uses conventional beamforming. They compute the optimal beamforming strategy at the relay node via convex optimization techniques with fixed powers, i.e., no power optimization is carried out. In [7], Wang and Zhang study the conventional MIMO-TWRC and propose a suboptimal method to compute the beamforming matrices. To the best of our knowledge, the MIMO-TWRC when the nodes use analog beamforming has not been considered yet in the literature. In this paper, we characterize the capacity region of the multi-antenna TWRC-AF, when all nodes use analog beamforming; what we call RF-TWRC. We show that the optimal beamforming strategy and the power allocation problem can be solved through efficient convex optimization techniques.

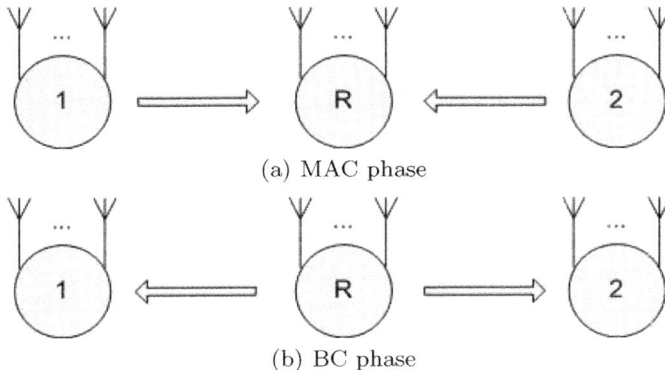

(a) MAC phase

(b) BC phase

Fig. 2. Two-way relay channel with amplify-and-forward strategy and two-phase protocol. In the MAC phase, the source nodes send their messages to the relay node; while in the BC phase, the relay node retransmits a linear composition of the received signal.

1.1 Notation

Bold upper and lower case letters denote matrices and vectors respectively; light-faced lower case letters denote scalar quantities. We use $\mathbf{A}^H$, $\mathbf{A}^*$ and $\mathbf{A}^T$ to denote Hermitian, conjugate and transpose of $\mathbf{A}$, respectively; $\mathrm{Tr}\,(\mathbf{A})$ denotes the trace of $\mathbf{A}$, and $\mathrm{rank}\,(\mathbf{A})$ denotes the rank of $\mathbf{A}$. For vectors, $\|\mathbf{a}\|$ denotes the Euclidean norm of $\mathbf{a}$; and for complex scalars, $|a|$ denotes the absolute value of a. The optimal solution of an optimization problem is indicated by $(\cdot)^{(*)}$.

2   System Model

We consider the TWRC depicted in Fig. 2, where two source nodes equipped with N_S antennas[1] establish a bidirectional communication through a relay node with N_R antennas. The two multi-antenna nodes and the relay perform beamforming in the RF domain, what it is called analog beamforming. We use the two-phase TWRC protocol which was also adopted in [6], [7]; and assume perfect channel state information at every node. In the MAC phase, both source nodes transmit simultaneously to the relay node. Due to the restrictions of the RF-MIMO architecture, the nodes are able to transmit a single data stream. Then, assuming flat-fading channels, the signal received at the relay node can be written as

$$\mathbf{y}_R = \mathbf{H}_1 \mathbf{v}_1 \sqrt{p_1} s_1 + \mathbf{H}_2 \mathbf{v}_2 \sqrt{p_2} s_2 + \mathbf{r}_R \ , \tag{1}$$

where $(\mathbf{v}_1, \mathbf{v}_2) \in \mathbb{C}^{N_S \times 1}$ are the unit-norm analog transmit beamformers of nodes 1 and 2, respectively; $(\mathbf{H}_1, \mathbf{H}_2) \in \mathbb{C}^{N_R \times N_S}$ are the channel matrices; and s_1 and s_2 are the symbols transmitted by nodes 1 and 2, respectively, which are assumed to be distributed as $CN(0,1)$; p_1 and p_2 are the transmit powers of each source

---

[1] The extension to source nodes with different number of antennas is straightforward.

node; and $\mathbf{r}_R \sim CN(0, \sigma^2 \mathbf{I})$ represents the noise at the relay node. Note that, in contrast to [6], we consider multiple antennas at the nodes and the relay. In the BC phase, the relay node performs the amplify-and-forward strategy to linearly process the received signal (1). The analog beamforming matrix is

$$\mathbf{B} = \mathbf{v}_R \mathbf{u}_R^H \tag{2}$$

where $\mathbf{u}_R$ and $\mathbf{v}_R$ are the $N_R \times 1$ receive and transmit beamformers, respectively. Notice that, unlike [6], the RF-MIMO architecture imposes a rank-one constraint in the relay beamforming matrix (2).

Without loss of generality, we can assume $\|\mathbf{v}_R\|^2 = 1$ and $\|\mathbf{v}_i\|^2 = 1$, $i = 1, 2$. Thus, the transmit power of the relay node is given by

$$p_R(p_1, p_2, \mathbf{u}_R, \mathbf{v}_1, \mathbf{v}_2) = p_{\text{ef}_1} + p_{\text{ef}_2} + \sigma^2 \|\mathbf{u}_R\|^2 \tag{3}$$

where we have defined the effective power of source node i $(i = 1, 2)$, p_{ef_i}, as

$$p_{\text{ef}_i} = p_i \left| \mathbf{u}_R^H \mathbf{H}_i \mathbf{v}_i \right|^2 , \qquad i = 1, 2 . \tag{4}$$

If channel reciprocity holds[2], i.e., the channels of source nodes 1 and 2 in the BC phase are $\mathbf{H}_1^T$ and $\mathbf{H}_2^T$, respectively; the signal received by the source nodes, after suppressing the self-interference, is

$$y_1 = \mathbf{u}_1^H \mathbf{H}_1^T \mathbf{v}_R \sqrt{p_{\text{ef}_2}} s_2 + \tilde{r}_1 , \tag{5}$$
$$y_2 = \mathbf{u}_2^H \mathbf{H}_2^T \mathbf{v}_R \sqrt{p_{\text{ef}_1}} s_1 + \tilde{r}_2 ,$$

where $\mathbf{u}_i \in \mathbb{C}^{N_s \times 1}$ and $\tilde{r}_i \sim CN\left(0, \sigma^2 \left[1 + \|\mathbf{u}_R\|^2 \left| \mathbf{u}_i^H \mathbf{H}_i^T \mathbf{v}_R \right|^2 \right] \right)$ are the unit-norm analog receive beamformer and the equivalent noise at node i. Thus, the achievable bidirectional rate pairs, denoted by R_{12} (link from node 1 to node 2, through the relay node) and R_{21} (link from node 2 to node 1, through the relay node), are given by

$$R_{12} \leq \frac{1}{2} \log_2 \left[1 + \frac{p_{\text{ef}_1} \left| \mathbf{u}_2^H \mathbf{H}_2^T \mathbf{v}_R \right|^2}{\sigma^2 \left(1 + \|\mathbf{u}_R\|^2 \left| \mathbf{u}_2^H \mathbf{H}_2^T \mathbf{v}_R \right|^2 \right)} \right] , \tag{6}$$

$$R_{21} \leq \frac{1}{2} \log_2 \left[1 + \frac{p_{\text{ef}_2} \left| \mathbf{u}_1^H \mathbf{H}_1^T \mathbf{v}_R \right|^2}{\sigma^2 \left(1 + \|\mathbf{u}_R\|^2 \left| \mathbf{u}_1^H \mathbf{H}_1^T \mathbf{v}_R \right|^2 \right)} \right] . \tag{7}$$

In the next section, we derive the optimal beamforming vectors to be applied at the source nodes and define the capacity region of the MIMO RF-TWRC. Finally, we propose an iterative algorithm based on convex optimization techniques to compute the boundary of the capacity region.

---

[2] This assumption is made only for simplicity. The results hold also if different transmit and receive channels are considered.

3   Capacity Region

3.1   Optimal Node Beamformers and Parametrization

For a fixed transmit beamforming at the relay node, $\mathbf{v}_R$, the BC phase reduces to a single-input multiple-output (SIMO) channel. Thus, the optimal strategy for both nodes is matched filtering with respect to their channels or maximum ratio combining (MRC) [10]. Therefore, the unit-norm optimal analog receive beamformers are

$$\mathbf{u}_i = \frac{\mathbf{H}_i^T \mathbf{v}_R}{\left\| \mathbf{H}_i^T \mathbf{v}_R \right\|}, \qquad i = 1, 2 \ . \tag{8}$$

Similarly, for a fixed receive beamforming at the relay node, $\mathbf{u}_R$, the MAC phase reduces to a multiple-input single-output (MISO) channel. In order to achieve the boundary of the capacity region, each source node must control its effective power, p_{ef_i} ($i = 1, 2$). According to (4), this power control can be done by varying either the source node power or its transmit beamformer, $\mathbf{v}_i$. Hence, if the source nodes perform maximum ratio transmission (MRT) [10], the effective power can be controlled solely by the source node power, p_i. Thus, MRT is always optimal. The MRT beamformers of the source nodes are given by

$$\mathbf{v}_i = \frac{\mathbf{H}_i^H \mathbf{u}_R}{\left\| \mathbf{H}_i^H \mathbf{u}_R \right\|}, \qquad i = 1, 2 \ . \tag{9}$$

In summary, the optimal Tx-Rx node beamformers can be written in terms of the Tx-Rx relay beamformers. Notice that, (8) and (9) implies that a feedback channel must exist between the nodes and the relay. According to that, the node beamformers can be computed at the relay node, and then sent them back to the nodes. With the optimal node beamformers, the rates in (6) and (7) can be rewritten as

$$R_{12} \le \frac{1}{2} \log_2 \left[1 + \frac{p_{\text{ef}_1} \left\| \mathbf{H}_2^T \mathbf{v}_R \right\|^2}{\sigma^2 \left(1 + \left\| \mathbf{u}_R \right\|^2 \left\| \mathbf{H}_2^T \mathbf{v}_R \right\|^2 \right)} \right] , \tag{10}$$

$$R_{21} \le \frac{1}{2} \log_2 \left[1 + \frac{p_{\text{ef}_2} \left\| \mathbf{H}_1^T \mathbf{v}_R \right\|^2}{\sigma^2 \left(1 + \left\| \mathbf{u}_R \right\|^2 \left\| \mathbf{H}_1^T \mathbf{v}_R \right\|^2 \right)} \right] , \tag{11}$$

where now $p_{\text{ef}_i} = p_i \left\| \mathbf{H}_i^H \mathbf{u}_R \right\|^2$ ($i = 1, 2$). Including the power constraints at the nodes and the relay, we can now define the capacity of the RF-TWRC as

$$C\left(P_1, P_2, P_R\right) \triangleq \bigcup_{p_1 \le P_1, p_2 \le P_2} \left[\bigcup_{\|\mathbf{v}_R\|^2 = 1, p_R(p_1, p_2, \mathbf{u}_R) \le P_R} \{R_{12}, R_{21}\} \right] , \tag{12}$$

where R_{12} and R_{21} are defined in (10) and (11), respectively. It is easy to see that the optimal relay beamforming vectors must lie in the subspace spanned by the columns of the channel matrices, i.e.,

$$u_R \in \text{span}\,(\mathbf{H})\ , \tag{13}$$
$$v_R \in \text{span}\,(\mathbf{H}^*)\ ,$$

where $\mathbf{H} = [\mathbf{H}_1, \mathbf{H}_2]$. Here, $\text{span}\,(\mathbf{A})$ denotes the subspace spanned by the columns of $\mathbf{A}$. Hence, taking the singular value decomposition (SVD) of $\mathbf{H}$, we can express the beamformers and the channel matrices in terms of a unitary basis as follows

$$\mathbf{u}_R = \mathbf{U}\mathbf{a}_r\ , \tag{14}$$
$$\mathbf{v}_R = \mathbf{U}^*\mathbf{a}_t\ ,$$
$$\mathbf{H}_1 = \mathbf{U}\mathbf{G}_1\ ,$$
$$\mathbf{H}_2 = \mathbf{U}\mathbf{G}_2\ ,$$

where $\mathbf{U} \in \mathbb{C}^{N_R \times \min(N_R, 2N_S)}$ is an orthogonal basis for the column range of $\mathbf{H}$. Vectors $\{\mathbf{a}_r, \mathbf{a}_t\} \in \mathbb{C}^{\min(N_R, 2N_S) \times 1}$ are the coefficients of the linear expansion of the beamformers in terms of the basis $\mathbf{U}$. As $\mathbf{U}$ is unitary, $\|\mathbf{v}_R\|^2 = 1$ implies $\|\mathbf{a}_t\|^2 = 1$. Note that, with this parametrization, the number of complex variables of $\mathbf{u}_R$ and $\mathbf{v}_R$ can be reduced from N_R to $2N_S$ if $N_R > 2N_S$.

With the parametrization given in (14), the rates achieved by the nodes are now given by

$$R_{12} \le \frac{1}{2} \log_2 \left[1 + \frac{p_{\text{ef}_1} \left\| \mathbf{G}_2^T \mathbf{a}_t \right\|^2}{\sigma^2 \left(1 + \|\mathbf{a}_r\|^2 \left\| \mathbf{G}_2^T \mathbf{a}_t \right\|^2 \right)} \right]\ , \tag{15}$$

$$R_{21} \le \frac{1}{2} \log_2 \left[1 + \frac{p_{\text{ef}_2} \left\| \mathbf{G}_1^T \mathbf{a}_t \right\|^2}{\sigma^2 \left(1 + \|\mathbf{a}_r\|^2 \left\| \mathbf{G}_1^T \mathbf{a}_t \right\|^2 \right)} \right]\ , \tag{16}$$

where now $p_{\text{ef}_i} = p_i \left\| \mathbf{G}_i^H \mathbf{a}_r \right\|^2$ ($i = 1, 2$). The capacity region is now given by

$$C\,(P_1, P_2, P_R) \triangleq \bigcup_{p_1 \le P_1, p_2 \le P_2} \left[\bigcup_{\|\mathbf{a}_t\|^2 = 1, p_R(p_1, p_2, \mathbf{a}_r) \le P_R} \{R_{12}, R_{21}\} \right]\ , \tag{17}$$

with R_{12} and R_{21} defined in (15) and (16), respectively.

3.2   Computing the Capacity Region

The capacity region in (17) can be characterized by solving a weighted sum-rate maximization problem (WSRmax). This approach assigns different weights to the source nodes in order to establish a priority between them. Hence, varying the weights, every point on the capacity boundary can be computed. However, the WSRmax problem is non-convex and it is very difficult to solve [11].

The authors of [12] proposed an alternative method to compute the boundary rate-tuples of the capacity region called *rate profile*, that was also applied in

[6] to the single-antenna TWRC. Applying this idea, the rate at each node can be expressed as a portion of the sum rate, i.e., $[R_{12}, R_{21}]^T = R_{\text{sum}} [\tau, 1 - \tau]^T$, where $[\tau, 1 - \tau]^T$ is the *rate profile* vector. Thus, for a fixed value of τ between 0 and 1, we can compute a boundary point of the capacity region by solving the following optimization problem

$$\underset{p_1, p_2, \mathbf{a}_t, \mathbf{a}_r, R_{\text{sum}}}{\text{maximize}} \quad R_{\text{sum}} , \tag{18}$$

$$\text{subject to} : \frac{1}{2} \log_2 \left[1 + \frac{p_1 \left\| \mathbf{G}_1^H \mathbf{a}_r \right\|^2 \left\| \mathbf{G}_2^T \mathbf{a}_t \right\|^2}{\sigma^2 \left(1 + \left\| \mathbf{a}_r \right\|^2 \left\| \mathbf{G}_2^T \mathbf{a}_t \right\|^2 \right)} \right] \geq \tau R_{\text{sum}} ,$$

$$\frac{1}{2} \log_2 \left[1 + \frac{p_2 \left\| \mathbf{G}_2^H \mathbf{a}_r \right\|^2 \left\| \mathbf{G}_1^T \mathbf{a}_t \right\|^2}{\sigma^2 \left(1 + \left\| \mathbf{a}_r \right\|^2 \left\| \mathbf{G}_1^T \mathbf{a}_t \right\|^2 \right)} \right] \geq (1 - \tau) R_{\text{sum}} ,$$

$$p_1 \left\| \mathbf{G}_1^H \mathbf{a}_r \right\|^2 + p_2 \left\| \mathbf{G}_2^H \mathbf{a}_r \right\|^2 + \sigma^2 \left\| \mathbf{a}_r \right\|^2 \leq P_R ,$$

$$p_1 \leq P_1 ,$$

$$p_2 \leq P_2 .$$

Since R_{12} and R_{21} grow monotonically with the signal-to-noise plus interference ratio (SINR), we can express the *rate profile* in terms of a *SINR profile*. Hence, for a fixed value of R_{sum}, the rate constraints in (18) are equivalent to target SINRs at the nodes, and thus the optimization problem (18) with fixed R_{sum} is equivalent to a power minimization problem with SINR constraints as follows

$$\underset{p_1, p_2, \mathbf{a}_t, \mathbf{a}_r}{\text{minimize}} \quad p_1 \left\| \mathbf{G}_1^H \mathbf{a}_r \right\|^2 + p_2 \left\| \mathbf{G}_2^H \mathbf{a}_r \right\|^2 + \sigma^2 \left\| \mathbf{a}_r \right\|^2 , \tag{19}$$

$$\text{subject to} : \frac{p_2 \left\| \mathbf{G}_2^H \mathbf{a}_r \right\|^2 \left\| \mathbf{G}_1^T \mathbf{a}_t \right\|^2}{\sigma^2 \left(1 + \left\| \mathbf{a}_r \right\|^2 \left\| \mathbf{G}_1^T \mathbf{a}_t \right\|^2 \right)} \geq \alpha \gamma_{\text{sum}} ,$$

$$\frac{p_1 \left\| \mathbf{G}_1^H \mathbf{a}_r \right\|^2 \left\| \mathbf{G}_2^T \mathbf{a}_t \right\|^2}{\sigma^2 \left(1 + \left\| \mathbf{a}_r \right\|^2 \left\| \mathbf{G}_2^T \mathbf{a}_t \right\|^2 \right)} \geq (1 - \alpha) \gamma_{\text{sum}} ,$$

$$p_1 \leq P_1 ,$$

$$p_2 \leq P_2 ,$$

for some $0 \leq \alpha \leq 1$. For a given γ_{sum}, the solution of the above problem provides a feasible point of the capacity region if and only if $p_R^{(*)} \leq P_R$, where $p_R^{(*)}$ is the optimal power value. It turns out that the boundary of the capacity region can be obtained by a bisection method over γ_{sum}, solving problem (19) in each step, as indicated in algorithm 1.

Algorithm 1:

- **Initialize** $\gamma_{sum}^{min} = 0$, $\gamma_{sum}^{max} = \gamma_{sum}^{UB}$.
- **Repeat**

1. $\gamma_{sum} = \frac{1}{2} \left(\gamma_{sum}^{min} + \gamma_{sum}^{max} \right)$.
2. Solve problem (19).
   (a) If the problem is feasible, $\gamma_{sum}^{min} = \gamma_{sum}$.
   (b) Otherwise, $\gamma_{sum}^{max} = \gamma_{sum}$.
- **Until** $\left(\gamma_{sum}^{max} - \gamma_{sum}^{min} \right) \leq \epsilon$

A reasonable value of γ_{sum}^{UB} can be obtained through the SINR upper bound derived in [6]. This upper bound is obtained considering the optimal beamforming of both links independently. In the analog beamforming case, the optimal strategy at the relay node is transmitting and receiving through the principal eigenvectors of the channels. Hence, a SINR upper bound for the RF-TWRC is

$$\gamma_{sum}^{UB} = P_R \sigma_1^2 \sigma_2^2 \left(\frac{p_2}{P_R \sigma_1^2 + P_2 \sigma_2^2 + \frac{1}{2}} + \frac{p_1}{P_R \sigma_2^2 + P_1 \sigma_1^2 + \frac{1}{2}} \right) , \qquad (20)$$

where σ_i is the strongest singular value of $\mathbf{G}_i$, $i = 1, 2$.

The optimization problem (19) is non-convex, but a solution can be found through a relaxed semidefinite programm (SDP), as we show in the next subsection.

3.3   Semidefinite Relaxation

Defining the Hermitian matrices $\mathbf{A}_r$ and $\mathbf{A}_t$ as

$$\mathbf{A}_r = \mathbf{a}_r \mathbf{a}_r^H , \qquad (21)$$
$$\mathbf{A}_t = \mathbf{a}_t \mathbf{a}_t^H ,$$

the optimization problem (19) can be written as

$$\underset{p_1, p_2, \mathbf{A}_t, \mathbf{A}_r}{\text{minimize}} \quad p_1 \text{Tr} \left(\mathbf{R}_1 \mathbf{A}_r \right) + p_2 \text{Tr} \left(\mathbf{R}_2 \mathbf{A}_r \right) + \sigma^2 \text{Tr} \left(\mathbf{A}_r \right) , \qquad (22)$$

subject to : $p_2 \text{Tr} \left(\mathbf{R}_2 \mathbf{A}_r \right) - (1 - \alpha) \gamma_{\text{sum}} \sigma^2 \text{Tr} \left(\mathbf{A}_r \right) \geq \dfrac{(1 - \alpha) \gamma_{\text{sum}} \sigma^2}{\text{Tr} \left(\mathbf{R}_1^* \mathbf{A}_t \right)}$,

$\qquad\qquad p_1 \text{Tr} \left(\mathbf{R}_1 \mathbf{A}_r \right) - \alpha \gamma_{\text{sum}} \sigma^2 \text{Tr} \left(\mathbf{A}_r \right) \geq \dfrac{\alpha \gamma_{\text{sum}} \sigma^2}{\text{Tr} \left(\mathbf{R}_2^* \mathbf{A}_t \right)}$,

$\qquad\qquad \text{Tr} \left(\mathbf{A}_t \right) = 1$,

$\qquad\qquad \mathbf{A}_t \succeq 0$,

$\qquad\qquad \mathbf{A}_r \succeq 0$,

$\qquad\qquad \text{rank} \left(\mathbf{A}_t \right) = 1$,

$\qquad\qquad \text{rank} \left(\mathbf{A}_r \right) = 1$,

$\qquad\qquad p_1 \leq P_1$,

$\qquad\qquad p_2 \leq P_2$,

where $\mathbf{R}_i = \mathbf{G}_i \mathbf{G}_i^H$. The above problem can be shown to be still non-convex due to several reasons. First, the cross products between $\mathbf{A}_r$ and the power variables (p_1 and p_2) make the first two constraints non-convex. However, we can get through it by optimizing the effective powers instead. Thus, problem (22) is equivalent to

$$\underset{p_{\text{ef}_1}, p_{\text{ef}_2}, \mathbf{A}_t, \mathbf{A}_r}{\text{minimize}} \quad p_{\text{ef}_1} + p_{\text{ef}_2} + \sigma^2 \text{Tr}\left(\mathbf{A}_r\right) , \tag{23}$$

$$\text{subject to}: p_{\text{ef}_2} - (1-\alpha)\,\gamma_{\text{sum}}\sigma^2 \text{Tr}\left(\mathbf{A}_r\right) \geq \frac{(1-\alpha)\,\gamma_{\text{sum}}\sigma^2}{\text{Tr}\left(\mathbf{R}_1^*\mathbf{A}_t\right)} ,$$

$$p_{\text{ef}_1} - \alpha\gamma_{\text{sum}}\sigma^2 \text{Tr}\left(\mathbf{A}_r\right) \geq \frac{\alpha\gamma_{\text{sum}}\sigma^2}{\text{Tr}\left(\mathbf{R}_2^*\mathbf{A}_t\right)} ,$$

$$\text{Tr}\left(\mathbf{A}_t\right) = 1 ,$$
$$\mathbf{A}_t \succeq 0 ,$$
$$\mathbf{A}_r \succeq 0 ,$$
$$\text{rank}\left(\mathbf{A}_t\right) = 1 ,$$
$$\text{rank}\left(\mathbf{A}_r\right) = 1 ,$$
$$p_{\text{ef}_1} \leq P_1 \text{Tr}\left(\mathbf{R}_1\mathbf{A}_r\right) ,$$
$$p_{\text{ef}_2} \leq P_2 \text{Tr}\left(\mathbf{R}_2\mathbf{A}_r\right) .$$

Note that the last two inequalities are the power constraints of the source nodes, according to the definition of the effective powers in (4). On the other hand, the rank-one constraints are non-convex, although we can find a solution relaxing these two constraints, what is called in the literature a relaxed SDP, as we show in the following.

$$\underset{p_{\text{ef}_1}, p_{\text{ef}_2}, \mathbf{A}_t, \mathbf{A}_r}{\text{minimize}} \quad p_{\text{ef}_1} + p_{\text{ef}_2} + \sigma^2 \text{Tr}\left(\mathbf{A}_r\right) , \tag{24}$$

$$\text{subject to}: p_{\text{ef}_2} - (1-\alpha)\,\gamma_{\text{sum}}\sigma^2 \text{Tr}\left(\mathbf{A}_r\right) \geq \frac{(1-\alpha)\,\gamma_{\text{sum}}\sigma^2}{\text{Tr}\left(\mathbf{R}_1^*\mathbf{A}_t\right)} ,$$

$$p_{\text{ef}_1} - \alpha\gamma_{\text{sum}}\sigma^2 \text{Tr}\left(\mathbf{A}_r\right) \geq \frac{\alpha\gamma_{\text{sum}}\sigma^2}{\text{Tr}\left(\mathbf{R}_2^*\mathbf{A}_t\right)} ,$$

$$\text{Tr}\left(\mathbf{A}_t\right) = 1 ,$$
$$\mathbf{A}_t \succeq 0 ,$$
$$\mathbf{A}_r \succeq 0 ,$$
$$p_{\text{ef}_1} \leq P_1 \text{Tr}\left(\mathbf{R}_1\mathbf{A}_r\right) ,$$
$$p_{\text{ef}_2} \leq P_2 \text{Tr}\left(\mathbf{R}_2\mathbf{A}_r\right) .$$

The above problem is convex and can be efficiently solved using convex optimization methods. Moreover, as we show in the Appendix following the lines in [13], [14]; when the rank of the optimal beamforming matrices, $\mathbf{A}_t^{(*)}$ and $\mathbf{A}_r^{(*)}$, is greater than one, we are able to find an optimal rank-one solution. Thus, the solution of (24) is also the optimal solution of the original problem (19). Given the solution of (24), the optimal power assigned to each source node, $p_1^{(*)}$ and $p_2^{(*)}$, are given by

$$p_1^{(*)} = \frac{p_{ef_1}^{(*)}}{\text{Tr}\left(\mathbf{R}_1 \mathbf{A}_r^{(*)}\right)} , \qquad (25)$$

$$p_2^{(*)} = \frac{p_{ef_2}^{(*)}}{\text{Tr}\left(\mathbf{R}_2 \mathbf{A}_r^{(*)}\right)} .$$

Note that, in contrast to [6], we are able to optimize the power of each source node and thus characterize completely the capacity region of the RF-TWRC, without resorting to an exhaustive search on the transmit powers.

4   Numerical Examples

In this section, we present some examples to illustrate the results of this paper, and to compare the achievable rates of the analog beamforming architecture with those achieved by conventional MIMO schemes. The elements of $\mathbf{H}_1$ and $\mathbf{H}_2$ are i.i.d. zero-mean circular complex Gaussian random variables with unit variance. We consider equal power constraints for the nodes and the relay, i.e., $P_R = P_1 = P_2 = P$, and define the signal-to-noise ratio as $\text{SNR} = 10\log_{10}\frac{P}{\sigma^2}$. Without loss of generality, we take $\sigma^2 = 1$. To get the boundary of the capacity region, we follow Algorithm 1 varying α between 0 and 1.

4.1   Conventional MIMO vs. Analog Beamforming

In this subsection we compare the achievable rate region when the nodes use analog beamforming against the conventional (baseband) MIMO beamforming. The capacity region with conventional MIMO is obtained through the algorithm proposed in [6]. This algorithm does not allow power optimization and requires the source nodes to be single antenna terminals. Thus, we focus on the following scenario: $N_s = 1$, $N_R = 4$ and fixed powers.

For convenience of the analysis, we consider one channel realization of $\mathbf{h}_1$ normalized by its own norm. Channel vector $\mathbf{h}_2$ is obtained such that $\|\mathbf{h}_2\| = 1$ and $\left|\mathbf{h}_1^H \mathbf{h}_2\right|^2 = \rho$, where ρ, $0 \le \rho \le 1$, is the squared cosine of the angle formed between $\mathbf{h}_1$ and $\mathbf{h}_2$.

Fig. 3(a) and Fig. 3(b) show the achievable rate region of both schemes for $\rho = 0.1$ and $\rho = 0.5$, respectively. The SNR is equal to 10 dB and the relay node is equipped with 4 antennas. As ρ increases, the channel vectors tend to be collinear and the gap between analog and conventional beamforming goes to 0.

4.2   Capacity Region of RF-TWRC

In this subsection we consider multi-antenna nodes and optimize the power transmitted by the source nodes. In Fig. 4 we show the capacity region of a RF-TWRC channel, when $N_R = 4$ and $N_S = 2$; and in Fig. 5 we show the power used by

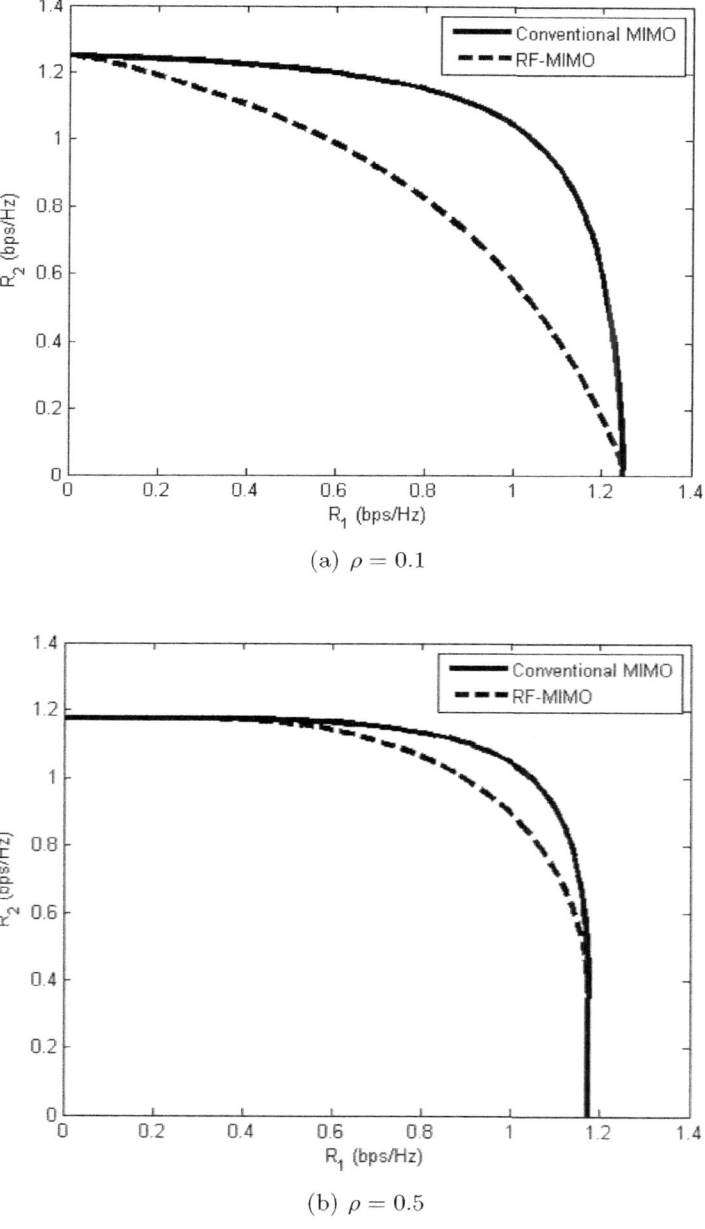

(a) $\rho = 0.1$

(b) $\rho = 0.5$

Fig. 3. Comparison between the achievable rates of conventional MIMO and RF-MIMO schemes when $N_s = 1$ and $N_R = 4$, for different values of ρ. The capacity of conventional MIMO-TWRC has been computed using the algorithm proposed in [6]. As ρ increases (i.e., more collinear channels), the gap between both regions tends to 0.

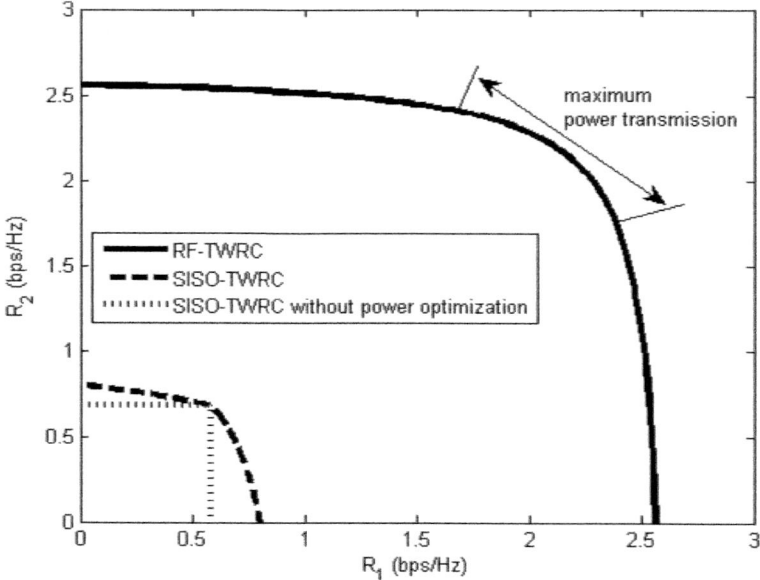

Fig. 4. Capacity region of the 4x2 RF-TWRC with a SNR of 10 dB. The dashed line depicts the achievable region of the SISO case with and without power optimization.

the source nodes. The SNR has a value of 10 dB and the energies of the channels are equal to unity. We observe that there are some points of the boundary that are achieved when the source nodes do not transmit at maximum power. This result can be easily explained in terms of the transmit power of the relay node, p_R. From (3), p_R is a function of the effective powers of both source nodes. As the relay node has a power constraint, P_R, the greater is the effective power of a node, the lower is the effective power of the other. To better understand this idea, consider one of the extreme points of the capacity boundary. If we fix $\alpha = 0$, i.e., full priority is assigned to node 1, then the optimal power of node 1, $p_1^{(*)}$, is equal to 0; while the optimal power of node 2, $p_2^{(*)}$, is maximized. As α increases, $p_1^{(*)}$ increases too, until it reaches its maximum value. Thus, maximum power transmission is optimal only in a subset of the capacity boundary.

Fig. 4 also shows the capacity region of the single antenna (SISO) case using the first antenna of each terminal, with and without power optimization (in the later case, both nodes always transmit at maximum power). We clearly observe the enlargement of the capacity region when the nodes are multi-antenna terminals performing analog beamforming. Moreover, optimizing the power of the source nodes noticeably improves the capacity region of the SISO-TWRC.

Similarly to the previous subsection, we also study the impact of the collinearity between the channels, that can be measured through the squared cosine of

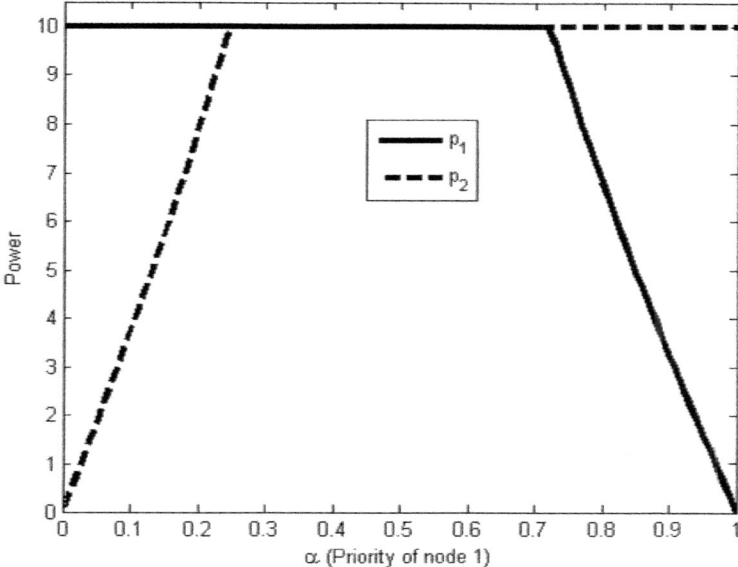

Fig. 5. Power allocation of the 4x2 RF-TWRC with a SNR of 10 dB. Note that only a subset of the boundary rate-tuples are achieved when both nodes are simultaneously transmitting at maximum power. Moreover, there is at least one node transmitting at maximum power at each boundary point.

the angle formed between them, ρ, that we compute using the Matlab function subspace(). For the convenience of the analysis, we normalize each channel by its own 2-norm. Fig. 6 shows the capacity of the RF-TWRC for different values of ρ, when $N_R = 4$, $N_S = 2$ and the SNR is 10 dB. As in the single-antenna case, the capacity region increases when the angle between the channels decreases.

4.3   Sum-Rate vs. SNR Analysis

In this subsection we evaluate the sum-rate capacity versus the SNR of the RF-TWRC through Monte Carlo simulation (specifically, we average the results of 100 channel realizations) and compare it with the conventional MIMO-TWRC and the SISO case, when the first antenna of each terminal is used. The sum-rate capacity is computed using exhaustive search over α, and we follow the algorithm proposed in [6] for the conventional MIMO-TWRC. Thus, we consider single antenna nodes, i.e., $N_S = 1$, and no power optimization, i.e., $p_1 = p_2 = P$. Fig. 7 shows the sum-rate capacity when the relay is equipped with $N_R = 4$ antennas. We observe that the RF-TWRC and the conventional MIMO-TWRC perform closely, as well as the enlargement in the sum-rate capacity with respect to the SISO case.

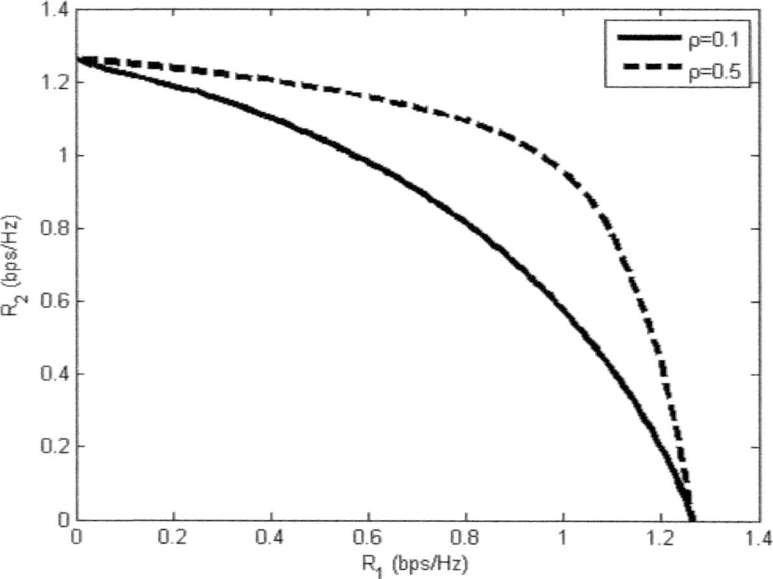

Fig. 6. Capacity region of the 4x2 RF-TWRC with a SNR of 10 dB, and different values of ρ. As ρ increases, the channels tend to be collinear and the capacity region increases.

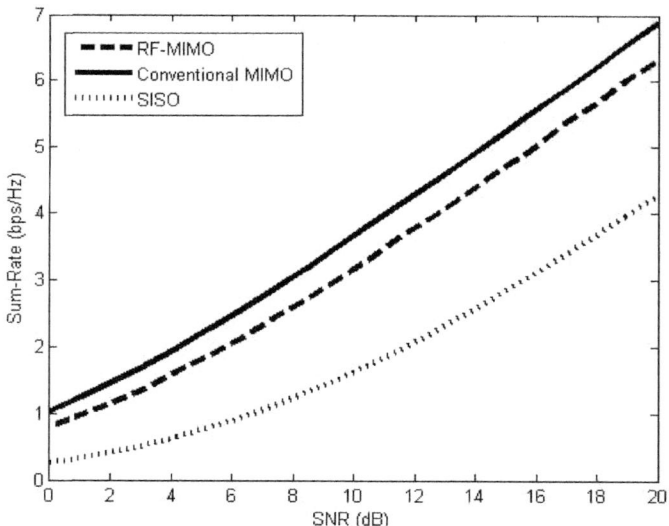

Fig. 7. Sum-rate capacity through Monte Carlo simulation (specifically, we average the results of 100 channel realizations) for the TWRC channel with $N_R = 4$ and $N_S = 1$, when no power optimization is performed.

5   Conclusions

RF-MIMO transceivers that apply beamforming in the analog domain are of practical interest due to the reduced system size, cost and power consumption in comparison with the conventional MIMO architectures. In this paper we have studied a basic TWRC-AF, when the two nodes and the relay are RF-MIMO terminals. Our main contribution has been to show that the optimal beamforming vectors and the power allocation can be efficiently computed using convex optimization techniques, through an iterative algorithm based on a bisection method. We have also shown that the capacity gap between analog beamforming and conventional MIMO schemes, when the source nodes are single antenna terminals and no power optimization is performed, goes towards 0 as the correlation coefficient between the channels increases.

A   Appendix I

Suppose that the optimal solution of (24), $\mathbf{A}_r^{(*)}$, has rank $r > 1$. If there exists an equivalent rank-one solution, the following must hold

$$\mathrm{Tr}\left(\mathbf{R}_1\tilde{\mathbf{A}}_r\right) = \mathrm{Tr}\left(\mathbf{R}_1\mathbf{A}_r^{(*)}\right) , \qquad (26)$$

$$\mathrm{Tr}\left(\mathbf{R}_2\tilde{\mathbf{A}}_r\right) = \mathrm{Tr}\left(\mathbf{R}_2\mathbf{A}_r^{(*)}\right) ,$$

$$\mathrm{Tr}\left(\tilde{\mathbf{A}}_r\right) = \mathrm{Tr}\left(\mathbf{A}_r^{(*)}\right) ,$$

where $\tilde{\mathbf{A}}_r$ is a rank-one matrix. Through the matrix decomposition theorem for Hermitian matrices [13], [14]; given the Hermitian matrices $\mathbf{R}_1$ and $\mathbf{R}_2$, there exists a decomposition of $\mathbf{A}_r^{(*)}$, $\mathbf{A}_r^{(*)} = \sum_{k=1}^r \mathbf{a}_r^{(k)}\left(\mathbf{a}_r^{(k)}\right)^H$, such that,

$$\mathrm{Tr}\left(\mathbf{R}_1\tilde{\mathbf{A}}_r^{(k)}\right) = \mathrm{Tr}\left(\mathbf{R}_1\mathbf{A}_r^{(*)}\right) , \qquad (27)$$

$$\mathrm{Tr}\left(\mathbf{R}_2\tilde{\mathbf{A}}_r^{(k)}\right) = \mathrm{Tr}\left(\mathbf{R}_2\mathbf{A}_r^{(*)}\right) ,$$

for all $k = 1, \ldots, r$; where $\tilde{\mathbf{A}}_r^{(k)} = r\mathbf{a}_r^{(k)}\left(\mathbf{a}_r^{(k)}\right)^H$ is a rank-one matrix. Thus, there exist r rank-one matrices that satisfy the first two conditions in (26). Taking into account that the trace of $\mathbf{A}_r^{(*)}$ and $\tilde{\mathbf{A}}_r^{(k)}$ are, respectively, given by

$$\mathrm{Tr}\left(\mathbf{A}_r^{(*)}\right) = \sum_{k=1}^r \left\|\mathbf{a}_r^{(k)}\right\|^2 , \qquad (28)$$

$$\mathrm{Tr}\left(\tilde{\mathbf{A}}_r^{(k)}\right) = r\left\|\mathbf{a}_r^{(k)}\right\|^2 ,$$

and assuming without loss of generality, $\left\|\mathbf{a}_r^{(1)}\right\|^2 \geq \left\|\mathbf{a}_r^{(2)}\right\|^2 \geq \ldots \geq \left\|\mathbf{a}_r^{(r)}\right\|^2$; it follows that

$$\mathrm{Tr}\left(\mathbf{A}_r^{(*)}\right) \geq \mathrm{Tr}\left(\tilde{\mathbf{A}}_r^{(r)}\right) . \qquad (29)$$

On the other hand, as $\mathbf{A}_r^{(*)}$ is an optimal solution of (24), the following must hold

$$\mathrm{Tr}\left(\mathbf{A}_r^{(*)}\right) \leq \mathrm{Tr}\left(\tilde{\mathbf{A}}_r^{(r)}\right) . \tag{30}$$

Thus, to satisfy (29) and (30), both traces must be equal, i.e.,

$$\mathrm{Tr}\left(\mathbf{A}_r^{(*)}\right) = \mathrm{Tr}\left(\tilde{\mathbf{A}}_r^{(r)}\right) . \tag{31}$$

Therefore, the rank-one matrix $\tilde{\mathbf{A}}_r^{(r)}$ is also optimal and equivalent to $\mathbf{A}_r^{(*)}$.

Similarly, suppose that the optimal solution of (24), $\mathbf{A}_t^{(*)}$, has rank $r > 1$. If there exists an equivalent rank-one solution, the following must hold

$$\mathrm{Tr}\left(\mathbf{R}_1^*\tilde{\mathbf{A}}_t\right) = \mathrm{Tr}\left(\mathbf{R}_1^*\mathbf{A}_t^{(*)}\right) , \tag{32}$$

$$\mathrm{Tr}\left(\mathbf{R}_2^*\tilde{\mathbf{A}}_t\right) = \mathrm{Tr}\left(\mathbf{R}_2^*\mathbf{A}_t^{(*)}\right) ,$$

$$\mathrm{Tr}\left(\tilde{\mathbf{A}}_t\right) = \mathrm{Tr}\left(\mathbf{A}_t^{(*)}\right) ,$$

where $\tilde{\mathbf{A}}_t$ is a rank-one matrix. The above conditions are equivalent to (26), and the same arguments can be invoked to prove the existence of an optimal rank-one matrix.

Acknowledgment. The research leading to these results has received funding from the European Community's Seventh Framework Programme (FP7/2007-2013) under grant agreement n 213952, MIMAX and by the Spanish Government (MICINN) under projects TEC2007-68020-C04-02/TCM (MULTIMIMO) and TEC2010-19545-C04-03 (COSIMA).

References

1. Eickhoff, R., Kraemer, R., Santamaría, I., González, L.: Developing energy-efficient MIMO radios. IEEE Vehicular Technology Magazine 4(1), 34–41 (2009)
2. Vía, J., Elvira, V., Santamaría, I., Eickhoff, R.: Analog antenna combining for maximum capacity under OFDM transmissions. In: Proceedings of the 2009 IEEE International Conference on Communications, Dresden, Germany, pp. 3129–3133 (2009)
3. Vía, J., Santamaría, I., Elvira, V., Eickhoff, R.: A general criterion for analog Tx-Rx beamforming under OFDM transmissions. IEEE Transactions on Signal Processing 58(4), 2155–2167 (2010)
4. Vía, J., Elvira, V., Santamaría, I., Eickhoff, R.: Minimum BER beamforming in the RF domain for OFDM transmissions and linear receivers. In: IEEE International Conference on Acoustics Speech and Signal Processing (ICASSP 2009), Taipei, Taiwan (March 2009)
5. Santamaría, I., Vía, J., Nazábal, A., Lameiro, C.: Capacity region of the multi-antenna gaussian broadcast channel with analog Tx-Rx beamforming. In: 5th International ICST Conference on Communications and Networking in China (CHINACOM), Beijing, China (August 2010)

6. Zhang, R., Liang, Y.-C., Chai, C.C., Cui, S.: Optimal beamforming for two-way multi-antenna relay channel with analogue network coding. IEEE Journal on Selected Areas in Communications 27(5), 699–712 (2009)
7. Wang, X., Zhang, X.-D.: Optimal beamforming in MIMO two-way relay channels. In: IEEE Global Communications Conference (GLOBECOM 2010), Miami, FL, USA (December 2010)
8. Wyrembelski, R.F., Oechtering, T.J., Boche, H.: MIMO bidirectional broadcast channels with common message. In: IEEE Global Communications Conference (GLOBECOM 2010), Miami, FL, USA (December 2010)
9. Oechtering, T., Wyrembelski, R., Boche, H.: Multiantenna bidirectional broadcast channels - optimal transmit strategies. IEEE Transactions on Signal Processing 57(5), 1948–1958 (2009)
10. Tse, D., Viswanath, P.: Fundamentals of Wireless Communications. Cambridge University Press (2004)
11. Boyd, S., Vandenberghe, L.: Convex Optimization. Cambridge University Press (2004)
12. Mohseni, M., Zhang, R., Cioffi, J.: Optimized transmission for fading multiple-access and broadcast channels with multiple antennas. IEEE Journal on Selected Areas in Communications 24(8), 1627–1639 (2006)
13. Huang, Y., Zhang, S.: Complex matrix decomposition and quadratic programming. Math. Oper. Res. 32, 758–768 (2007)
14. Palomar, D., Eldar, Y.: Convex Optimization in Signal Processing and Communications. Cambridge University Press (2010)

Analytical Foundation for Energy Efficiency Optimisation in Cellular Networks with Elastic Traffic

Zhijiat Chong and Eduard Jorswieck

Technische Universität Dresden, Institut für Nachrichtentechnik,
Lehrstuhl Theoretische Nachrichtentechnik, 01062 Dresden, Germany
{zhijiat.chong,eduard.jorswieck}@tu-dresden.de

Abstract. Lately, energy efficiency (EE) in mobile communications is receiving growing attention as its increasing energy consumption raises concern over climate effects. As a stepping stone to improve the situation, we provide some analytical tools for optimising the EE of a base station through power control, focusing on elastic traffic in the downlink scenario. Under certain assumptions, the problem is formulated such that any rate maximisation algorithm can be incorporated to achieve the optimum EE. Furthermore, when formulated as a function of one sum power variable, optimality is illustrated graphically and Pareto optimality is discussed. Finally, an EE function for a base station in a cellular network is proposed that captures essential factors of power consumption.

Keywords: energy efficiency, optimisation, cellular network, mobile communications.

1 Introduction

Following the increasing popularity of mobile cellular communications worldwide, its rising energy consumption has started to draw the attention of both regulatory bodies and network operators. It is not only a concern due to operational costs, but also due to future energy availability and environmental preservation. Currently, about 0.2% of the global CO_2 emissions are contributed by the mobile telecommunication industry [1]. Energy efficiency (EE) will become an inevitable concern in network design and architecture. Due to restrictions on power supply on mobile units, their EE is highly optimised to provide customer satisfaction. However, the power consumption of base stations (BS), which constitutes the major portion of a cellular network [2,3], has been neglected until lately [4]. Without considering users' phones, the power consumption of a typical mobile communications network is estimated to be about 40 MW [5].

One of the first steps for enhancing EE is to adequately model the communication system and measure its EE. The accurate evaluation of the EE requires the operational power of the entire access network at different levels to be taken into account [6]. However, its prohibitive complexity makes idealisation unavoidable. There is much research done in EE optimisation for specific system models.

J. Del Ser et al. (Eds.): MOBILIGHT 2011, LNICST 81, pp. 18–29, 2012.

Algorithms to achieve EE in OFDM systems and frequency-selective channels were presented in [7] and [8], respectively. Game theory was applied in [9] to analyse the EE of CDMA systems. In [10], the technique of switching between SIMO and MIMO to achieve EE in the uplink was investigated.

Our contribution is a framework that provides solutions to a general system model. In this work, we provide some general analytical tools and insights into optimising the EE at the BS in a cell of a network through power control, especially with downlink data transfer at the link and network level in mind. Nevertheless, the general results presented here can be adapted to include the uplink as well. Furthermore, we explore the case of elastic traffic, where the delay constraints are relaxed. This applies especially to data transfer through the Internet, which is increasingly being accessed through mobile devices [11].

In Section 2 we discuss the EE metric in a general sense. We also show that if the data rate function is a concave function of the power, we may incorporate any rate maximisation algorithm into the programme to find the optimum EE. This is formulated as a problem of a single total power variable, from which we obtain some useful insights concerning Pareto optimality w.r.t. EE and through-put. In Section 3 we present an EE function, which the results in the previous section can be applied to. This function captures essential factors for elastic data transmission while remaining general. Section 4 concludes the paper.

2   Energy Efficiency

2.1   Energy Efficiency Metric

In a general sense, efficiency can be seen as the ratio of goods produced to the resources consumed. Here, we focus on the EE in the physical and medium access control layers where goods are effective data transmitted measured in the information unit (bits or nats[1]) and the resources are the total energy (Joule) consumed for transmitting data. For static power configuration, EE can be expressed as the ratio of the sum rate r to the total expended power P_s that achieves this rate[2]:

$$EE = \frac{r(\boldsymbol{p})}{P_s(\boldsymbol{p})} \quad \left[\frac{\text{nat}}{\text{Joule}}\right]. \tag{1}$$

We characterise a *rate function* as being non-negative and $r(\boldsymbol{0}) = 0$. The power function P_s can be modelled as an affine function consisting of two terms, namely the constant power scalar p_C and the total input transmission power scalar $\sum_i^n p_i$, where p_i are the components of the non-negative vector $\boldsymbol{p} \in \mathbb{R}_+^n$ that represent input transmission powers e.g. in different subcarriers (as in [7]) or to different users. This constant or *base power* can be understood as the requirement for enabling transmission and is spent independently of the transmission power, which is justified considering the power consumption model of BSs [2]. It includes baseband processing, transceiver circuits, and other auxiliary components like

---

[1] A more convenient unit for analytical calculations.

[2] Unless otherwise indicated, *rate* refers to the *sum rate* hereafter.

climate control (see [12,13] for more details). This constant may vary depending on factors like the number of antennas, the traffic load and the computational efficiency. In Section 3 we show that these variations can be taken into account by considering them as separate constants in different traffic conditions.

If the powers are functions of time, e.g. adapted according to time-varying parameters (e.g. the channel coefficients), EE can be expressed as

$$EE = \frac{\int_0^T r(\boldsymbol{p}(t))\, dt}{\int_0^T P_s(\boldsymbol{p}(t))\, dt} = \frac{\mathbb{E}_T\left[r(\boldsymbol{p}(t))\right]}{\mathbb{E}_T\left[P_s(\boldsymbol{p}(t))\right]} \tag{2}$$

which is the ratio of the total amount of information units transmitted during a period T, $T\mathbb{E}_T\left[r(\boldsymbol{p}(t))\right]$, to the total energy expended $T\mathbb{E}_T\left[P_s(\boldsymbol{p}(t))\right]$, where $\mathbb{E}_T\left[X(t)\right] = \int_0^T X(t)\, dt$ is the mean value of X averaged over the period T, and $\boldsymbol{p}(t)$ is a vector function of time t. Under the assumption of ergodicity and stationarity (of the channel distribution), $\mathbb{E}_\infty\left[x\right] = \lim_{T\to\infty} \frac{1}{T}\int_0^T x(t)\, dt$ is the expected value of x averaged over time.

2.2   Optimal Energy Efficiency

Our aim is to maximise (1) over $\boldsymbol{p}$ or (2) over $\boldsymbol{p}(t)$. Due to the structure of (1) and (2), this belongs to a class of nonconvex problems called fractional programming [14,15]. Depending on the objective function EE, we can determine certain properties with regard to optimality. If it is a semistrictly quasiconcave function[3] of the transmit power vector, any local maximum is also the global maximum [16]. This reduces the problem to finding any local maximum. Additionally, if the objective function is pseudoconcave[4], any stationary point ($\nabla EE = 0$) is the local as well as the global optimum. The Karush-Kuhn-Tucker (KKT) conditions are then both necessary and sufficient for optimality, which does not necessarily apply to semistrictly quasiconcave functions.

For a function $f(x)/g(x)$, it is shown in [14] that if the numerator is concave and the denominator convex, then the function is semistrictly quasiconcave. If both the numerator and the denominator are differentiable in their domain, the function is pseudoconcave, implying that the system of equations given by the KKT conditions yields the global maximum. If in addition either the numerator or denominator is strictly concave or strictly convex, respectively, the function is strictly pseudoconcave, which implies that the global maximum, if it exists, is unique. Clearly, since the sum power is an affine function of its power components, the denominator fulfils the condition of being convex in $\boldsymbol{p}$. If the rate function $r(\boldsymbol{p})$ is concave, we have a semistrictly quasiconcave EE function. If it is differentiable, it is even pseudoconcave. Further, note that the positively

---

[3] A function $f : X \subset \mathbb{R}^n \to \mathbb{R}$ is *semistrictly quasiconcave* on X if $f(\lambda x_1 + (1-\lambda)x_2) > \min\{f(x_1), f(x_2)\}$ for all $x_1, x_2 \in X$, $f(x_1) \neq f(x_2)$, $\lambda \in (0,1)$ [16].

[4] A differentiable function $f : X \to \mathbb{R}$, where $X \subset \mathbb{R}^n$ is an open set, is *pseudoconcave* on X if $f(x_1) > f(x_2) \Rightarrow (x_1 - x_2)\nabla f(x_2) > 0$ for all $x_1, x_2 \in X$ [16].

weighted sum of (strictly) concave functions is (strictly) concave. This can be extended to integrals, i.e. the integral of a positively weighted (strictly) concave function is (strictly) concave [17].

For the case of static power configuration, we can formulate the programming problem as follows

$$\max_{\boldsymbol{p}} \ \frac{r(\boldsymbol{p})}{P_s(\boldsymbol{p})}. \tag{3}$$

If $r(\boldsymbol{p})$ is concave and differentiable, finding any feasible solution that satisfies the KKT conditions will yield the global optimum. Assume a convex constraint set of sum and individual power, and the optimal point $\boldsymbol{p}^*$ is found in it. Thus, we obtain the stationarity condition by setting the first derivative of (3) to zero, which after rearrangement yields

$$\frac{\frac{\partial}{\partial p_i} r(\boldsymbol{p})\Big|_{\boldsymbol{p}=\boldsymbol{p}^*}}{\frac{\partial}{\partial p_i} P_s(\boldsymbol{p})\Big|_{\boldsymbol{p}=\boldsymbol{p}^*}} = \frac{r(\boldsymbol{p}^*)}{P_s(\boldsymbol{p}^*)} = EE^*(\boldsymbol{p}^*) \tag{4}$$

for all $i = 1, ..., n$. The denominator on the left is one, if P_s is an unweighted sum of powers, $p_C + \sum p_i$. This means that at the vector $\boldsymbol{p}^*$ where the gradient of the rate equals its corresponding EE, we have the global optimum. If the optimal point is not within the constraint set, Lagrangian multipliers are utilised to obtain the solution.

In the case of time-varying powers, we obtain a similar result. Consider a case where the powers are adapted according to the channel coefficients $\boldsymbol{\alpha}$, which vary with time. We assume that the system has perfect channel information. The programming problem is finding the functional $\boldsymbol{p}(\boldsymbol{\alpha}) = (p_1(\boldsymbol{\alpha}), p_2(\boldsymbol{\alpha}), ..., p_n(\boldsymbol{\alpha}))$ that maximises the EE as follows:

$$\max_{\{\forall i, \, p_i(\boldsymbol{\alpha}) \geq 0\}} \frac{\int_H r(\boldsymbol{p}(\boldsymbol{\alpha})) f(\boldsymbol{\alpha}) \, d\boldsymbol{\alpha}}{\int_H P_s(\boldsymbol{p}(\boldsymbol{\alpha})) f(\boldsymbol{\alpha}) \, d\boldsymbol{\alpha}}, \tag{5}$$

where $f(\boldsymbol{\alpha})$ is the probability density function (PDF) of $\boldsymbol{\alpha}$, and H is the set of all values $\boldsymbol{\alpha}$ can have. We assume that P_s is an affine function of $p_1(\boldsymbol{\alpha}), p_2(\boldsymbol{\alpha}), ..., p_n(\boldsymbol{\alpha})$. Consider the case when the function r is concave and differentiable. Then by treating each $p_i(\boldsymbol{\alpha})$ as an infinitesimal vector where $p_i(\hat{\boldsymbol{\alpha}})$ for a particular value $\hat{\boldsymbol{\alpha}}$ and i is a component of it, we derive the objective function with respect to each component and obtain the following stationarity condition:

$$\frac{\frac{\partial}{\partial p_i(\hat{\boldsymbol{\alpha}})} r(\boldsymbol{p}(\boldsymbol{\alpha}))\Big|_{\boldsymbol{p}(\boldsymbol{\alpha})=\boldsymbol{p}^*(\boldsymbol{\alpha})}}{\frac{\partial}{\partial p_i(\hat{\boldsymbol{\alpha}})} P_s(\boldsymbol{p}(\boldsymbol{\alpha}))\Big|_{\boldsymbol{p}(\boldsymbol{\alpha})=\boldsymbol{p}^*(\boldsymbol{\alpha})}} = \frac{\int_H r(\boldsymbol{p}^*(\boldsymbol{\alpha})) f(\boldsymbol{\alpha}) \, d\boldsymbol{\alpha}}{\int_H P_s(\boldsymbol{p}^*(\boldsymbol{\alpha})) f(\boldsymbol{\alpha}) \, d\boldsymbol{\alpha}}, \tag{6}$$

for all is and all $\hat{\boldsymbol{\alpha}} \in H$, where $\boldsymbol{p}^*(\boldsymbol{\alpha})$ is the functional that maximises the problem above. Again, the denominator on the left is unity, if P_s is an unweighted sum of powers, $p_C + \sum p_i(\boldsymbol{\alpha})$. This means that when the differential increment of the rate function w.r.t. every component of $p(\alpha)$ for all $\hat{\alpha}$ is equal to its corresponding EE, we obtain the optimal function $p(\alpha)$. This is treated in depth in [18].

Of course perfect channel information is hardly available in real systems. Moreover the energy needed for the information feedback also needs to be accounted for, so that the true EE may be reflected. However, the solution to this problem provides us with theoretical knowledge of the upper bound of EE.

2.3   Equivalent Problems

Although we will focus on the EE maximisation problem shown in (3), we mention the variety of problem formulations that yield the same solution. For example, the problem can be formulated as the minimisation of the inverse of EE over power, $\frac{1}{EE} = P_s(p)/R(p)$. Since EE is non-negative, the solution p^* to $\max . EE(p)$ is identical to that of $\min . \frac{1}{EE(p)}$. The optimal value of one problem is just the reciprocal of the other. This is evident from the fact that if $EE(p^*) \geq EE(p)$ for any p in the constraint set, simply inverting it yields $\frac{1}{EE(p)} \geq \frac{1}{EE(p^*)}$.

In [8], the cost function, energy consumption per bit, corresponding to the dissipated power divided by the throughput is formulated as a function of rates r in subcarrier channels $\frac{1}{EE(r)} = E_a = P_s(r)/R(r)$ where $R(r) = \sum r_i$ here is the sum of all rates produced in the subcarriers. The cost function is to be minimised over r. This formulation is only possible if the transmit power can be written as a function of the individual rate, $p_i(r_i)$. In this case we have a bijective power function where each power allocation corresponds to a unique rate distribution (e.g. across the subcarriers). Because of the bijective property we know that the solution to the minimisation problem is obtained when the rate distribution yields the optimal power allocation solved by $\min . \frac{1}{EE(p)}$. Therefore, this optimisation problem is equivalent to the previous one.

2.4   Energy Efficiency Optimisation Using Rate Maximisation

The EE optimisation problem can be nested into inner and outer optimisation problems, namely to maximise the rate given the sum transmit power P and then to maximise this over the single variable P in the EE function. This allows the utilisation of any rate maximising algorithm for EE optimisation. Let us formulate the programming problem as follows:

$$\max . EE(P) = \max_P . \frac{\left(\max_{\sum_i p_i = P} R(p)\right)}{p_C + P} \tag{7}$$

$$= \max_P . \frac{R_{\max}(P)}{p_C + P}. \tag{8}$$

Based on the discussion in Section 2.2, a semistrictly quasiconcave objective function ensures that the local maximum is also the global maximum, which can then be found efficiently, e.g. using the bisection method or interior-point methods. This requires $R_{\max}$ to be concave in P. Using Corollary 1, which is attained through Lemma 5, we conclude that the objective function of (7) is semistrictly quasiconcave in P, if $R(\boldsymbol{p})$ is concave in $\boldsymbol{p}$. It is also pseudoconcave if $R_{\max}$ remains differentiable after the maximisation operation.

Corollary 1. *If $R(\boldsymbol{p})$ is concave in $\boldsymbol{p}$, $R_{\max}(P) = \max\left\{R(\boldsymbol{p}) : \mathbf{1}^T\boldsymbol{p} = P\right\}$ is concave in P.*

Note that this result is applicable if we additionally impose individual power constraints, provided that the set $\left\{\boldsymbol{p} : \mathbf{1}^T\boldsymbol{p} = P, 0 \leq p_i \leq p_{i,max} \,\forall i\right\}$, which is convex, is feasible.

2.5   Visualisation of Optimal EE Involving One Power Variable

We showed how the problem can be presented as having only one power variable, making it easier to handle. In this section we illustrate some properties of optimising EE with one power variable. Let us look at a simple example to obtain some insight. Consider

$$EE(P) = \frac{R(P)}{p_C + P} = \frac{B\log\left(1 + \frac{\alpha}{B}P\right)}{p_C + P}, \tag{9}$$

where the rate $R(P)$ is only dependent on a single power variable P and is modelled here as a logarithmic function with parameter α that represents the channel gain, and B the bandwidth. Since $B\log\left(1 + \frac{\alpha}{B}P\right)$ is a strictly concave function in P, (9) is a strictly pseudoconcave function. Using (4) the maximum EE is found where

$$\left.\frac{dR(P)}{dP}\right|_{P=P^*} = \frac{B\log\left(1 + \frac{\alpha}{B}P^*\right)}{p_C + P^*} \tag{10}$$

is fulfilled, P^* being the optimal point. This can be solved with the help of the Lambert-W function[5] $W(\cdot)$ such that

$$P^* = \frac{1}{\alpha}\left[\frac{\alpha p_C - B}{W\left((\alpha p_C - B)/B\exp(1)\right)} - B\right]. \tag{11}$$

This example is sketched in Fig. 1. Using this figure we illustrate that the solution can also be obtained graphically for any concave rate function $R(P)$. The figure shows the rate as a function of the total power $P_s = p_C + P$. The abscissa shows the magnitude of the total expended power whereas the ordinate shows the corresponding rate. p_C is the minimum power required for data transmission. The EE at any point of P_s is given as $EE = \tan\theta(P_s) = \frac{R(P)}{P_s}$. Since $R(P)$ is

---

[5] i.e. the inverse function of $x = W\exp(W)$ [19].

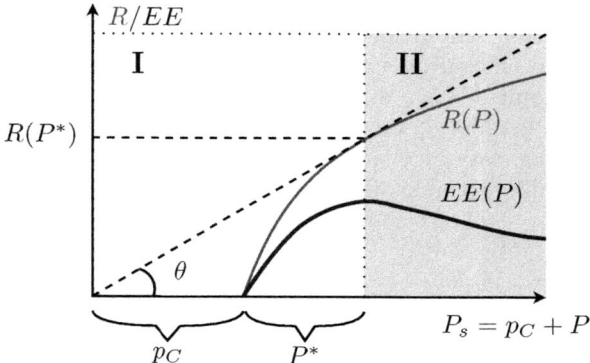

Fig. 1. Visualisation of the optimal EE. The optimal EE is obtained where the transmission power is P^*.

concave, we know that the optimal point is where (10) is fulfilled. Rearranging (10), we obtain

$$(p_C + P^*) \left.\frac{dR(P)}{dP}\right|_{P=P^*} = R(P^*), \tag{12}$$

which can be interpreted as follows. The linear function (l.h.s.) which has the slope identical to the slope of $R(P)$ at point P has to intersect the rate function (r.h.s.) in order that the optimality criterion is fulfilled. This means that the tangent of the rate function at P^* has to form a line that passes the origin of P_s. This is shown as the dashed line in Fig. 1. This is similar to the approach in [20], where the global maximiser is identified graphically for resource management problems.

Generally, maximising EE is equivalent to finding the point P along $R(P)$ that maximises θ, even when $R(P)$ is not concave, e.g. when cross-channel interference exists.

Lemma 2. *Given a concave rate function $R(P)$, the optimal point P^* of $EE = \frac{R(P)}{p_C+P}$ increases with p_C whereas the optimal value $EE^* = \frac{R(P^*)}{p_C+P^*}$ decreases with p_C.*

Proof. Intuitively, this can be observed graphically in Fig. 1. As p_C is increased, the point of contact P^* of the tangent line with the rate function increases and the angle θ decreases. To show it analytically, consider the following. It is evident that for any given P, EE decreases with an increasing p_C. This implies that the optimal value of EE also decreases with p_C. Recall the stationarity condition

$$\left.\frac{dR(P)}{dP}\right|_{P=P^*} = \frac{R(P^*)}{p_C+P^*} = EE^*.$$

Since $R(P)$ is a concave function, its derivative is a non-increasing function of P. It follows that a lower optimal EE^* would intersect with $\frac{dR}{dP}$ at a higher P^*. $\square$

Corollary 3. *If* $p_C = 0$, *the highest possible* $EE^* = \left.\frac{dR(P)}{dP}\right|_{P=P^*}$ *is achieved, where* $P^* = 0$. $\left.\frac{dR(P)}{dP}\right|_{P=0}$ *can be used as an upper bound for the* EE^*.

We say that R_1 dominates R_2 if $R_1(P) > R_2(P)$ for all $P > 0$, where R_1 and R_2 are two different rate functions. It is easy to see that if R_1 dominates R_2, their corresponding EE also follow the same order, such that $EE_1 = \frac{R_1(P)}{p_C+P} > \frac{R_2(P)}{p_C+P} = EE_2$ for all P. The same applies to the optimum EE.

Corollary 4. *If* R_1 *dominates* R_2, $EE_1^* > EE_2^*$.

2.6 Pareto Optimality. Trade-Off between Rate and Energy Efficiency.

It is beneficial if we can achieve both high rates and high EE. This can be considered as a multiobjective optimisation problem. An efficient operating point is where neither the rate nor the EE can be further improved without the decline of the other. Such a point is also said to be *Pareto-optimal*.

Any point in region I (not shaded) in Fig. 1 is not Pareto-optimal since both the rate and the EE can simultaneously be increased by applying a higher power. The points in region II (shaded) are Pareto-optimal because increasing either the rate or the EE causes the other to decline. This is where there is a trade-off between the rate and the EE.

Regulatory bodies usually impose constraints on radiation power to prevent potential damage on people exposed to electromagnetic waves used in communications, and to mitigate interference among links using the same bandwidth, which is also a cause of inefficiency. As a result, we may assume a given sum transmit power constraint $P_{\max}$. If $P_{\max} < P^*$ then maximising the EE is the same as maximising the rate using $P_{\max}$. Further improvement is possible with new solutions to reduce p_C or new designs that yields a better (or more dominant) $R(P)$ function, e.g. by increasing the power amplifier efficiency, but not through power control. If $P_{\max} > P^*$ EE can be improved by reducing the sum transmit power. If a sum rate constraint $R_{\mathrm{con}} > R_{\max}(P^*)$ exists such that $R_{\max}(P) \geq R_{\mathrm{con}}$, it will be fulfilled with equality when maximising the EE, while the optimal EE cannot be achieved. If $R_{\mathrm{con}} \leq R_{\max}(P^*)$ both the rate constraint and EE optimality can be simultaneously achievable.

3   Energy Efficiency Function for One Cell

In this section, we present a model for EE optimisation considering different traffic conditions. Imagine the downlink scenario of a cell in a mobile cellular network. During the day the links between the BS and the mobile stations experience higher interference from cells due to greater traffic and thus have statistically lower *signal to interference plus noise ratios* (SINR) than at night.

Assume we have the statistical information of these two traffic conditions, denoted by $i = 1$ for day-time and $i = 2$ for night-time. This stochastic information is captured as PDFs f_i of the normalised SINR α_i, which is the SINR at a unit transmit power. This can be written as a vector α_i to include the normalised SINR of different components (e.g. subcarriers) of the transmission. Note that the dimension of vector α_i may vary for different is. The PDF yields the probability that a certain channel realisation α_i occurs during traffic condition i. Assume that conditions $i = 1, 2$ applies to 60% and 40% of a day's cycle, respectively. We describe this using weights $w_1 = 0.6$ and $w_2 = 0.4$. One may choose to have different schemes for day-time and night-time traffic (e.g. MIMO and SISO as studied in [10] for the uplink) to lower the total base power during low load period. The variations in the base power can be modelled using separate constants $p_{C,i}$.

We generalise this example and describe the EE as follows:

$$EE = \frac{\sum_i w_i \beta_i \int_0^\infty R_i\left(\alpha_i, P_i, \gamma_i, \epsilon_i, G\right) f_i\left(\alpha_i\right) d\alpha_i}{\sum_i w_i\left(p_{C,i} + P_i\right)}. \tag{13}$$

i index for different traffic conditions or schemes at different time intervals.

w_i time fraction for traffic condition i such that $\sum_i w_i = 1$ and $w_i \geq 0, \forall i$.

R_i rate produced using scheme employed in i, a concave rate function of input power components p_i with the following parameters.

     α_i SINR,

     P_i sum input power such that $\mathbf{1}^T p_i = P_i$[6],

     γ_i SNR gap for scheme i,

     G additional imposed constraints such as individual power constraints.

     ϵ_i characterises the power amplifier efficiency, $\epsilon_i \in [0, 1]$, i.e. the output to input power ratio where $\epsilon_i p_i$ is the actual transmit power (more on PA efficiency in [11]).

f_i PDF of the SINR α_i for traffic condition i.

β_i proportion of the rate used for data transfer (without pilot signals for channel estimation etc.), $\beta_i \in [0, 1]$.

$p_{C,i}$ base power required during interval i.

Note again that if R_i for all i is concave, EE is semistrictly quasiconcave (pseudoconcave if EE is differentiable), ensuring efficiency in optimisation. The programming problem can be written in the form described in (8) as:

$$\max . EE\left(\mathcal{P}\right) = \max_{\mathcal{P}} . \frac{\mathcal{R}\left(\mathcal{P}\right)}{\mathcal{P}_C + \mathcal{P}}, \quad \text{where} \tag{14}$$

---

[6] The vector $\mathbf{1} = (1, 1, ..., 1)^T$.

$$\mathcal{R}\left(\mathcal{P}\right) = \max_{\sum_i w_i P_i = \mathcal{P}} \sum_i w_i \beta_i \int_0^\infty R_i\left(\boldsymbol{\alpha}_i, P_i, \gamma_i, \epsilon_i, G\right) f_i\left(\boldsymbol{\alpha}_i\right) d\boldsymbol{\alpha}_i,$$

$$\text{and} \quad \mathcal{P}_C = \sum_i w_i p_{C,i}.$$

Examples of R_i are $R_i\left(P_i\right) = \left\{\sum_{n=1}^N \log\left(1 + \gamma_i \epsilon_i \alpha_{i,n} p_{i,n}\right) : \sum_n p_{i,n} = P_i\right\}$ for OFDM systems with N subcarriers and $R_i\left(P_i\right) = \{\log\det\left(\mathbf{I} + \gamma_i \mathbf{HQH}^H\right)$: $\mathrm{tr}\left(\mathbf{Q}\right) = P_i\}$ for MIMO systems. A rate maximising function can also be chosen such as $R_i\left(P_i, \boldsymbol{\alpha_i}\right) = \max. \left\{\sum_{n=1}^N \log\left(1 + \gamma_i \epsilon_i \alpha_{i,n} p_{i,n}\right) : \sum_n p_{i,n} = P_i\right\}$ or one w.r.t. a scheduler that selects best users, as in [21]. The integral $\int_0^\infty \{\ldots\} d\boldsymbol{\alpha}_i$ implies an integration over all vector components, i.e. $\int_0^\infty \ldots \int_0^\infty \{\ldots\} d\alpha_{i,1}\ldots d\alpha_{i,N}$. If the SINRs are quantised (e.g. due to channel measurements or signal feedback), the integral is to be replaced by a sum, and $d\alpha_{i,n}$ by the quantisation interval. The variable P_i can be further generalised as a function of $\boldsymbol{\alpha}_i$, such that $\sum_i w_i \int_0^\infty P_i\left(\boldsymbol{\alpha}_i\right) f_i\left(\boldsymbol{\alpha}_i\right) d\boldsymbol{\alpha}_i = \mathcal{P}$.

A great energy saving potential lies in deactivating certain components or putting them to sleep modes during inactivity [6]. This can be modelled by assigning a certain $p_{C,i}$ at which $P_i = 0$ with some time fraction w_i. In the elastic traffic scenario, it is even possible to collect enough data so that there can be continuous transmission at a later period while energy can be saved during the data collection period by shutting down inactive components. Additionally, it is more efficient to transmit during the best traffic condition (e.g. at night), that is, when the power is most effectively used due to low interference. Further improvements can be achieved by adjusting the weights w_i. This has to be done with caution since f_i may be altered by changing the time interval. Pilot and overhead signals that make channel information available at the receiver and transmitter, which takes up a fraction of the transmission, are considered in β_i.

4   Conclusion

An accurate evaluation of the EE requires adequate modelling of the whole network. Here, we present some general analytical tools for optimising EE at the BS, assuming elastic traffic. The optimisation problem can be formulated utilising rate maximisation algorithms. As a function of a single variable, insights into EE were illustrated graphically. An EE function that captures essential factors in one cell of a network is introduced and discussed.

Since this is also an idealised model, further work is needed to refine it, e.g. to describe the non-linearity of power amplifier efficiency, which has a significant influence on the EE. A model for inelastic traffic is also necessary, where more stringent quality of service demands are considered. Other figures of merit like fairness and coverage can still be incorporated into the model. The relationship between the base power and system model parameters, such as number of antennas and subcarriers, would also be a future task that enables EE optimisation over these parameters in addition to power control.

Acknowledgement. The authors would like to thank our project partners and colleagues from TU Dresden for helpful discussions, especially Christian Isheden for pointing out Schaible's works and his valuable comments. This work is supported by the German Federal State of Saxony in the excellence cluster Cool Silicon within the frame of the project Cool Cellular under grant number 14056/2367.

Appendix

Lemma 5. *If f is jointly concave in $(x, y) \in \overline{C}$, $g(x) = \max\limits_{y \in C(x)} f(x, y)$ is also concave where $C(x)$ is a convex set for any $x \in \mathbf{dom}\, g$ and $\overline{C} = \{(x, y) : \forall y \in C(x), \forall x \in \mathbf{dom}\, g\}$ is a convex set for all $x \in \mathbf{dom}\, g$. The domain g is defined as $\mathbf{dom}\, g = \{x : (x, y) \in \mathbf{dom}\, f \text{ for some } y \in C(x)\}$. Variables x and y can be vectors or matrices of identical dimensions.*

Proof. The function $g(x)$ is concave if

$$g(\lambda x_1 + (1 - \lambda) x_2) \geq \lambda g(x_1) + (1 - \lambda) g(x_2)$$

for any $\lambda \in [0, 1]$ and any $x_1, x_2 \in \mathbf{dom}\, g$. We show that the following holds:

$$
\begin{aligned}
& g(\lambda x_1 + (1 - \lambda) x_2) \\
& = \max_{y} \{f(\lambda x_1 + (1 - \lambda) x_2, y) : y \in C(\lambda x_1 + (1 - \lambda) x_2)\} \\
\text{(a)} \quad & = \max_{y_1, y_2} \{f(\lambda x_1 + (1 - \lambda) x_2, \lambda y_1 + (1 - \lambda) y_2) \\
& \quad : \lambda y_1 + (1 - \lambda) y_2 \in C(\lambda x_1 + (1 - \lambda) x_2)\} \\
\text{(b)} \quad & \geq \max_{y_1, y_2} \{f(\lambda x_1 + (1 - \lambda) x_2, \lambda y_1 + (1 - \lambda) y_2) : y_1 \in C(x_1), y_2 \in C(x_2)\} \\
\text{(c)} \quad & \geq \max_{y_1, y_2} \{\lambda f(x_1, y_1) + (1 - \lambda) f(x_2, y_2) : y_1 \in C(x_1), y_2 \in C(x_2)\} \\
\text{(d)} \quad & = \lambda \max_{y_1} \{f(x_1, y_1) : y_1 \in C(x_1)\} + (1 - \lambda) \max_{y_2} \{f(x_2, y_2) : y_2 \in C(x_2)\} \\
& = \lambda g(x_1) + (1 - \lambda) g(x_2).
\end{aligned}
$$

The variable y is substituted by $\lambda y_1 + (1 - \lambda) y_2$ in (a). We arrive at the inequality in (b) because $\{(y_1, y_2) : y_1 \in C(x_1), y_2 \in C(x_2)\}$ forms a subset of $\{(y_1, y_2) : \lambda y_1 + (1 - \lambda) y_2 \in C(\lambda x_1 + (1 - \lambda) x_2)\}$. The reason for this is understood by considering the following. Assume that $(x_1, y_1) \in \overline{C}$ and $(x_2, y_2) \in \overline{C}$. Because $\overline{C}$ is a convex set, $(\lambda x_1 + (1 - \lambda) x_2, \lambda y_1 + (1 - \lambda) y_2) \in \overline{C}$ also. By definition of $\overline{C}$, $\lambda y_1 + (1 - \lambda) y_2 \in C(\lambda x_1 + (1 - \lambda) x_2)$ is automatically satisfied under these assumptions. Since we now have a more restricted constraint set for y_1 and y_2 in the assumptions, a smaller maximum value may be induced. Using the property of joint concavity of f, we derive the inequality in (c). Since each summand has independent variables and the constraints can be separately considered, step (d) is derived. $\qquad \square$

References

1. Ericsson, Sustainable energy use in mobile communications (2007)
2. Chen, T., Zhang, H., Zhao, Z., Chen, X.: Towards Green Wireless Access Networks. In: 5th Int. ICST Conference on Communications and Networking (2010)
3. Karl, H.: An Overview of Energy-Efficiency Techniques for Mobile Communication Systems. Telecommunication Networks Group, Technical University Berlin, Germany, Tech. Rep. TKN-03-017 (2003)
4. Auer, G., et al.: Enablers for Energy Efficient Wireless Networks. In: IEEE 72nd Vehicular Technology Conference Fall (2010)
5. He, J., et al.: Energy Efficient Architectures and Techniques for Green Radio Access Networks. In: 5th Int. ICST Conference on Communications and Networking (2010)
6. Correia, L., et al.: Challenges and enabling technologies for energy aware mobile radio networks. IEEE Communications Magazine 48, 66–72 (2010)
7. Miao, G., Himayat, N., Li, Y., Bormann, D.: Energy-efficient design in wireless OFDMA. In: IEEE ICC (2008)
8. Isheden, C., Fettweis, G.P.: Energy-Efficient Multi-Carrier Link Adaptation with Sum Rate-Dependent Circuit Power. In: IEEE GLOBECOM (December 2010)
9. Betz, S., Poor, H.: Energy Efficient Communications in CDMA Networks: A Game Theoretic Analysis Considering Operating Costs. IEEE Transactions on Signal Processing 56(10), 5181–5190 (2008)
10. Kim, H., Chae, C.-B., de Veciana, G., Heath, R.: A cross-layer approach to energy efficiency for adaptive MIMO systems exploiting spare capacity. IEEE Transactions on Wireless Communications 8(8), 4264–4275 (2009)
11. Chen, T., Kim, H., Yang, Y.: Energy efficiency metrics for green wireless communications. In: Int. Conference on Wireless Communications and Signal Processing (2010)
12. Energy Efficiency of Wireless Access Network Equipment, ETSI TS102706 (2009)
13. Arnold, O., Richter, F., Fettweis, G., Blume, O.: Power consumption modeling of different base station types in heterogeneous cellular networks. In: Proc. of 19th Future Network & MobileSummit (2010)
14. Schaible, S.: Fractional programming. Zeitschrift für Operations Research 27(1), 39–54 (1983)
15. Schaible, S.: Minimization of ratios. Journal of Optimization Theory and Applications 19(2), 347–352 (1976)
16. Giorgi, G., Guerraggio, A., Thierfelder, J.: Mathematics of optimization: smooth and nonsmooth case. Elsevier Science (2004)
17. Boyd, S., Vandenberghe, L.: Convex optimization. Cambridge University Press (2004)
18. Chong, Z., Jorswieck, E.A.: Energy-efficient Power Control for MIMO Time-varying Channels. In: IEEE Online Green Communications Conference (2011)
19. Corless, R., Gonnet, G., Hare, D., Jeffrey, D., Knuth, D.: On the LambertW function. Advances in Computational Mathematics 5, 329–359 (1996)
20. Rodriguez, V.: An analytical foundation for resource management in wireless communication. In: IEEE GLOBECOM 2003, vol. 2, pp. 898–902 (2003)
21. Chong, Z., Jorswieck, E.A.: Energy efficiency in Random Opportunistic Beamforming. In: IEEE 73rd Vehicular Technology Conference (2011)

A Note on the Wiener Filter
for Vector Random Processes

Jesús Gutiérrez-Gutiérrez[1], Adam Podhorski[1],
Iñaki Iglesias[2], and Javier Del Ser[3]

[1] CEIT and Tecnun (University of Navarra),
Manuel de Lardizábal 15, 20018 San Sebastián, Spain
{jgutierrez,apodhorski}@ceit.es
[2] Department of Electrical and Computer Engineering,
University of Delaware, 302 Evans Hall, Newark, DE 19716 USA
iglesias@udel.edu
[3] Tecnalia Research & Innovation
P. Tecnológico, Ed. 202, 48170 Zamudio, Spain
javier.delser@tecnalia.com

Abstract. In the present paper we compute the geometric minimum mean square error for the vector linear estimation problem. We do this by proving that the vector linear estimator that minimizes the mean square error (MSE) also minimizes the geometric MSE.

Keywords: minimum mean square error (MMSE), vector linear estimation, vector linear interpolation, vector linear prediction.

1 Introduction

The linear estimation of a random vector from another random vector, which, at first sight, may seem to be a very particular case of vector linear estimation, represents, in fact, a very general framework, which includes, as it will be shown in the appendix, vector linear prediction, vector linear interpolation, multiple-input multiple-output (MIMO) linear equalization and MIMO decision-feedback equalization (DFE).

As we are dealing with random vectors, the definition of mean square error (MSE) is not unique. Here we will consider two possible definitions, which will be called MSE and geometric MSE (GMSE), respectively. We will name the former simply MSE because it is the definition commonly used [1,2]. The latter was introduced in [3] for MIMO DFE. The interest in using one or the other error measure depends on the problem considered.

In this paper we will prove that the vector linear estimator that minimizes the MSE also minimizes the GMSE. This fact was proved in [4] for MIMO DFE, and it was assumed true, but not proved, in [5] for MIMO linear prediction.

J. Del Ser et al. (Eds.): MOBILIGHT 2011, LNICST 81, pp. 30–36, 2012.

2   General Framework

2.1   Vector Linear Estimation Problem Considered

Let $\mathbf{x}$ and $\mathbf{y}$ be two complex random vectors of dimensions N and M, respectively. We assume that the correlation matrix of $\mathbf{y}$ is invertible, and we estimate $\mathbf{x}$ from $\mathbf{y}$ in the following manner:

$$\hat{\mathbf{x}} = W\mathbf{y}, \tag{1}$$

where $W \in \mathbb{C}^{N \times M}$ and $\mathbb{C}^{N \times M}$ is the set of all $N \times M$ complex matrices. We denote by $\mathbf{e}$ the error vector $\mathbf{x} - \hat{\mathbf{x}}$, and by R its correlation matrix, that is,

$$
\begin{aligned}
R &= \mathrm{E}\left[\mathbf{e}\mathbf{e}^*\right] \\
&= \mathrm{E}\left[(\mathbf{x} - W\mathbf{y})(\mathbf{x}^* - \mathbf{y}^* W^*)\right] \\
&= \mathrm{E}\left[\mathbf{x}\mathbf{x}^*\right] - \mathrm{E}\left[\mathbf{x}\mathbf{y}^*\right] W^* - W\mathrm{E}\left[\mathbf{y}\mathbf{x}^*\right] + W\mathrm{E}\left[\mathbf{y}\mathbf{y}^*\right] W^*,
\end{aligned} \tag{2}
$$

where $*$ indicates complex conjugate transpose. MSE and GMSE are defined as the trace and the determinant of R, respectively.

2.2   Principle of Orthogonality

We now compute the matrices W and R associated with the linear estimator $\hat{\mathbf{x}}$ that satisfies the following condition that is called principle of orthogonality:

$$0_{N \times M} = \mathrm{E}\left[\mathbf{e}\mathbf{y}^*\right], \tag{3}$$

where $0_{N \times M}$ is the $N \times M$ zero matrix. We denote these two matrices by W_0 and R_0. From (3) we have

$$0_{N \times M} = \mathrm{E}\left[\mathbf{x}\mathbf{y}^*\right] - W_0 \mathrm{E}\left[\mathbf{y}\mathbf{y}^*\right], \tag{4}$$

and consequently,

$$W_0 = \mathrm{E}\left[\mathbf{x}\mathbf{y}^*\right]\left(\mathrm{E}\left[\mathbf{y}\mathbf{y}^*\right]\right)^{-1}. \tag{5}$$

Combining (2) and (4) we obtain

$$R_0 = \mathrm{E}\left[\mathbf{x}\mathbf{x}^*\right] - W_0 \mathrm{E}\left[\mathbf{y}\mathbf{x}^*\right], \tag{6}$$

and applying (5) yields

$$R_0 = \mathrm{E}\left[\mathbf{x}\mathbf{x}^*\right] - \mathrm{E}\left[\mathbf{x}\mathbf{y}^*\right]\left(\mathrm{E}\left[\mathbf{y}\mathbf{y}^*\right]\right)^{-1}\mathrm{E}\left[\mathbf{y}\mathbf{x}^*\right]. \tag{7}$$

Observe that (4) and (6) can be written as

$$
\begin{cases}
R_0 = \begin{pmatrix} I_N & -W_0 \end{pmatrix} \begin{pmatrix} \mathrm{E}\left[\mathbf{x}\mathbf{x}^*\right] \\ \mathrm{E}\left[\mathbf{y}\mathbf{x}^*\right] \end{pmatrix}, \\[2ex]
0_{N \times M} = \begin{pmatrix} I_N & -W_0 \end{pmatrix} \begin{pmatrix} \mathrm{E}\left[\mathbf{x}\mathbf{y}^*\right] \\ \mathrm{E}\left[\mathbf{y}\mathbf{y}^*\right] \end{pmatrix},
\end{cases} \tag{8}
$$

where I_N is the $N \times N$ identity matrix. The system of two equations (8) is equivalent to the following single equation:

$$(R_0 \quad 0_{N \times M}) = (I_N \quad -W_0) \, \mathrm{E}[\mathbf{z}\mathbf{z}^*], \tag{9}$$

with $\mathbf{z} = \begin{pmatrix} \mathbf{x} \\ \mathbf{y} \end{pmatrix}$. As in the theory of linear prediction [1], we can call equations (4) and (9) normal equation and augmented normal equation, respectively.

2.3   Relation between R and R_0

Since R and R_0 are Hermitian[1], $S = R - R_0$ is Hermitian. In this subsection we prove that S is also positive semidefinite. Using (2) and (7) we have

$$S = \mathrm{E}\,[\mathbf{x}\mathbf{y}^*] \, (\mathrm{E}\,[\mathbf{y}\mathbf{y}^*])^{-1} \mathrm{E}\,[\mathbf{y}\mathbf{x}^*] - \mathrm{E}\,[\mathbf{x}\mathbf{y}^*]\, W^* - W \mathrm{E}\,[\mathbf{y}\mathbf{x}^*] + W \mathrm{E}\,[\mathbf{y}\mathbf{y}^*]\, W^*.$$

Thus

$$S = B \mathrm{E}\,[\mathbf{y}\mathbf{y}^*]\, B^*, \tag{10}$$

with

$$B = \mathrm{E}\,[\mathbf{x}\mathbf{y}^*] \, (\mathrm{E}\,[\mathbf{y}\mathbf{y}^*])^{-1} - W.$$

From the expression given in (10) for the matrix S, we can now show that S is positive semidefinite:

$$v^* S v = \mathrm{E}\,[v^* B \mathbf{y}\mathbf{y}^* B^* v] = \mathrm{E}\,\left[|v^* B \mathbf{y}|^2\right] \geq 0 \quad \forall v \in \mathbb{C}^{N \times 1}.$$

2.4   MMSE

If A is an $N \times N$ diagonalizable matrix, then $\mathrm{tr}(A) = \sum_{k=1}^{N} \lambda_k(A)$, where tr denotes trace and $\lambda_1(A), \ldots, \lambda_N(A)$ are the eigenvalues of A counted with their multiplicities. Since S is a Hermitian positive semidefinite matrix, S is diagonalizable and all its eigenvalues are non-negative, and consequently, $\mathrm{tr}(S) \geq 0$. Therefore using that $\mathrm{tr}(S) = \mathrm{tr}(R) - \mathrm{tr}(R_0)$ we conclude that

$$\mathrm{tr}(R) \geq \mathrm{tr}(R_0). \tag{11}$$

Thus we have proved that the minimum MSE (MMSE) is

$$\mathrm{MMSE} = \mathrm{tr}\,(R_0).$$

2.5   Geometric MMSE

In this subsection we assume that the Hermitian positive semidefinite matrix R_0 is invertible (or equivalently, positive definite) and we show that (11) is also true when trace is replaced by determinant.

---

[1] Any correlation matrix is Hermitian and positive semidefinite.

Let $R_0 = V \operatorname{diag}(\lambda_1(R_0), \ldots, \lambda_N(R_0)) V^*$ be an eigenvalue decomposition of the Hermitian matrix R_0. Since R_0 is positive definite, all its eigenvalues are positive, and consequently, we can define the following two Hermitian matrices:

$$R_0^{\frac{1}{2}} := V \operatorname{diag}\left((\lambda_1(R_0))^{\frac{1}{2}}, \ldots, (\lambda_N(R_0))^{\frac{1}{2}}\right) V^*,$$

and

$$R_0^{-\frac{1}{2}} := V \operatorname{diag}\left((\lambda_1(R_0))^{-\frac{1}{2}}, \ldots, (\lambda_N(R_0))^{-\frac{1}{2}}\right) V^* = \left(R_0^{\frac{1}{2}}\right)^{-1}.$$

Hence the determinant of the matrix R can be expressed as

$$\begin{aligned}
\det(R) &= \det(R_0 + S) \\
&= \det\left(R_0^{\frac{1}{2}}(R_0 + S) R_0^{-\frac{1}{2}}\right) \\
&= \det\left(R_0 + R_0^{\frac{1}{2}} S R_0^{-\frac{1}{2}}\right) \\
&= \det\left(R_0\left(I_N + R_0^{-\frac{1}{2}} S R_0^{-\frac{1}{2}}\right)\right) \\
&= \det(R_0) \det\left(I_N + R_0^{-\frac{1}{2}} S \left(R_0^{-\frac{1}{2}}\right)^*\right).
\end{aligned}$$

Therefore if $R_0^{-\frac{1}{2}} S \left(R_0^{-\frac{1}{2}}\right)^* = U D U^*$ is an eigenvalue decomposition of the Hermitian matrix $R_0^{-\frac{1}{2}} S \left(R_0^{-\frac{1}{2}}\right)^*$, then

$$\begin{aligned}
\frac{\det(R)}{\det(R_0)} &= \det(I_N + U D U^*) \\
&= \det(U U^* + U D U^*) \\
&= \det(U(I_N + D) U^*) \\
&= \det(I_N + D) \\
&= \prod_{k=1}^{N}\left(1 + \lambda_k\left(R_0^{-\frac{1}{2}} S \left(R_0^{-\frac{1}{2}}\right)^*\right)\right). \tag{12}
\end{aligned}$$

Since S is positive semidefinite, $R_0^{-\frac{1}{2}} S \left(R_0^{-\frac{1}{2}}\right)^*$ is also positive semidefinite:

$$v^* R_0^{-\frac{1}{2}} S \left(R_0^{-\frac{1}{2}}\right)^* v = \left(\left(R_0^{-\frac{1}{2}}\right)^* v\right)^* S \left(R_0^{-\frac{1}{2}}\right)^* v \geq 0 \qquad \forall v \in \mathbb{C}^{N \times 1}.$$

Consequently, all the eigenvalues of $R_0^{-\frac{1}{2}} S \left(R_0^{-\frac{1}{2}}\right)^*$ are non-negative, and from (12) we obtain

$$\frac{\det(R)}{\det(R_0)} \geq 1.$$

Finally, since $\det(R_0) = \prod_{k=1}^{N} \lambda_k(R_0) > 0$ we conclude that

$$\det(R) \geq \det(R_0).$$

Thus we have proved that the geometric MMSE (GMMSE) is

$$\text{GMMSE} = \det(R_0).$$

Acknowledgments. The work of Jesús Gutiérrez-Gutiérrez and Javier Del Ser was supported in part by the Basque Government through the MIMONET project (PC2009-27B), and by the Spanish Ministry of Science and Innovation through the projects COSIMA (TEC2010-19545-C04-02), COMONSENS (CSD2008-00010), and a Torres-Quevedo grant (PTQ-09-01-00740).

Appendix A: Some Specific Cases

A.1 Vector Linear Interpolation and Prediction

Consider $m, d \in \mathbb{N}$, with $1 \leq d \leq m$. Let $\{\mathbf{x}_n; n \in \mathbb{Z}\}$ be a complex N-dimensional random process. We estimate $\mathbf{x}_d$ from $\mathbf{x}_1, \ldots, \mathbf{x}_{d-1}, \mathbf{x}_{d+1}, \ldots, \mathbf{x}_m$ in the following manner:

$$\hat{\mathbf{x}}_d = \sum_{\substack{1 \leq k \leq m \\ k \neq d}} W_k \mathbf{x}_k, \tag{13}$$

where $W_k \in \mathbb{C}^{N \times N}$.

The estimator given in (13) is called linear interpolator when $1 < d < m$, forward linear predictor when $d = m$, and backward linear predictor when $d = 1$.

The estimation scheme (13) can be obtained as a particular case of (1) by taking:

$$\begin{cases} \hat{\mathbf{x}} = \hat{\mathbf{x}}_d, \\ \mathbf{y} = \begin{pmatrix} \mathbf{x}_1 \\ \vdots \\ \mathbf{x}_{d-1} \\ \mathbf{x}_{d+1} \\ \vdots \\ \mathbf{x}_m \end{pmatrix}, \\ W = (W_1 \quad \cdots \quad W_{d-1} \quad W_{d+1} \quad \cdots \quad W_m). \end{cases}$$

A.2 Vector Wiener Filtering

Let $\{\mathbf{x}_n; n \in \mathbb{Z}\}$ and $\{\mathbf{y}_n; n \in \mathbb{Z}\}$ be two complex vector random processes of dimensions N and M, respectively. Given $m \in \mathbb{N}$ we estimate $\mathbf{x}_n$ from $\mathbf{y}_n, \mathbf{y}_{n-1}, \ldots, \mathbf{y}_{n-m+1}$ in the following manner:

$$\hat{\mathbf{x}}_n = \sum_{k=1}^{m} W_k \mathbf{y}_{n-k+1}, \tag{14}$$

where $W_k \in \mathbb{C}^{N \times M}$.

The filter given in (14) is called the vector Wiener filter when it minimizes the MSE. We have proved in this paper that the vector Wiener filter can also be defined as the filter that minimizes the GMSE.

The estimation scheme (14) can be obtained as a particular case of (1) by taking:

$$\begin{cases} \hat{\mathbf{x}} = \hat{\mathbf{x}}_n, \\ \mathbf{y} = \begin{pmatrix} \mathbf{y}_n \\ \vdots \\ \mathbf{y}_{n-m+1} \end{pmatrix}, \\ W = (W_1 \quad \cdots \quad W_m). \end{cases}$$

Observe that (1) can also be obtained as a particular case of (14) by taking:

$$\begin{cases} \hat{\mathbf{x}}_n = \hat{\mathbf{x}}, \\ m = 1, \\ W_1 = W, \\ \mathbf{y}_n = \mathbf{y}. \end{cases}$$

That is, the estimation schemes (1) and (14) are equivalent.

A.3 MIMO DFE

We consider a MIMO communication system given by

$$\mathbf{y}_n = \sum_{k=0}^{p} H_k \mathbf{x}_{n-k} + \mathbf{s}_n,$$

where p is a non-negative integer, $H_k \in \mathbb{C}^{M \times N}$, the input $\{\mathbf{x}_n; n \in \mathbb{Z}\}$ is a discrete complex vector random process of dimension N, and the noise $\{\mathbf{s}_n; n \in \mathbb{Z}\}$ is a complex vector random process of dimension M.

We estimate $\mathbf{x}_{n-d}$ from m output vectors $\mathbf{y}_n, \mathbf{y}_{n-1}, \ldots, \mathbf{y}_{n-m+1}$, and l input vectors $\mathbf{x}_{n-1-d}, \mathbf{x}_{n-2-d}, \ldots, \mathbf{x}_{n-l-d}$ in the following manner:

$$\hat{\mathbf{x}}_{n-d} = \sum_{k=1}^{m} W_k \mathbf{y}_{n-k+1} + \sum_{k=1}^{l} V_k \mathbf{x}_{n-k-d}, \tag{15}$$

where $W_k \in \mathbb{C}^{N \times M}$, $V_k \in \mathbb{C}^{N \times N}$, d is the decision delay, and $m, l \in \mathbb{N}$ should satisfy $d + l + 1 = m + p$.

The estimator given in (15) is called MIMO decision-feedback equalizer except when $V_k = 0$ with $1 \leq k \leq l$. In that other case it is called MIMO linear equalizer.

The estimation scheme (15) can be obtained as a particular case of (1) by taking:

$$\begin{cases} \hat{\mathbf{x}} = \hat{\mathbf{x}}_{n-d}, \\[2mm] \mathbf{y} = \begin{pmatrix} \mathbf{y}_n \\ \vdots \\ \mathbf{y}_{n-m+1} \\ \mathbf{x}_{n-1-d} \\ \vdots \\ \mathbf{x}_{n-l-d} \end{pmatrix}, \\[2mm] W = (W_1 \quad \dots \quad W_m \quad V_1 \quad \dots \quad V_l). \end{cases}$$

References

1. Vaidyanathan, P.P.: The Theory of Linear Prediction. Morgan & Claypool (2008)
2. Benesty, J., Huang, Y., Chen, J.: Wiener and adaptive filters. In: Benesty, J., Sondhi, M.M., Huang, Y. (eds.) Springer Handbook of Speech Processing. Springer, Heidelberg (2008)
3. Yang, J., Roy, S.: Joint transmitter-receiver optimization for multi-input multi-output systems with decision feedback. IEEE Trans. Inf. Theory 40(5), 1334–1347 (1994)
4. Al-Dhahir, N., Sayed, A.H.: The finite-length multi-input multi-output MMSE-DFE. IEEE Trans. Signal Process. 48(10), 2921–2936 (2000)
5. Gutiérrez-Gutiérrez, J., Crespo, P.M.: Asymptotically equivalent sequences of matrices and Hermitian block Toeplitz matrices with continuous symbols: applications to MIMO systems. IEEE Trans. Inf. Theory 54(12), 5671–5680 (2008)

Energy Efficient Network Coding-based Cooperative ARQ Scheme for Wireless Networks

Angelos Antonopoulos and Christos Verikoukis

Telecommunications Technological Centre of Catalonia
Parc Mediterrani de la Tecnologia (PMT) - Building B4
Av. Carl Friedrich Gauss 7, 08860
Castelldefels (Barcelona), Spain
{aantonopoulos,cveri}@cttc.es

Abstract. In this paper we present an energy consumption model for a network coding-based cooperative ARQ (Automatic Repeat reQuest) scheme for wireless networks. Applying network coding techniques, we achieve a significant enhancement in the energy efficiency of the network without compromising the Quality of Service in terms of throughput and packet delay. In order to evaluate the proposed solution, we compare our protocol with simple cooperative ARQ protocols, where the retransmissions take place via relays without any network coding capabilities. Finally, both analytical and simulation results are used in order to verify our conclusions.

Keywords: Network Coding, Cooperative Networks, Automatic Repeat reQuest (ARQ), Energy Efficiency.

1 Introduction

Last years there is a trend towards designing energy efficient protocols, since power consumption is one of the most critical issues in mobile communications. The purpose of these new protocols is twofold: i) to extend the battery's life in the portable devices and ii) to efficiently use the environment's natural resources. Therefore, "green"[1] communications have become one of the hottest topics in the research community.

Network Coding [1],[2] is a technique that has been introduced in order to benefit the wireless communication in terms of throughput. However, minimizing the total number of transmissions, inherently implies savings in the energy consumption. The impact of network coding in "green" communications has been already studied in the literature, especially in broadcast and multi-cast scenarios [3]-[6].

The recent research work that investigates the energy aspect of network coding applications, focuses mostly on the network layer. In [7], an algorithm that generates a multipath route is presented. The simulation results confirm that

---

[1] "Green" refers to all environment-aware methods.

J. Del Ser et al. (Eds.): MOBILIGHT 2011, LNICST 81, pp. 37–47, 2012.

the energy consumption is lower compared to traditional simple path routes. Cui et al. [8] introduce CORP by using a suboptimal scheduling algorithm that exploits network coding opportunities, thus achieving a significant power saving over pure routing.

Towards this direction, and since network coding affects the MAC layer of the network as well, this paper presents an energy consumption model for a network coding-based scheme designed for cooperative wireless networks. Our main objective is to investigate the energy efficiency of network coding-based protocols and determine key ideas in order to optimize the design of such schemes, without degrading the provided Quality of Service (QoS).

The rest of the paper is organized as follows. Section 2 briefly describes the network coding-based cooperative ARQ (NCC-ARQ) scheme. In Section 3 we present our proposed energy consumption model for NCC-ARQ. The simulation set up and the results (both numerical and simulation) are provided in Section 4. Finally, Section 5 concludes the paper.

2   Network Coding-based Cooperative ARQ Scheme

2.1   Introduction

NCC-ARQ [9] is one of the fundamental works that implement network coding in cooperative ARQ schemes. In our previous work, it has been shown that NCC-ARQ improves the total throughput of the system, while at the same time the average packet delay is reduced. In this section we will briefly review the basic aspects of the protocol in order to make the paper self-standing and facilitate the comprehension of our proposed energy consumption model.

2.2   Protocol Description

When NCC-ARQ is applied in the network, all the nodes should operate in promiscuous mode in order to be able to capture all ongoing transmissions and cooperate if required. In addition, they should keep a copy of any received data packet (regardless of its destination address) until it is acknowledged by the destination station.

Whenever a data packet is received with errors at the destination node, a cooperation phase can be initiated. The error control could be performed by checking a cyclic redundancy code (CRC) attached to the header of the packet or any other equivalent mechanism. The cooperation phase is initiated by the destination station by transmitting a Call for Cooperation (CFC) message to the best relay in terms of channel conditions (i.e. SINR) after sensing the channel idle for a Short Interframe Space (SIFS) period. This message has the form of a control packet and higher priority over regular data traffic, since data transmissions in IEEE 802.11 take place after a longer period of silence (DIFS). Furthermore, in the special but not rare case of bidirectional traffic, when the destination station has a data packet for the source station, it transmits this packet piggybacked with the CFC message.

Upon the reception of the CFC, the helper node gets ready to forward its information. Since the relay has already stored the packets that destined both to the destination (the cooperative packet) and to the source (the piggybacked packet), it creates a new coded packet by combining the two existing data packets, using the XOR method. In this point we have to state that NCC-ARQ is backwards compatible with IEEE 802.11 Standard, as it uses the same frame structure and follows the same principles with the standard. However, there have been some modifications that are necessary in order for the protocol to exploit efficiently the advantages of using both cooperative and network coding techniques:

1. There is no expected ACK packet associated to each transmitted packet that is sent piggybacked with the CFC message.
2. In case of bidirectional traffic, the packet that is destined back to the source is sent along with the Call for Cooperation packet, without taking part in the contention phase.
3. There are ACK packets for the multicast transmission of the coded packet in order to provide a reliable communication scheme.

Once the source and the destination receive the network coded packet from the relay, they are able to decode it and extract the respective original data packets. Subsequently, they acknowledge the received data packet by transmitting the respective ACK, thus terminating the cooperation phase. In case that the received coded packets can not be decoded after a certain maximum cooperation timeout due to transmission errors, the relay is obliged to forward again the network coded packet. Figure 1 graphically demonstrates the general idea of our proposed NCC-ARQ scheme.

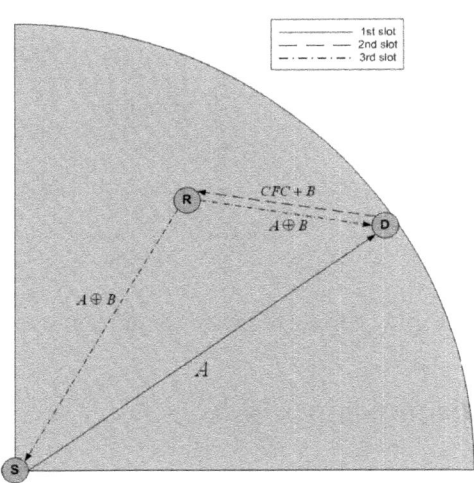

Fig. 1. General idea of NCC-ARQ scheme

2.3 Operational Example

In this subsection we provide a simple example in order to clarify the operation of the protocol. A basic network topology with 3 stations is considered, all of them in the transmission range of each other. A source station (S) transmits a data packet (A) to a destination station (D) that has also a packet (B) destined to the source station. There is also one relay (R) that has been chosen as the most appropriate helper node in terms of Signal to Interference-Noise Ratio (SINR) and supports this particular bidirectional communication. The whole procedure is depicted in Figure 2 and explained as follows:

1. At instant t_1, station S wins the contention phase and sends the data packet A to station D.
2. Upon reception, at instant t_2, station D fails to demodulate the packet A, thus transmitting a CFC packet to R along with the data packet B, destined to the station S.
3. At instant t_3, the relay R transmits the coded packet $A \oplus B$ to the nodes S and D simultaneously.
4. At instant t_4 the station D sends back an ACK packet since it is able to decode properly the XOR-ed packet and retrieve the original packet A.
5. At instant t_5 the node S acknowledges the packet B since it is able to decode properly the coded packet $A \oplus B$.

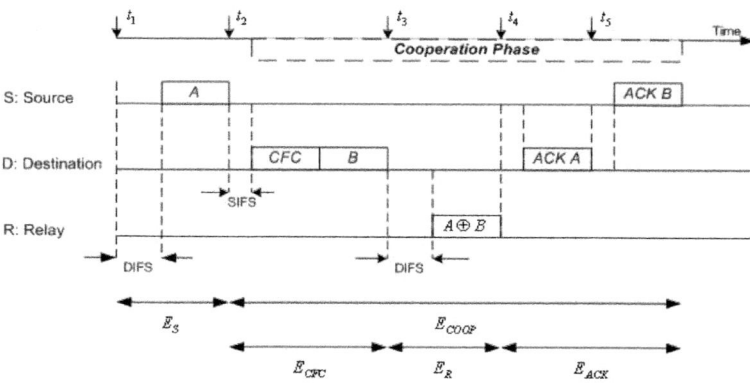

Fig. 2. NCC-ARQ example of operation

3 Energy Consumption Model

As we have already mentioned, the formed network consists of three nodes: the source, the destination and the relay. Considering the operation of NCC-ARQ protocol, we derive a closed-form expression that describes the power consumption in the network:

$$E_{TOTAL} = E_S + E_{COOP} \qquad (1)$$

where E_S and E_{COOP} are the energy consumptions during the initial transmission from the source and the cooperative phase, respectively. The term E_{COOP} could be further expressed as:

$$E_{COOP} = E_{CFC} + E[r] \cdot E_R + E_{ACK} \qquad (2)$$

where E_{CFC} represents the energy consumption during the beginning of the co-operation, E_R is the energy waste during the relay's transmission, while E_{ACK} is the energy that is consumed during the transmission of the acknowledgement (ACK) packets. The energy consumption during the particular time intervals is graphically demonstrated in Figure 2. $E[r]$ is the average number of the re-transmissions that are required in order to properly decode the X-OR packets at the destination nodes. It depends on the channel conditions and specifically on the packet error rate (PER) between the relay and the destination nodes. Lower values of PER imply higher probability for successful decoding of the packets at the destination nodes. This relationship could be mathematically expressed as:

$$E[r] = 1/(1 - PER_{R \to D}) \qquad (3)$$

In order to clarify the equation (1), we try to compute each term analytically. We consider three different modes:

1. **Transmission mode**, when the node is transmitting data/control packets
2. **Reception mode**, when the node is receiving data/control packets
3. **Idle mode**, when the node is sensing the medium, without performing any action.

The power levels associated to each mode are P_T, P_R and P_I respectively. Furthermore, the relationship between energy and power is given by $E = P \cdot T$, where the terms E, P and T represent the energy, the power and the time, respectively. Therefore, considering the network's topology, we have:

$$E_S = 3 \cdot P_I \cdot T_{DIFS} + P_T \cdot T_A + 2 \cdot P_R \cdot T_A \qquad (4)$$

$$E_{CFC} = 3 \cdot P_I \cdot T_{SIFS} + P_T \cdot T_{CFC} + 2 \cdot P_R \cdot T_{CFC} + P_T \cdot T_B + 2 \cdot P_R \cdot T_B \qquad (5)$$

$$E_R = 3 \cdot P_I \cdot T_{DIFS} + P_T \cdot T_{A \oplus B} + 2 \cdot P_R \cdot T_{A \oplus B} \qquad (6)$$

$$E_{ACK} = 2 \cdot 3 \cdot P_I \cdot T_{SIFS} + 2 \cdot P_T \cdot T_{ACK} + 2 \cdot 2 \cdot P_R \cdot T_{ACK} \qquad (7)$$

The above equations (4)-(7) are based on the following ideas:

– All stations remain idle during the SIFS and DIFS times.
– When a station transmits a packet (control or data), the rest stations are in promiscuous mode, thus capturing the packets.

4   Performance Results

In order to evaluate the performance of the NCC-ARQ we have developed an event-driven C++ simulator that executes the rules of the protocol. In this section we present the simulation set up and results of our experiments.

4.1   Simulation Scenario

We simulate an 802.11g network formed by a pair of transmitter-receiver (the two nodes are both transmitting and receiving data) and a relay node that facilitates the communication, all of them in the transmission range of each other. Furthermore, the relay node is able to perform network coding to its buffered packets before relaying them. In order to focus the analysis of the impact of both network coding and cooperative communication on energy consumption, the following assumptions have been made:

1. The traffic is bidirectional, i.e. the destination node has always a packet destined back to the source node.
2. Original transmissions from source to destination are always received with errors, thus initiating a cooperative phase.
3. The packet error rate (PER) - and consequently the required number of retransmissions that have to be made by the relay until the packets received correctly - is known in advance.

The configuration parameters of the stations in the network are summarized in Table 1 considering the IEEE 802.11g PHY layer [10]. Ebert et al. [11] have measured the power consumption of a wireless interface during the transmission and reception phase. Based on their work, we have chosen the following power levels for our scenarios: $P_T = 1900mW$, $P_R = P_I = 1340mW$. The value of P_T has been selected as an average value of transmission consumed power, since it varies according to the Radio Frequency (RF) power level.

Table 1. System Parameters

Parameter	Value	Parameter	Value
MAC Header	34 bytes	PHY Header	96 μsec
DIFS	50 μsec	SIFS	10 μsec
CFC, ACK	14 bytes	P_T	1900 mW
P_R	1340 mW	P_I	1340 mW

In order to evaluate the energy performance of our proposed solution, we consider various scenarios with different SNR values between the original source and the destination. The control packets are transmitted always at the rate of 6 Mb/s, while the transmission rate for the data packets is 6, 24 and 54 Mb/s for low, medium and high SNR values, respectively. On the other hand, given that the relays are usually placed close to the destination, we assume that the transmission rates for all scenarios are 6 and 54 Mb/s for control and data packets, respectively. The three different scenarios are summarized in Table 2. Furthermore, the network operates under saturated conditions, which means that the nodes have always packets to send in their buffers.

Table 2. Simulation Scenarios

SNR (S-D)	Source Control Rate	Source Data Rate	Relay Control Rate	Relay Data Rate
Low	6Mb/s	6Mb/s	6Mb/s	54Mb/s
Medium	6Mb/s	24Mb/s	6Mb/s	54Mb/s
High	6Mb/s	54Mb/s	6Mb/s	54Mb/s

In this point, we have to describe the transmission procedure in simple cooperative schemes (C-ARQ), where the bidirectional communication takes place in two steps. In the first step, node S sends the packet to D and, upon the erroneous reception, D transmits the CFC packet, thus triggering the relay to retransmit the packet. In the second step, node D transmits its own packet to S and the same procedure as in the first step is repeated, thus consuming valuable network resources.

In order to evaluate the performance of NCC-ARQ compared to simple cooperative schemes in terms of energy, we use as metric the energy efficiency of a protocol, which was introduced in [12]. The energy efficiency (denoted by η) is defined as:

$$\eta = \frac{total\ amount\ of\ useful\ data\ delivered\ (bits)}{total\ enery\ consumed\ (Joule)} \tag{8}$$

Before proceeding to the simulation results, it is worth mentioning that the definition in equation (8) inherently implies that network coding benefits the energy efficiency of a protocol, as the number of the delivered bits increases by combining different data packets.

4.2   Simulation Results

As it has been already mentioned, it is of paramount importance to save energy without decreasing the performance of the network. The simulation results that are plotted in Figures 3 and 4 show that using our proposed scheme, we are able to enhance the QoS of the network, compared to simple cooperative schemes. Figure 3 depicts the system's aggregated throughput for the three described scenarios in both NCC-ARQ and C-ARQ protocols. We can see that in all cases our proposed solution outperforms the simple cooperative ARQ scheme, since the throughput is greatly increased. On the other hand, regarding the packet delay, we notice that using the NCC-ARQ scheme, the total packet delay is decreased, as it is shown in Figure 4.

Figures 5 and 6 show that our analysis verifies the simulation results with regard to the energy performance. Comparing our proposed network coding-based scheme with simple cooperative protocols for different number of retransmissions (and consequently different PER between the relay and the destinations), we observe that our scheme is more energy efficient than non-network-coding-based schemes, since more bits are delivered over the same amount of consumed energy. Keeping constant the data packet length (1500 bytes), the energy efficiency of NCC-ARQ is decreased as the number of relay retransmissions grows. However, the difference with simple cooperative schemes remains steadily over 80% (Figure 5).

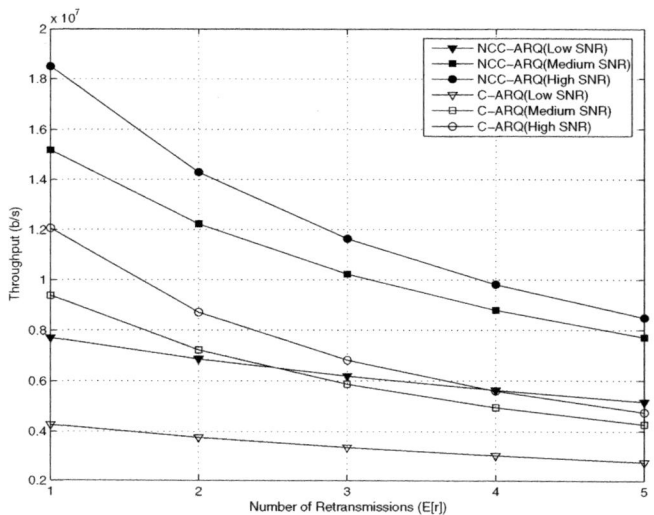

Fig. 3. System's Throughput (NCC-ARQ vs C-ARQ) (Packet Payload=1500 bytes)

Specifically, we can observe that NCC-ARQ outperforms the classic cooperative ARQ schemes under the same conditions (i.e. when both schemes operate under similar SNR values). However, it is worth noticing that NCC-ARQ outperforms C-ARQ even for worse SNR scenarios. To make it clear, observing the energy performance of the NCC-ARQ for the *medium SNR scenario* we can see that it clearly outperforms the C-ARQ under the *high SNR scenario*. Furthermore, comparing the *low SNR scenario* for NCC-ARQ with the *medium SNR scenario* for C-ARQ, we see that NCC-ARQ acts more efficiently in terms of energy when the channel conditions between the relay and the destinations are not good (i.e. when the packets need to be retransmitted three or more times).

Figure 6 shows the energy efficiency for both network coding-based and simple cooperative ARQ schemes using data packets of different size. The simulations have been conducted considering the *low SNR scenario* (i.e. low data rate between the source and the destination). In all cases we consider that one retransmission is needed ($E[r] = 1$) in order for the packets to be correctly received by the respective destinations. As it was expected, it can be observed that the bigger the data payload, the higher the energy efficiency of the protocol (up to 70%), since more bits are delivered in one transmission cycle. However, it has to be pointed that the gain we get in case of NCC-ARQ is significantly higher than the other schemes, since the delivered bytes in each transmission cycle are doubled because of the network coding techniques. It is also worth noticing that

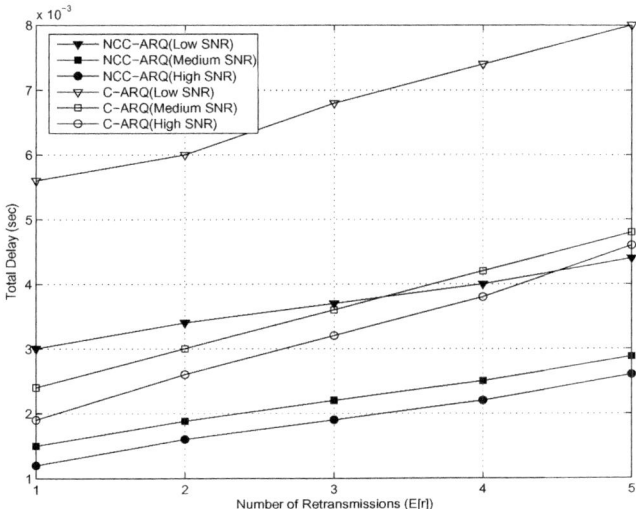

Fig. 4. Packet Delay (NCC-ARQ vs C-ARQ) (Packet Payload=1500 bytes)

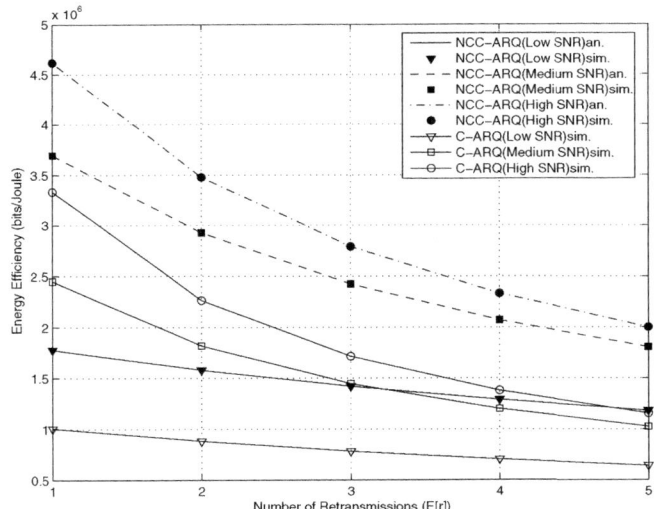

Fig. 5. Energy Efficiency vs Number of Retransmissions (NCC-ARQ vs C-ARQ) (Packet Payload=1500 bytes)

our protocol outperforms simple cooperative protocols in all cases, thus proving a better and more efficient approach in terms of energy management.

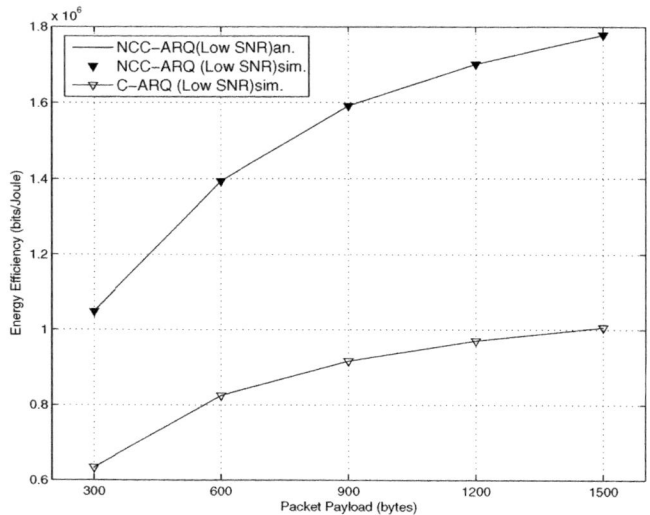

Fig. 6. Energy Efficiency vs Packet Payload (NCC-ARQ vs C-ARQ) (E[r]=1)

5   Conclusion

In this paper, an energy consumption model for a network coding-based cooperative ARQ (NCC-ARQ) scheme is presented. Compared to simple cooperative schemes, our protocol achieves significant enhancement in terms of power consumption, since it has been proven to be up to 80% more energy efficient, without degrading the offered Quality of Service.

Furthermore, we have shown that the energy efficiency is decreased with respect to the number of retransmissions, while we achieve better results by using bigger data packets. In order to optimize the energy management in the network, MAC schemes have to be combined with energy-aware routing solutions, while sleep mode for inactive nodes should be considered as an extra option. Our future research will be focused on such issues.

Acknowledgments. This work has been funded by the Research Projects GREENET (PITN-GA-2010-264759), CO2GREEN (TEC2010-20823) and R2D2 (CP6-013).

References

1. Ahlswede, R., Cai, N., Li, S.-Y.R., Yeung, R.W.: Network information flow. IEEE Transactions on Information Theory 46(4), 1204–1216 (2000)
2. Li, S.-Y.R., Yeung, R.W., Cai, N.: Linear network coding. IEEE Transactions on Information Theory 49(2), 371–381 (2003)

3. Mansouri, H.S., Pakravan, M.R.: Reliable and energy efficient single source broadcasting using network coding in wireless Ad-hoc networks. In: IEEE International Conference on Telecommunications and Malaysia International Conference on Communications (ICT-MICC), May 14-17, pp. 81–85 (2007)
4. Wu, Y., Chou, P.A., Kung, S.-Y.: Minimum-energy multicast in mobile ad hoc networks using network coding. IEEE Transactions on Communications 53(11), 1906–1918 (2005)
5. Hosseinmardi, H., Lahouti, F.: Online multicasting using network coding in energy constrained wireless ad hoc networks. In: 3rd International Symposium on Wireless Pervasive Computing (ISWPC), May 7-9, pp. 545–549 (2008)
6. Kim, S., Ho, T., Effros, M.: Network coding with periodic recomputation for minimum energy multicasting in mobile ad-hoc networks. In: 46th Annual Allerton Conference on Communication, Control, and Computing, September 23-26, pp. 154–161 (2008)
7. Tao, X., Zhang, C., Lu, J.: Network Coding for Energy Efficient Wireless Multimedia Transmission in Ad Hoc Network. In: International Conference on Communication Technology (ICCT), November 27-30, pp. 1–5 (2006)
8. Cui, T., Chen, L., Ho, T.: Energy Efficient Opportunistic Network Coding for Wireless Networks. In: The 27th Conference on Computer Communications. IEEE INFOCOM 2008, April 13-18, pp. 361–365 (2008)
9. Antonopoulos, A., Verikoukis, C.: Network Coding-based Cooperative ARQ Scheme. In: Proc. of IEEE International Conference in Communications (ICC 2011), Kyoto, Japan, June 5-9 (2011)
10. IEEE 802.11g WG, Part 11: Wireless LAN Medium Access Control (MAC) and Physical Layer (PHY) specifications - Amendment 4: Further Higher Data Rate Extension in the 2.4 GHz Band (June 2003)
11. Ebert, J.-P., Aier, S., Kofahl, G., Becker, A., Burns, B., Wolisz, A.: Measurement and Simulation of the Energy Consumption of a WLAN Interface, Technical University of Berlin,Telecommunication Networks Group, Tech. Rep. TKN-02-010 (June 2002)
12. Zorzi, M., Rao, R.R.: Energy-constrained error control for wireless channels. IEEE Personal Communications 4(6), 27–33 (1997)

A Real-Time FPGA-Based Implementation of a High-Performance MIMO-OFDM Mobile WiMAX Transmitter[*]

Oriol Font-Bach[1], Nikolaos Bartzoudis[1],
Antonio Pascual-Iserte[1,2], and David López Bueno[1]

[1] Centre Tecnològic de Telecomunicacions de Catalunya (CTTC),
Parc Mediterrani de la Tecnologia (PMT), Av. Carl Friedrich Gauss 7,
08860 Castelldefels, Barcelona, Spain
{ofont,nbartzoudis,dlopez}@cttc.cat
[2] Department of Signal Theory and Communications,
Universitat Politècnica de Catalunya (UPC), Campus Nord, Jordi Girona 1-3,
08034 Barcelona, Spain
antonio.pascual@upc.edu

Abstract. The Multiple Input Multiple Output (MIMO)-Orthogonal Frequency Division Multiplexing (OFDM) is considered a key technology in modern wireless-access communication systems. The IEEE 802.16e standard, also denoted as mobile WiMAX, utilizes the MIMO-OFDM technology and it was one of the first initiatives towards the roadmap of fourth generation systems. This paper presents the PHY-layer design, implementation and validation of a high-performance real-time 2x2 MIMO mobile WiMAX transmitter that accounts for low-level deployment issues and signal impairments. The focus is mainly laid on the impact of the selected high bandwidth, which scales the implementation complexity of the baseband signal processing algorithms. The latter also requires an advanced pipelined memory architecture to timely address the datapath operations that involve high memory utilization. We present in this paper a first evaluation of the extracted results that demonstrate the performance of the system using a 2x2 MIMO channel emulation.

Keywords: MIMO, testbeds, IEEE 802.16e, real-time systems, space-time coding, FPGAs, DSP.

1 Introduction

The most innovating PHYsical layer (PHY) algorithms for Multiple Input Multiple Output (MIMO) systems supporting Orthogonal Frequency-Division Multiplexing (OFDM) are usually deployed using testbeds that operate in off-line

[*] This work was partially supported by the European Commission under projects NEWCOM++ (216715) and BuNGee (248267); by the Catalan Government under grants 2009 SGR 891 and 2010 VALOR 198; and by the Spanish Government under project TEC2008-06327-C03 (MULTI-ADAPTIVE) and Torres Quevedo grants PTQ-08-01-06441, PTQ06-02-0540, PTQ06-2-0553.

J. Del Ser et al. (Eds.): MOBILIGHT 2011, LNICST 81, pp. 48–66, 2012.
© Institute for Computer Sciences, Social Informatics and Telecommunications Engineering 2012

mode. The PHY layer algorithms of such experimental MIMO testbeds are commonly designed with the help of system-modelling tools such as Matlab, whereas the rest of the testbed comprises Commercial-Off-The-Shelf (COTS) equipment that are responsible for the Radio Frequency (RF) signal up and down conversion and the signal synthesis and acquisition (i.e., Digital to Analog Conversion, DAC, and Analog to Digital Conversion, ADC). Although this implementation approach enables the rapid prototyping, it is lacking an insight to the implementation complexity and limitations that characterize the deployment of a real-time MIMO-OFDM system. Moreover, the focus of the offline baseband implementations is limited to the analysis of static testing scenarios (i.e., real-time system adaptivity is not feasible), which is commonly constrained to processing short data frames. Hence, offline MIMO testbeds are not suitable for the implementation of high capacity systems whose response has to be adapted according to dynamically changeable parameters or features. This consequently renders the MIMO testbeds that operate in offline mode unsuitable for exploring systems employing Adaptive Modulation and Coding (AMC) or closed-loop communication schemes. On the other hand, real-time testbeds allow the validation and analysis of such systems in realistic environments accounting for both hardware limitations and software or low-level programming constraints. The massive parallelism required for the baseband signal processing in real-time MIMO-OFDM testbeds, makes the Field Programmable Gate Array (FPGA) devices an appropriate candidate for implementing such systems.

This paper presents the PHY-layer design, implementation and validation of a high performance real-time MIMO-OFDM transmitter that was deployed using FPGA technology (based on the IEEE 802.16e-2005[1] standard). The transmitter is able to host two transmit antennas, features a 20 MHz bandwidth and uses matrix A encoding based on Alamouti's Space-Time Block Code (STBC)[2] on a per carrier basis. The PHY-layer algorithms of the transmitter were modelled in Matlab, designed in VHDL and implemented using a real-time FPGA platform. The GEDOMIS® (GEneric hardware DemOnstrator for MIMO Systems) testbed has facilitated the functional validation of the transmitter under realistic operating conditions. GEDOMIS® is a multi-antenna wireless communication testbed that enables the prototyping and evaluation of MIMO PHY algorithms. In the past, it has been used for the PHY-layer implementation of a real-time MIMO-OFDM mobile WiMAX receiver [3]. The transmitter in that case was emulated by using two Vector Signal Generators (VSG; containing an arbitrary waveform generator), each one of which was loaded with a Matlab-based vector file. These files are the output of a Matlab model that emulates the baseband algorithms of the MIMO-OFDM mobile WiMAX transmitter.

The principal issue that this paper aims to address is the absence in the literature of high bandwidth real-time implementations of MIMO-OFDM mobile WiMAX systems that account for low-level implementation issues and signal impairments. Thus, the work presented herein underlines the complexity of such an undertaking in the design, verification and debugging stages. At the same time, our work offers an insight into the critical design issues that should be accounted in similar real-time FPGA-based deployments. Particular focus is given on the impact of the selected high bandwidth in the implementation of the baseband signal processing algorithms in terms of memory utilization and the interfacing of the baseband processing components with the DACs. In fact, the 20 MHz channel bandwidth is one of the key changes that is introduced in the WiMAX 2 standard (IEEE 802.16m), to provide compliance with the prerequisites of fourth generation (4G) wireless communication systems.

2   Related Work

A great number of MIMO testbeds that were encountered in the literature are exclusively assembled by COTS equipment [4,5,6]. Nevertheless, these otherwise excellent development and testing environments do not allow the in-depth evaluation of the PHY-layer algorithms. This is due to the fact that their baseband part is a black box that cannot be modified or extended, limiting by this way their overall scope (i.e., conducting measurement campaigns and performance benchmarking). The PHY-layer implementation of our MIMO-OFDM mobile WiMAX transmitter can be easily expanded to include future WiMAX features or apply experimental PHY-layer concepts.

In the case of MIMO testbeds having custom PHY-layer implementations, it is important to make a distinction between the baseband signal processing deployments that operate in real-time and those that feature off-line functionality. The latter are software-based (e.g., Matlab) baseband implementations [7,8,9] which, as already mentioned before, are not suitable for validating certain testing scenarios that require adaptive signal processing. Also, the assessment of their deployment feasibility is impaired by the absence of real-world baseband signal processing implementation. On the other side, our MIMO-OFDM mobile WiMAX transmitter goes beyond the modelling of PHY-layer algorithms in Matlab; that is for, apart from the challenging task of mapping the algorithms to custom VHDL code, the system-deployment included the board-level code integration (e.g., interpolation filters, DAC ICs programming) and the real-time debugging under realistic laboratory conditions. The latter implied the use of heterogeneous equipment with critical system integration features such as a MIMO-enabled channel emulator.

Various authors have presented papers regarding real-time PHY-layer implementations of WLAN and LTE MIMO tranmsitters [15,16,17,18]. However, we have encountered relative few mobile WiMAX-related PHY-layer implementations available in the literature. In most of the cases the work described in such papers mainly focuses on partial FPGA implementations of specific mobile WiMAX PHY-layer algorithms, that either lack real-time operation or do not evaluate mobile MIMO channels (i.e., absence of either field testing results or laboratory results using a channel emulator). In [10] a low bandwidth FPGA-based 2x2 MIMO fixed WiMAX testbed is used to carry out a channel measurement campaign in order to analyse its capacity. The authors in [11] implemented in a FPGA a 2x2 MIMO Fixed Sphere Decoder (FSD), while the rest of the

PHY-layer was developed in a Matlab model, allowing by this way rapid co-simulations with different FFT sizes. Another FPGA-based implementation of a sphere detector and a channel matrix pre-processor for a 4x4 MIMO system is presented in [12]. In [13] a 2x2 MIMO system with 10 MHz bandwidth is implemented using a cell processor, with the aim to provide a single chip baseband implementation of the IEEE 802.16e OFDMA PHY for a base station transceiver. This system cannot be accurately validated, since it is not accounting for an appropriately applied channel model. Finally, a 2x2 MIMO system targeting a FPGA implementation is presented in [14]. The performance of the baseband transceiver implementation cannot be assessed in the absence of important system characteristics since it mainly presents partial simulation or emulation results.

3 Design of the Real-Time Transmitter

3.1 System Features

The IEEE 802.16e-2005 standard specifies a system for combined fixed and mobile broadband wireless access. The specified PHY supports scalable OFDMA architectures, AMC, various subchannelization permutation techniques and MIMO-aided transmit/receive diversity. However, our system is only adopting a fixed subset of this flexible configuration. In short, the main system parameters are as follows: 2 transmit and 2 receive antennas, RF frequency at 2.595 GHz, Intermediate Frequency (IF) at 67.2 MHz, fixed channel bandwidth of 20 MHz and QPSK modulation. It has to be noted that although this paper presents a MIMO-OFDM mobile WiMAX transmitter that supports only the Partial Usage of Subchannels (PUSC) permutation scheme, the AMC permutation scheme has also been implemented and validated for an equivalent SISO system, but it is not presented herein. It is also useful to mention that channel coding has not been included in our system. The WiMAX signal comprises various frames, which encapsulate user data, separated by equivalent silence periods. The OFDM frame used in our testbed is composed of a single burst with a fixed predefined format (i.e., the FCH and DL-MAP are not used).

3.2 Design, Implementation and Validation Methodology

The challenging inter-operability issues related to setting-up, calibrating and configuring a high-capacity real-time MIMO testbed require an efficient design, implementation and system-validation strategy. Hence, we present in this section a development-flow that was iteratively improved during the deployment of the present system.

The first vital step was to define the specifications of the transmitted signal and the respective channel models. After the completion of this stage, the initial selection of the MIMO signal processing algorithms was conducted, together with the definition of the basic system architecture. Another important

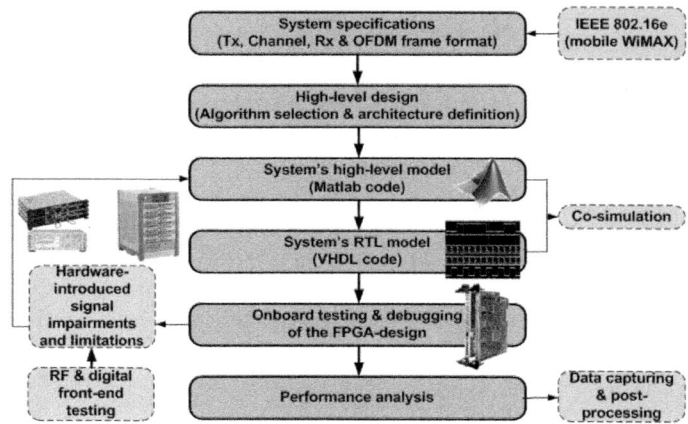

Fig. 1. Proposed development flow for a real-time MIMO testbed

step involved the estimation of the computational complexity of the PHY-layer algorithms (i.e., having as a reference the target FPGA device). The result of this estimation was used to re-adjust the scalability of certain system features.

A: Emulated transmitter (high-level system modelling)
This first evaluation stage was significant because it allowed us to start developing a precise Matlab model of the system, where the fixed-point logic was emulated by applying specific quantizations. This model was a vehicle to rapidly verify the behaviour of the designed architecture. The selected algorithms were optimized according to their estimated computational complexity. The GEDOMIS® testbed facilitated the rapid validation of the Matlab model. In more detail, the output of the baseband Matlab model of the MIMO-OFDM mobile WiMAX transmitter was stored in two files and uploaded in two VSGs, appropriately configured for MIMO signal transmission. The VSGs were then used to play back in real-time the MIMO signal and up-convert it to the desired RF frequency (i.e., 2.595 GHz). Next, the MIMO signal was fed to the channel emulator and finally it was demodulated using a multichannel oscilloscope that is remotely accessed by a certified mobile WiMAX software package (i.e., Agilent VSA with the IEEE 802.16e-2005 option). After validating the conformance with the mobile WiMAX standard, the receiver's data acquisition board was used to capture again the received signal (i.e., higher ADC resolution compared to the oscilloscope data-captures).

B: Real-time FPGA-based MIMO-OFDM signal transmission
Once the Matlab model satisfied the defined set of system requirements, every processing block of this model was mapped to RTL code (i.e., VHDL model). A very important stage of the design-validation flow was the Matlab/VHDL co-simulation. The latter offers a reliable comparison method that proves the

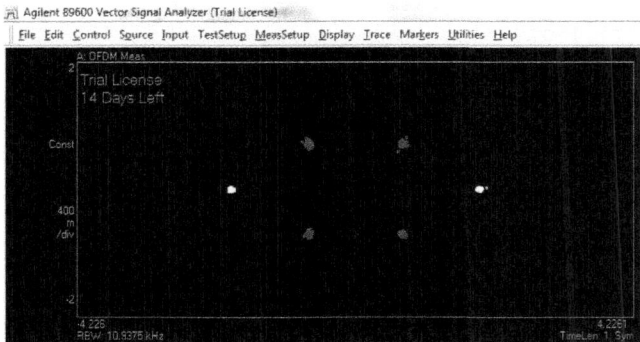

Fig. 2. Demodulation of the MIMO signal transmitted by CTTC's real-time FPGA-based MIMO-OFDM mobile WiMAX transmitter, using Agilent's VSA signal analysis software

functional and behavioural validity of the RTL-model and provides an assessment of the fixed-point implementation's precision.

The following stage included the implementation of the VHDL design targeting a specific FPGA-based processing platform. This implied the interfacing of the MIMO-OFDM mobile WiMAX transmitter with various on-board processing components using VHDL code (e.g., DACs, SDRAMs). Once the FPGA binary executable was produced, we were able to test the MIMO-OFDM mobile WiMAX transmitter using the entire equipment set-up of the GEDOMIS® testbed. A similar mobile WiMAX standard conformity test was conducted following the steps described before. However, this time the Matlab model was replaced by the real-time FPGA-based implementation of the equivalent baseband algorithms (a snap-shot of the demodulated signal is seen in figure 2). It has to be underlined that the same MIMO-OFDM mobile WiMAX signal generated by our FPGA-based transmitter, it was demodulated using the respective receiver, which we have developed previously [3] (figure 9 shows the demodulated data in a QPSK constellation). The real-world FPGA-based implementation of the MIMO-OFDM mobile WiMAX transmitter was debugged on the fly through vendor-specific tools and traces were captured to provide the final refinement of the system model.

C: Performance analysis
A final step of the presented design flow included the performance validation of the transmitter. Large buffers at the receiver side were used to capture data. This facilitated a detailed off-line analysis and characterization of the system performance. The extracted real-world performance metrics can be compared to the ones achieved by the Matlab/VHDL co-simulations, providing a quantification of the losses introduced by the hardware and by the RTL-implementation.

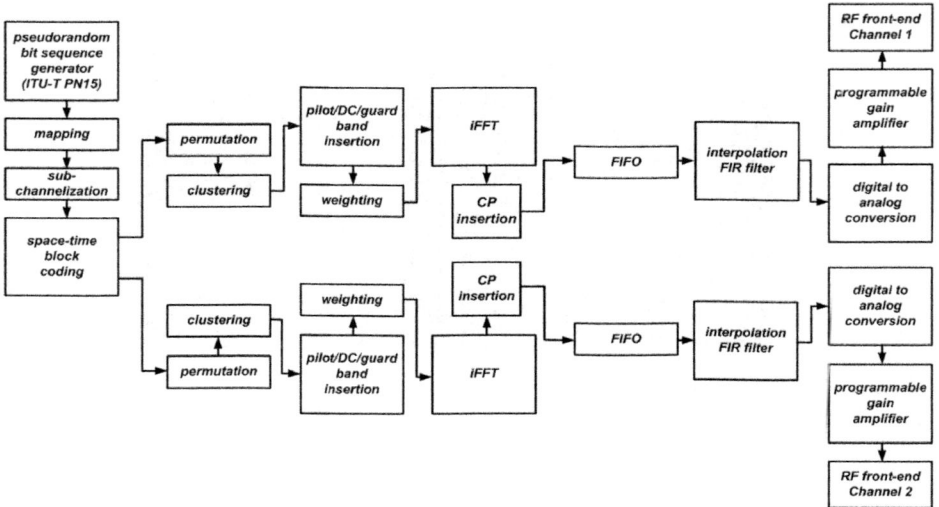

Fig. 3. General architecture of the real-time mobile WiMAX MIMO transmitter

3.3   General Architecture of the Transmitter

The general architecture of the system is depicted in figure 3. The first processing block of the real-time mobile WiMAX MIMO transmitter is a PseudoRandom Binary Sequence (PRBS) generator (i.e., based on the ITU-T PN15 specification [19]). Since our OFDM-frame is constructed by 46 data-symbols (using the PUSC scheme), each containing 2880 bits, the generated sequence will be composed of 132480 pseudo-random bits. The modulator then maps the generated sequence into QPSK constellation points.

At this stage, it is important to distinguish the operations related to the mobile WiMAX standard with those signal processing operations that are commonly encountered in MIMO-OFDM transmitters. The first type of signal processing operations define OFDM symbols that make use of the PUSC scheme according to the IEEE 802.16e standard:

- *Subchannelization*: the subchannels are created at the output of the modulator by distributing the sequence of QPSK symbols in two consecutive OFDM symbols in an interleaved fashion. This means that an adjacent group of 24 complex values is allocated in the first symbol and another one is allocated in the second symbol; this process is recursively repeated until the whole capacity of both symbols is filled up. The subchannelization process continues with the next two symbols, until the whole OFDM frame is constructed. The subchannelization process is shown in figure 4a.
- *Permutation*: logical structures of 12 adjacent subcarriers, named clusters, are created by scrambling the outputs of the STBC processing component. Additionally, the subchannels are grouped in 6 larger structures, namely Major Groups (MG) that have to be taken into account during this permutation

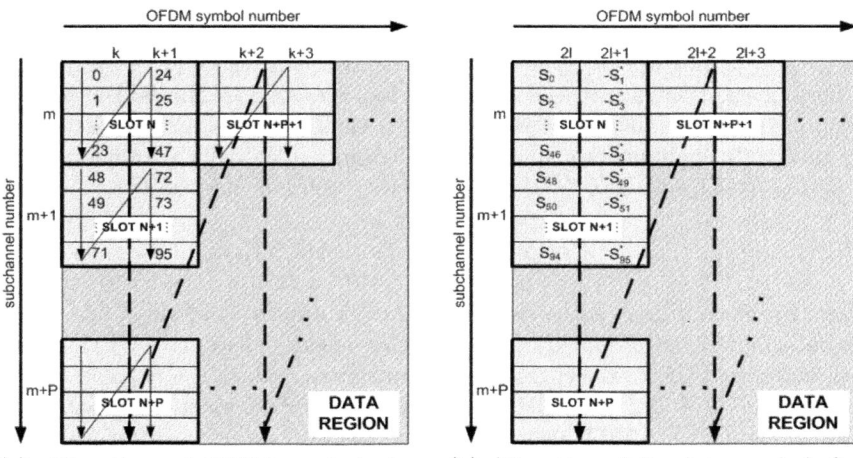

(a) Allocation of QPSK symbols to subcarriers and subchannels.

(b) Allocation of the data symbols for transmit antenna 1 according to matrix A (i.e., similar for transmit antenna 2).

Fig. 4. Mobile WiMAX standard-related operations applied on PUSC structured OFDM symbols

process. The mobile WiMAX standard defines the following permutation formula for the PUSC-structured OFDM symbols:

$$scr(k, s) = N_{sch} \cdot n_k + permbase[(s + n_k)modN_{sch}], \qquad (1)$$

where scr is the index of the subcarrier within the OFDM symbol, $n_k = (k + 13s)mod24$, N_{subch} is the number of subchannels within the MG (i.e., it can be either 8 or 12, depending on the parity of the MG's index), $permbase$ is a predefined subchannel permutation mask (i.e., defined for both lengths of the MG), s is the subchannel index within the MG, and k is the subcarrier index within the subchannel (i.e., up to 24).

– *Clustering*: the previously created clusters are permuted according to a predefined renumbering sequence (i.e., interchanging of adjacent subcarrier groups).
– *Pilot/DC/guard-band insertion*: two pilot subcarriers are inserted to each cluster in given positions that depend on the parity of the OFDM symbol's index. The DC is inserted in the position 1024 of each OFDM symbol. Finally, two sets of null-carriers (with fixed-length) are inserted in the beginning and in the end of each symbol of the OFDM frame.
– *Weighting*: an additional subcarrier-randomization is applied (i.e., some subcarriers will be inverted), using the PRBS generator specified by the WiMAX standard, which applies a homogeneous distribution of the transmitted power.

The remaining of the signal processing operations at the transmitter are commonly found in MIMO-OFDM implementations. These include the inverse FFT that transforms the frequency-domain signal to a time-domain signal, and the Cyclic Prefix (CP) insertion, which complements each OFDM symbol in the frame (i.e., a portion of each symbol is repeated in the end of each symbol). A fundamental feature of the system is implemented in the STBC processing component, which allocates the carriers for each transmit antenna according to the matrix A configuration specified in the mobile WiMAX standard, as shown in figure 4b. This matrix for the case of two antennas is given by:

$$A = \begin{bmatrix} S_N & -S_{N+1}^* \\ S_{N+1} & S_N^* \end{bmatrix}, \tag{2}$$

where $S_1, S_2, ..., S_{66240}$ is the single stream of complex symbols at the output of the subchannelization block; the columns represent consecutive OFDM symbols (i.e., time) and the rows represent the transmit antennas (i.e., spatial streams).

After the insertion of the CP, several OFDM symbols must be stored in a large First In First Out (FIFO) memory in order to provide an uninterrupted flow of data to the DAC device. Although it is not shown in figure 3, dedicated-logic has been designed to control the synchronous communication between the different processing components. In other words, this logic is responsible for counting the length of each inter-frame silence period (during which null-carriers will be transmitted) and trigger the FIFO to provide the stored symbols in a strict timing manner. An additional OFDM symbol (i.e., preamble) must be inserted to facilitate the synchronization process at the receiver. The preamble is calculated off-line, stored in a RAM memory and read just before the first FIFO outputs are available.

3.4   Intelligent Memory System

The real-time implementation of the MIMO-OFDM mobile WiMAX transmitter requires a new bit to be generated at each clock cycle (i.e., a new QPSK symbol is generated each two clock cycles). At the same time, both the generic MIMO-OFDM processing stages and the operations related to the WiMAX standard are applied to the generated symbols. The implementation complexity of all these signal processing operations (e.g., symbols need to be permuted, inverted and grouped in logical structures) and the high bandwidth that was selected (i.e., 20 MHz, which corresponds to an FFT size of 2048 points), imposed a pipelined memory-structure with high throughput. The latter involved the storage of large data-sets in numerous intermediate stages that require low-latency interfacing. This was achieved by designing a generic memory structure for each intermediate stage in the processing chain, comprising an advanced memory controller and an adaptive memory block as shown in figure 5.

The operation of the transmitter involves the storing of large sequences of complex values that comprise two 16-bit words (i.e., real and imaginary parts - in-phase and quadrature, I&Q, components). The length of this sequence varies according to the operations applied in each processing stage (i.e., additional sub-carriers are inserted in various stages of the baseband processing chain). There-fore, the first step towards the design of a generic memory structure was the

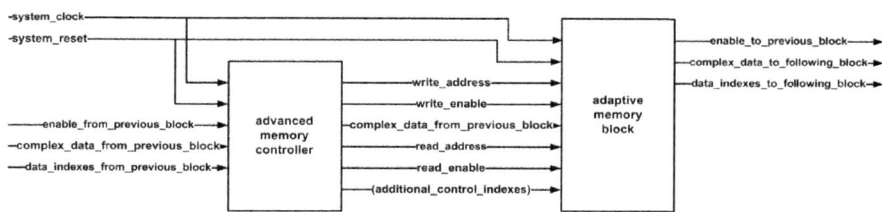

Fig. 5. Generic memory structure for each signal processing stage

implementation of adaptive memory blocks. These include a mechanism that enables the adjustable instantiation and grouping of numerous FPGA RAM-block primitives as a single memory entity, offering by this way a simplified interface for seamless memory-access. Another important benefit of this mechanism is that it allows the simultaneous read and write operations of the OFDM-symbols, minimizing the memory-access latency and optimizing the usage of the embedded RAM blocks of the target FPGA device. The adaptive memory block includes additional functionality that enables the concurrent execution of complementary baseband operations with the storage and retrieval of the data sequences (featuring minimal latency). This top-up functionality is triggered by the advanced memory controllers using dedicated control signalling.

The operations related to the mobile WiMAX standard are basically devoted to reordering the data at each processing stage. The advanced memory controllers are therefore abstracting away the complexity of the required access to the embedded-memory elements. This is achieved by providing transparent memory read and write operations that allow the processing elements of the transmitter to continue performing data-manipulation operations. Moreover, the synchronous control of the FPGA RAM blocks composing each adaptive memory block is not introducing any impairment to the distinct processing stages. Thus, the main function of the advanced memory controllers is to optimize the read and write operations by implementing the required scrambling-operations with the minimum possible latency. During the write operations the symbol-reordering is applied with no latency by storing each incoming QPSK symbol in the indicated position. The latter is calculated on-the-fly according to the description given by the mobile WiMAX standard for the respective operation. For instance, the value of the index received with each symbol's complex value will be introduced together with additional control indexes (that are automatically generated), to a dedicated logic-component that implements the permutation formula. The result of this operation indicates the position where the given complex value needs to be stored (i.e., position within the OFDM symbol). Once a whole data stream (i.e., 1440 QPSK symbols for a PUSC-structured OFDM symbol) has been stored, the read operation will be triggered simultaneously with the next write operation.

It has to be noted that the read operations take into account the interrelations between the different OFDM symbols in a frame (i.e., in the PUSC permutation scheme, the pilot-tone distribution is applied in groups of two OFDM symbols). This in more detail means that the read operations are used to apply additional functions to the QPSK symbol sequence, before reaching the following stage. These functions include additional scrambling, zero latency subcarrier insertion (i.e., pilots, DC, guard-band), subcarrier weighting, insertion of the CP after the inverse FFT (i.e., allowing cyclic memory reading) and cyclic shift of the data-symbols inserted in the FIFO (i.e., the DC carrier is to be found in position 0). As previously mentioned, additional dedicated logic, memory-initialization values or predefined masks have been included within the adaptive memory blocks, accommodating by this the latter functionality.

4   Implementation and Integration

4.1   Testbed Setup

A graphic-overview of the GEDOMIS® testbed setup featuring a 2x2 MIMO con-figuration is shown in figure 6. The whole transceiver is implemented in a high-end signal processing development platform equipped with Lyrtech's DAC, ADC and FPGA/DSP boards. The VHS-DAC board features 8 phase synchronous channels with 14-bit DAC devices. The board also includes a Xilinx Virtex-4 FPGA device that hosts the complete PHY-layer of the real-time MIMO-OFDM mobile WiMAX transmitter. Similarly, the VHS-ADC board, features 8 channels with 14-bit ADC devices and a Xilinx Virtex-4 FPGA device that partially hosts the PHY-layer implementation of our MIMO-OFDM mobile WiMAX receiver. The signal acquisition board also includes a digitally controlled Programmable Gain Amplifier (PGA) for each channel, facilitating by this the development of automatic gain control algorithms. Finally, the SignalMaster Quad board has four TSM320C6416 DSPs and two Xilinx Virtex-4 FPGAs. One of these FPGAs is hosting the remaining signal processing blocks of the receiver.

The IF outputs generated by the DAC components of the MIMO-OFDM mo-bile WiMAX transmitter are fed in two ESG4438C VSG instruments, which up-convert the signal and provide the RF output centered at 2.595 GHz. The Elektrobit Propsim C8 radio channel emulator, allows accurate multi-channel emulation of custom or standardized models at the laboratory. This is feasible by adding complex and time-varying effects of multipath and Doppler-shifts in the digital domain (e.g., adjusting the tap amplitude, delay spread, operation frequency and mobile speed). For the 2x2 MIMO scenario considered in this paper, four uncorrelated multipath fading channels were created (with different distribution seeds), using either the ITU Vehicular A or the ITU Pedestrian B channel model [20]. The receiver's RF front-end from Mercury Computer Sys-tems (Echotek Series RF 3000 Tuners) comprises 1 Direct Digital Synthesizer (DDS) and 4 receiver modules. This instrument features high spectral purity and dynamic range and applies phase-coherent down-conversion of the RF signal to the system's IF frequency. Finally, two uncorrelated broadband RF noise gen-erators provide extremely flat white noise at IF and facilitate the measurement campaign and the assessment of system's performance under variable SNR con-ditions (e.g., the two noise sources were calibrated and balanced for every 2dB attenuation step).

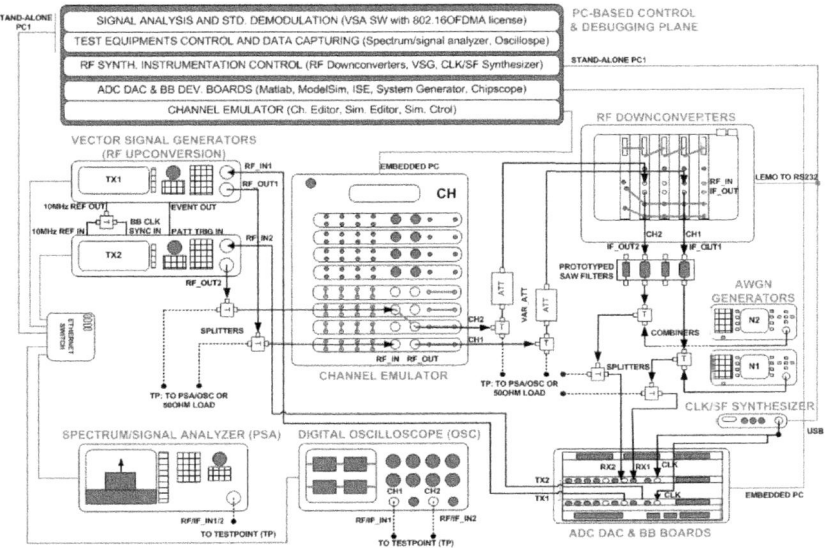

Fig. 6. The GEDOMIS® testbed setup

4.2 Baseband and DAC Circuitry Integration

Once the baseband RTL design of the MIMO-OFDM mobile WiMAX transmitter was verified (VHDL versus Matlab co-simulations), a detailed study was conducted to provision its integration with the DAC circuitry of the target boards.

In order to produce the global synthesized IF (i.e., 67.2 MHz) and take the maximum advantage of the filters contained in the DAC devices of Lyrtech's VHS-DAC board, an appropriate filter-interpolation strategy was selected. Hence, considering that a total x16 interpolation is required for each of the two I&Q data-streams of the transmitter, it was selected to apply a x2 interpolation in the FPGA domain and a x8 interpolation in each dual DAC device.

In more detail, the I&Q 16-bit data at the output of the FIFOs shown in figure 3 are fed to four separate filters (hosted in the FPGA device), which apply an interpolation by two. The filters are clocked at 44.8 MHz (i.e., the baseband clock is at 22.4 MHz). The use of two separate filters for the I&Q data components of each antenna was made to avoid the multiplexing of data, which requires a double clock rate from the one currently used at baseband (i.e. increasing the input clock from 44.8 MHz to 89.6 MHz). This decision was made after realizing that what was gained in terms of FPGA slice area (by using two instead of four interpolation filters clocked at 89.6 MHz), was resulting in a complicated place-and-route process, which is prone to timing errors (i.e., the FPGA routing algorithms are stressed to the limits for achieving the desirable clock rate). Manual placing of the implemented logic to the FPGA device area is another design option that is equally complicated when considering large FPGA design with numerous digital processing components.

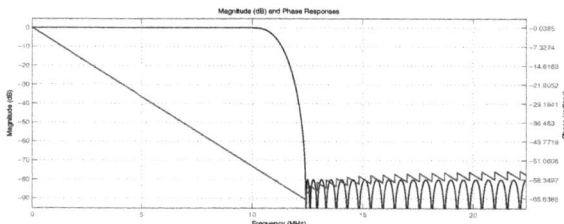

Fig. 7. The magnitude and phase response of the x2-interpolation pre-DAC interpolation filters used in the transmitter

Each dual DAC device, from the other side, is using 3 filters that are applying a x8 interpolation, which results in a 358.4 MHz clock (i.e., x8 FMIX QMC CMIX [21]). Taking into account the pass-bands of the different interpolation filters within the DAC, the frequency response of the inverse-sinc filter, and the frequency bands of the signals and their aliases, the coarse mixer was set to 89.6 MHz (equal to Fs/4, where Fs 358.4 MHz, i.e., the sampling frequency of the DAC) and the fine mixer was set to -22.4 MHz. The registers corresponding to the NCO frequency, the fine mixer gain and the coarse mixer mode-configuration (i.e., cm_mode(3:0) = 1000) were programmed through a C-based application. The latter allows the read/write access of several internal registers of the DAC 5687 chips on-the-fly.

The choice of the interpolation filters was a significant design, implementation and system-integration decision, which in fact has to be carefully considered every time a designer of MIMO-OFDM transmitters intends to interface the baseband design with a DAC device. The filter interpolation is an excessively processing-consuming implementation for FPGA devices, having a very hard to define trade-off between of FPGA resources utilization (embedded memory blocks and DSP48s versus generic FPGA slices) and maximum achievable clock rate. The heavily populated FPGA designs with multiple clock domains are making the place-and-route process of the FPGA implementation-flow, a particularly hard task. Therefore, a key system-design objective is to migrate the challenging implementation of the interpolating filters from the FPGA domain to high performance versatile DAC devices. In our case for instance, it would have been impossible to implement a higher order interpolation in the available FPGA area (perhaps, this is feasible using the largest Xilinx Virtex-6 devices).

The filters were designed using the FDAtool of Matlab having a high rejection response (i.e., 80 dB). This in turn implied a high number of filter coefficients, which ultimately is translated to high computational cost. The filter coefficients are featuring a symmetric response (i.e., 76 coefficients) and the produced file was used to configure the four respective Xilinx IP cores (i.e., Finite Impulse Response (FIR) filter). Distributed arithmetic logic has been selected for implementing the internal filter computational architecture, because the multiply and accumulate filter implementation option is requiring a high number of embedded FPGA RAM blocks (i.e., RAMB16s). This implementation decision was taken

Table 1. The main FPGA utilization metric corresponding to the PHY-layer implementation of the real-time MIMO-OFDM mobile WiMAX transmitter

Device utilization summary			
Logic Utilization	**Used**	**Available**	**Utilization**
Number of Slice Flip Flops	76,598	135,168	56%
Number of 4 input LUTs	69,189	135,168	51%
Logic Distribution			
Number of occupied Slices	52,650	67,584	77%
Total Number of 4 input LUTs	80,466	135,168	59%
Number of FIFO16/RAMB16s	285	288	98%
Number of DSP48s	81	96	84%
Number of DCM_ADVs	8	12	66%
Total equivalent gate count for design	20,565,470		

considering the total number of RAMB16s, DPS48s and slices available in the Virtex-4 LX160 device and the internal processing architecture of the MIMO-OFDM mobile WiMAX transmitter. The latter has several intermediate stages that require large data-set storage (i.e., high RAMB16s utilization). Finally, the 28-bit output of the I and Q filters had to be appropriately truncated to allow the utilization of the full dynamic range of the baseband part of the transmitter. This was achieved by running simulations of the whole MIMO mobile WiMAX transmitter. The truncated 14 bits are then interfaced with the DAC device of the Lyrtech board (the board firmware utilizes a 2 bit back-off margin).

4.3   FPGA Resource Utilization

The design of the transmitter has passed through numerous optimization stages to improve the performance and minimize the FPGA resource utilization, when targeting the Virtex-4 LX-160 device and using the Xilinx ISE 9.2 (i.e., mandatory software version for maintaining compatibility with the target board's firmware). The FPGA implementation was proved to be quite challenging since the chosen arithmetic-logic implementation of the four interpolation filters residing in the FPGA domain is making exclusive use of FPGA slices. Therefore, a study of the appropriate synthesis, mapping and place-and-route options was conducted trying to balance the implementation trade-offs, in terms of design speed (e.g., meet timing requirements) and FPGA area utilization. The place-and-route process resulted in significantly different FPGA implementation times, since the ISE 9.2 software tool is applying radically different execution routines.

As it is seen in table 1 the MIMO-OFDM mobile WiMAX transmitter is practically leaving very limited resources to the given Virtex-4 LX160 FPGA device for future extensions of the design. This could be partially solved with the migration of the whole design to the new Xilinx ISE 12 design suite. We have conducted some preliminary tests which show a reasonable reduction of the FPGA resource utilization. However, it has to be noted that the implementation

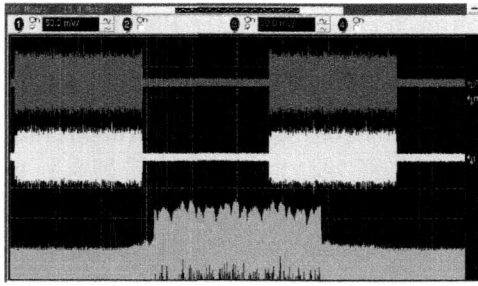
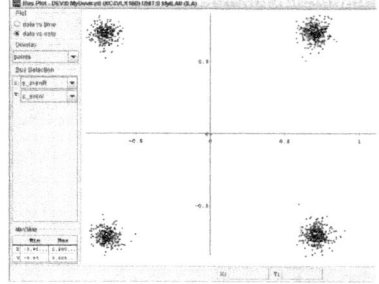

Fig. 8. Resulting MIMO time-signal, along with its spectrum (Agilent 80804B Infiniium Oscilloscope)

Fig. 9. Visualization of the demodulated data at the receiver side (Xilinx Chipscope Pro)

feasibility-analysis of this option is tightly coupled with an equivalent migration to a new FPGA board (i.e., out of scope for this paper).

5   Experimental Validation and Results

The resulting data at the output of the PHY-layer of the mobile WiMAX receiver is captured and visualized in real-time through vendor-specific tools and the Chipscope Pro software from Xilinx, using two different hardware configurations. In the first case, the emulated transmitter presented in [3] is used, while in the second case it is used the real-time FPGA-based MIMO-OFDM mobile WiMAX transmitter presented in this paper (see figures 8 and 9).

It is important to underline that the same VSGs are used to playback in real-time the Matlab-generated MIMO mobile WiMAX signal (i.e., PHY-layer emulation and baseband to RF signal up-conversion), and to up-convert to RF the IF signal generated by the real-time FPGA-based MIMO-OFDM mobile WiMAX transmitter. Therefore, the RF front-end performance is common for both testing scenarios. In other words, the deviations that may be observed when comparing the performance of both deployments have their origins in the baseband signal processing implementation (e.g., fixed-point arithmetic operations). Other indicative factors that could also result in differences during the result-comparison are the DAC features and the connector losses.

In order to analyse the performance of the system, the real-time channel emulator was configured with two static MIMO channel models (i.e., no mobility is emulated). These random channel models have the tap and delay spread characteristics, defined by the ITU Vehicular A and the ITU Pedestrian B specifications. The accuracy of the received constellation points is validated at the receiver by calculating the Error Vector Magnitude (EVM), as shown in figure 10. Additional assessment of the system's performance is provided by means of the raw Bit Error Rate (BER) calculation (i.e., no channel coding was used), as it can be seen in figure 11. Both performance metrics are calculated for a

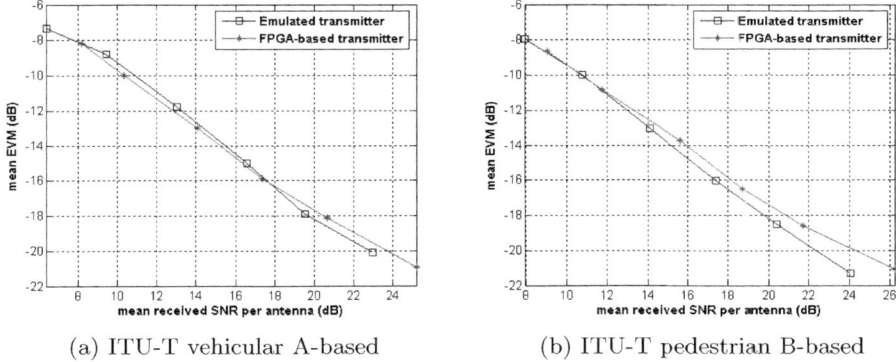

(a) ITU-T vehicular A-based   (b) ITU-T pedestrian B-based

Fig. 10. System performance under static channel conditions: EVM-SNR curves

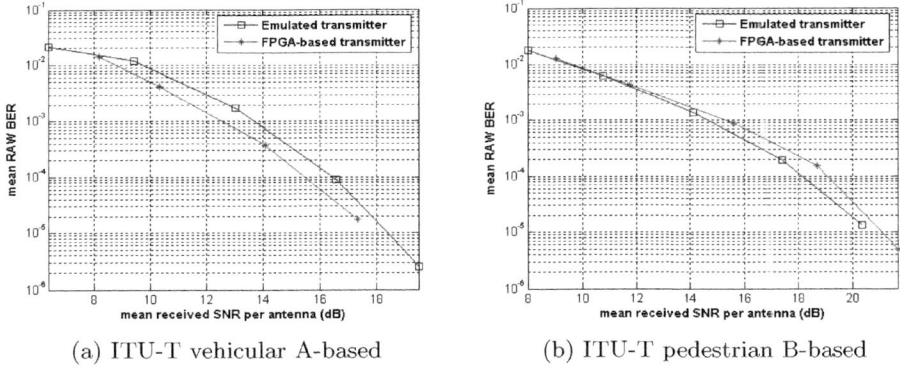

(a) ITU-T vehicular A-based   (b) ITU-T pedestrian B-based

Fig. 11. System performance under static channel conditions: BER-SNR curves

set of suitable Signal-to-Noise Ratio (SNR) values that cover a wide range of signal reception conditions (i.e., pre-defined white noise gain attenuation steps are used to generate uncorrelated white noise signals, which are added to each of the received signals). The real-time captures that have been performed are covering several entire frames, allowing by this way an averaged (and thus more robust) calculation of the presented metrics for each analysed scenario.

As it may be observed in the related figures, the performance for both deployments is quite similar. This not only proves the functional validity of the FPGA development, but also demonstrates the high implementation precision of the baseband signal processing algorithms. The minimal performance differences are due to the factors described next. First of all, the MIMO-OFDM mobile WiMAX transmitter is only validated under static channel conditions; this means that the performance comparison between the emulated transmitter and its FPGA-based counterpart is only analysed for a specific channel frequency response for each

Fig. 12. The MIMO-OFDM mobile WiMAX transmitter during a live demonstration (European Nanoelectronics Forum 2010)

channel model, which could represent a different performance condition among models at a given time (e.g., very low or very high attenuation). This frequency response would vary over time in mobility testing scenarios. In that case, the two system performance metrics are extracted by averaging numerous measurements corresponding to different channel seeds. This important testing scenario will be presented as a future work. Second, despite the fact that we tried to re-produce the exact testing conditions in both scenarios (e.g., transmitted power, noise levels), it is unavoidable that the hardware set-up features certain differences (e.g., number of RF/IF cables, DAC features, intermediate gain values) that influence the overall system's performance.

A limited setup of the full MIMO-OFDM mobile WiMAX transceiver is shown in figure 12. The real-time operation of the transmitter was demonstrated with success during the European Nanoelectronics Forum 2010.

6   Conclusions and Future Work

This paper presented the design, implementation and validation of the PHY-layer of a high-performance real-time MIMO mobile WiMAX transmitter featuring the Alamouti's STBC-based scheme. The system has been successfully validated under realistic channel conditions in static scenarios. The performance of the system has been analysed, by comparing the FPGA-based deployment to an emulated version of the transmitter (i.e., ideal baseband implementation). The presented work covers the most relevant issues to be taken into account, when confronting the complex task of designing, implementing and validating a real-time MIMO-OFDM system. The authors' contribution is mainly found in building and validating a realistic MIMO system based on the mobile WiMAX standard, having a high channel bandwidth of 20 MHz. These system specifications have scaled up the implementation complexity, but have offered at the same time an added-value to the whole undertaking.

The modular design of the system will be extended to include more advanced MIMO-OFDM schemes in the future. For instance, a real-time MIMO-OFDM closed-loop scheme could enhance the scalability and MIMO configuration options of the entire testbed. This will be implemented by using a feedback channel from the receiver to the transmitter. This real-time feedback link will contain quantized information for the actual channel conditions, which will enable an adaptive signal transmission. Finally, a broader analysis will be conducted in the future, to cover the performance of the system under the presence of mobility at high-speeds (e.g., ITU-T vehicular A up to 120 Km/h).

References

1. Mobile WiMAX air interface, P802.16Rev2/D9 (Revision of IEEE Std 802.16-2004 and consolidates material from IEEE Std 802.16e-2005, IEEE Std 802.16-2004/Cor1-2005)
2. Alamouti, S.M.: A simple transmit diversity technique for wireless communications. IEEE Journal on Selected Areas in Communications 16(8), 1451–1458 (1998)
3. Font-Bach, O., Bartzoudis, N., Pascual-Iserte, A., López Bueno, D.: A real-time MIMO-OFDM mobile WiMAX receiver: Architecture, design and FPGA implementation. Elsevier Journal of Computer Networks (2011) (in press), doi:10.1016/j.comnet.2011.02.018
4. De Bruyne, J., Joseph, W., Verloock, L., Olivier, C., De Ketelaere, W., Martens, L.: Field Measurements and Performance Analysis of an 802.16 System in a Suburban Environment. IEEE Transactions on Wireless Communications 8(3), 1424–1434 (2009)
5. Pelechrinis, K., Broustis, I., Salonidis, T., Krishnamurthy, S.V., Mohapatra, P.: Design and Deployment Considerations for High Performance MIMO Testbeds. In: 4th International Wireless Internet Conference (WICON), Hawaii (2008)
6. Imperatore, P., Salvadori, E., Chlamtac, I.: Path Loss Measurements at 3.5 GHz: A Trial Test WiMAX Based in Rural Environment. In: 3rd International ICST Conference on Testbeds and Research Infrastructure for the Development of Networks and Communities (TridentCom), Orlando (2007)
7. Wang, G., Yin, B., Amiri, K., Sun, Y., Wu, M., Cavallaro, J.R.: FPGA Prototyping of a High Data Rate LTE Uplink Baseband Receiver. In: 2009 Asilomar Conference on Signals, Systems and Computers (ASILOMAR), Monterey (2009)
8. Pérez-Iglesias, H.J., García-Naya, J.A., Dapena, A., Castedo, L., Zarzoso, V.: Blind Channel Identification in Alamouti Coded Systems: A Comparative Study of Eigendecomposition Methods in Indoor Transmissions at 2.4 GHz. European Transactions on Telecomunications. Special Issue: European Wireless 2007 19(3), 751–759 (2008)
9. Ramírez, D., Santamaría, I., Pérez, J., Vía, J., García-Naya, J.A., Fernández-Caramés, T.M., Pérez-Iglesias, H.J., González-López, M., Castedo, L., Torres-Royo, J.M.: A comparative study of STBC transmissions at 2.4 GHz over indoor channels using a 2 x 2 MIMO testbed. Wireless Communications and Mobile Computing 8(9), 1149–1164 (2008)
10. Gil Jiménez, V.P., Fernández-Getino García, M.J., García Armada, A., Torres, R.P., García Fernández, J.J., Sánchez-Fernández, M.P., Domingo, M., Fernández, O.: MIMO-OFDM Testbed, Channel Measurements, and System Considerations

for Outdoor-Indoor WiMAX. EURASIP Journal on Wireless Communications and Networking 2010 (2010)

11. Khairy, M.S., Abdallah, M.M., Habib, S.E.-D.: Efficient FPGA Implementation of MIMO Decoder for Mobile WiMAX System. In: IEEE International Conference on Communications 2009 (ICC), Dresden (2009)

12. Dick, C., Trajkovic, M., Denic, S., Vuletic, D., Rao, R., Harris, F., Amiri, K.: FPGA implementation of a near-ML sphere detector for 802.16e broadband wireless systems. In: SDR Technical Conference and Product Exposition 2009 (SDR), Arlington (2009)

13. Wang, Q., Fan, D., Lin, Y.H., Chen, J., Zhu, Z.: Design of BS transceiver for IEEE 802.16E OFDMA mode. In: IEEE International Conference on Acoustics, Speech and Signal Processing 2008 (ICASSP), Las Vegas (2008)

14. Wu, Y.-J., Lin, J.-M., Yu, H.-Y., Ma, H.-P.: A baseband testbed for uplink mobile MIMO WiMAX communications. In: IEEE International Symposium on Circuits and Systems 2009 (ISCAS), Taipei (2009)

15. Chen, J., Zhu, W., Daneshrad, B., Bhatia, J., Kim, H.S., Mohammed, K., Sasi, S., Shah, A.: A Real Time 4x4 MIMO-OFDM SDR for Wireless Networking Research. In: 15th European Signal Processing Conference (EUSIPCO), Poznań (2007)

16. Haene, S., Perels, D., Burg, A.: A Real-Time 4-Stream MIMO-OFDM Transceiver: System Design, FPGA Implementation, and Characterization. IEEE Journal on Selected Areas in Communications 26(6), 877–889 (2008)

17. Ibing, A., Kühling, D., Kuszak, M., Helmolt, C.v., Jungnickel, V.: Flexible demonstrator platform for cooperative joint transmission and detection in next generation wireless MIMO-OFDM networks. In: 4th International ICST Conference on Testbeds and Research Infrastructure for the Development of Networks and Communities (TridentCom), Innsbruck (2008)

18. Jungnickel, V., Schellmann, M., Thiele, L., Wirth, T., Haustein, T., Koch, O., Zirwas, W., Schulz, E.: Interference-Aware Scheduling in the Multiuser MIMO-OFDM Downlink. IEEE Communications Magazine 47(6), 56–66 (2009)

19. General requirements for instrumentation for performance measurements on digital transmission equipment, Rec. ITU-T O.150 (1996)

20. Guidelines for Evaluation of Radio Transmission Technologies for IMT-2000, Rec. ITU-R M.1225 (1997)

21. 16-bit, 500 MSPS 2x-8x Interpolating Dual-Channel Digital-to-Analog Converter (DAC), Texas Instruments,
http://focus.ti.com/docs/prod/folders/print/dac5687.html

Delay Modeling for 3G Mobile Multimedia Services QoE Estimation

Ianire Taboada, Fidel Liberal, and Jose Oscar Fajardo

University of the Basque Country, Spain
ianire.taboada@ehu.es
http://det.bi.ehu.es/NQAS

Abstract. In this paper, we consider mobile multimedia services delivered over a UMTS network. In this convergent scenario, low level error recovery mechanisms of the access network entail that almost all packet losses are caused by frames arriving at the receiver later than the playout time. Our aim is to find a simplified yet realistic expression for the statistics of the end–to–end delay in order to characterize related application level losses and infer resulting Quality of Experience (QoE). We present a simple and easy to implement model based on empirically obtained parameters that quantitatively reflects the statistical properties of delay. We apply the model to estimate the behavior of application level losses and delay which are the key factors while forecasting QoE in VoIP. Obtained results prove the suitability of the model to be integrated in cross–layer adaptation mechanisms and its utility for dimensioning VoIP playout buffers is also tested.

Keywords: UTRAN, SR–ARQ, delay, QoE, VoIP.

1 Introduction

Nowadays the utilization of mobile communications for multimedia services over converged networks is becoming more frequent. For the reproduction of these streaming services, packets in the reception buffer are read at deterministically–spaced time intervals, so that those packets that do not respect the timing constraints imposed by applications (i.e. arrive later than expected playout time) are discarded. Therefore, at the playout time of these discarded packets either "nothing" is reproduced or just copies of previous samples, which negatively affects user Quality of Experience (QoE). So, these application level losses happen even in allegedly lossless networks, even under not–severe degradations, whenever the end–to–end (e2e) delay of a packet is above a playout time related threshold (implemented with a buffer). Non–interactive multimedia services usually allocate long buffers in order to cope with these losses, so that it is the jitter and not the delay the indicator to consider. Unfortunately, VoIP services can not use so long buffers due to associated initial delay (see [1]) since it would affect the interactivity of the conversation. Even with dynamic adaptive buffers the tradeoff between delay and losses imposes a maximum size for the playout buffer.

J. Del Ser et al. (Eds.): MOBILIGHT 2011, LNICST 81, pp. 67–80, 2012.

Then, our analysis aims at determining the probability that the e2e delay is above the threshold related to player buffer length, $Pr\{d_{e2e} > d_{max}\}$, which will provide us with application level loss rate. As a result, we need the Complementary Cumulative Distribution Function (CCDF) of the e2e delay to be evaluated at that threshold point instead. The scenario considered in this work includes a traditional 3G cellular access (UMTS) and a wired CN (Internet). However, the core part of the UMTS network (i.e. SGSN/GGSN nodes) and the Internet are assumed to introduce a constant delay, so the delay caused by low level error recovery mechanisms in the last mile radio part is the only contributor to the total variable e2e delay.

We have considered UMTS and not a HSPA/LTE/WiMAX scenario because it is the most widespread commercial network nowadays. Furthermore, unlike variants such as HSPA/HSPA+ where the final QoS (Quality of Service) will depend on the number of users and the scheduler, classical 3G connections provide dedicated channels (DCH), so that QoS constraints can be fulfilled. Additionally, the analysis of delay will be focused on ARQ (Automatic Repeat reQuest) based low level error recovery mechanisms that are quite similar to those in Beyond–3G (B3G) technologies.

Hence, the main objective of this work is to obtain an expression of the e2e delay CCDF under different reception and UMTS network conditions. To obtain the analytical expression of the delay statistics in the UTRAN (UMTS Radio Access Network), an in-depth analysis of retransmission techniques in UMTS has been carried out together with empirical fitting of simulation data. The impact of particular network and service parameters such as Block Error Rate (BLER), packet size, bitrate and average burst length (ABL) has been taken into account. Estimation of the delay calculated with the obtained model has been later used for predicting QoE on mobile VoIP service. Such closed–form solution will allow researchers to model network effects in cross–layer adaptation systems and design decision making algorithms accordingly.

The rest of the article is organized as follows. Section 2 describes retransmission schemes for UMTS Acknowledged Mode (AM). In Section 3 we propose using empirical data and fitting techniques in order to overcome the drawbacks of classical ARQ analysing schemes. Resulting expressions and fitted coefficients will be calculated just once. Later they will be included in a delay estimation system that will interpolate the obtained model to every possible input situation. Finally, Section 4 will show the capability of the proposed method for forecasting QoE for VoIP calls and Section 5 will summarize main contributions.

2   Preliminary Analysis of Delay for UMTS AM Mode

In this Section we will describe the analysis of the additional delay caused by retransmissions due to level 2 error recovery mechanisms in UMTS networks. Concerning the transmission in the UTRAN, different ARQ techniques are used in order to face the high error probability in the wireless medium, resulting in erroneous packets retransmissions. These mechanisms introduce a random delay at RLC (Radio Link Control) level, which is translated to a random delay at the application layer.

In UMTS networks, Selective Reject (SR) ARQ (see for example [2]) schemes are used instead of the typical ARQ ones (like Stop & Wait and Go-Back-N), due to their superior throughput performance. The basic operation of SR-ARQ consists on performing selective retransmission of frames, this is, only erroneous frames are retransmitted. These retransmissions happen when the sender receives a "NACK" (or Bitmap ACK with erroneous packets) belonging to an erroneous frame or when a timer reaches its time-out.

As defined in 3GPP Specifications [3], the RLC layer operation mode responsible for handling the frames recovery by means of SR-ARQ is called AM (Acknowledged Mode). As specified in this standard, error correction is done at MAC PDU (Medium Access Control Protocol Data Unit) or Transport Block (TB) level. In typical configurations these blocks have a payload capacity of 40 bytes, in which data from superior layers can be inserted, plus the corresponding bytes of RLC+MAC headers. In this way, upper level packets are segmentated in blocks of 40 bytes that are later on individually transmitted. TBs are transmitted at time intervals called TTI (Transmission Time Interval). One or several PDUs can be transmitted in each TTI, and it is the MAC layer which decides how many PDUs must be transmitted in each TTI (a typical example of bearer configuration is: 12 TBs transmitted per TTI and 10 ms TTI duration).

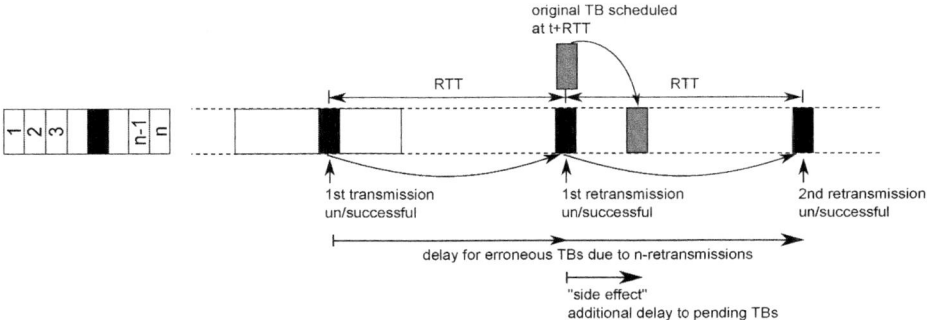

Fig. 1. ARQ retransmission from the sender point of view

Normally a RLC level transmitter uses both a transmission and a retransmission buffer. When a PDU is scheduled for submission, it is placed in the transmission buffer, and once transmitted to the channel, it is moved to the retransmission buffer. The PDUs in the retransmission buffer are deleted or retransmitted depending on the answer sent by the receiver. Retransmission buffer has priority over the transmission one; this is, upon the reception of erroneous frame notification, those to be retransmitted must be sent before the one originally scheduled in the transmission buffer, which has to wait to the next TTI (considering that there is no TB available in the RLC frame). Furthermore, since some feedback from the receiver is needed for the sender to detect the loss/error, possible retransmissions will arrive at the receiver delayed $n \cdot RTT_u$ ms., considering this UTRAN Round Trip Time (we will use the term RTT_u from now on) as the time needed for the "NACK" to go back to the sender plus the time needed for the retransmitted AMD (Acknowledged Model Data) PDU to arrive at the receiver (i.e. 100ms for typical UMTS commercial configurations). The overall processs is depicted in Fig. 1.

In the receiver, AMD PDUs remain in the reception buffer until a complete RLC SDU (Service Data Unit) is received. However, if a retransmitted RLC PDU is once again lost, retransmissions are tried until either the SDU Discard Timer expires or the number of retransmission exceeds a threshold (see different retransmission modes in [3] and [4]). Once all the PDUs associated to a RLC SDU are correctly received, they are reassembled and passed to superior layers. When all the PDUs of a RLC SDU are correctly received, resulting SDU is moved forward to the upper layer regardless the status of the previous RLC SDUs. This reception buffer policy is called not-in-order delivery. On the other hand, if SDU order is respected and SDUs with lower sequence number are required to be correctly received before sending data to higher layers, the buffer policy is in-order delivery.

Finally, due to the characteristics of the propagation of radio signals over the air, errors usually occur in bursts, and a correlation exists among them also at TB level in UTRAN. In [7] this behavior is modeled by means of modified Markov chains. For a dynamic scene (with mobility), it is concluded that the behavior of error bursts at TB level is similar to those of burst at TTI level, and this is modeled with a modified Gilbert–Elliot model (see [5] and [6]) at TTI level. On the contrary, for the static case a lower granularity analysis at TB level is required resulting in a more complex Markov model. In this context, burst length concept is defined as the number of consecutive TTIs or TBs with errors. These advanced models represent with higher fidelity the behavior of the wireless link, far beyond typical ones based on independent losses.

As a conclusion of this preliminary analysis note here that in our study we will consider UMTS RLC AM mode, SR-ARQ retransmission scheme, with 12 TB/TTI configuration, not–in–order delivery and complex bursty air error model in mobility based on modifications of Gilbert–Elliot's one in [7]. These considerations are not trivial since many studies do not state them explicitly and usually yield inaccurate simulation results. A common example is the UMTS module for ns-2 of the Eurane project [8] that does not support not–in–order delivery model, so that the evolution of the delay is not the expected one.

3   UMTS Delay Modeling

Once the behavior of the AN (Access Network) considered in our study has been analyzed we focus on modeling SR-ARQ.

In [9] a simple UTRAN SR-ARQ delay modeling is carried out based on combinatorics both for in–order and out–of–order delivery. A discrete PDF

(PMF-Probability Mass Function) evaluated only at RTT_u multiples is obtained. However, obtained expressions are too simplified solutions. On the one hand, they do not consider the typical situation where TBs are scheduled for transmission at the same time slot that a retransmission request. In this case, the slot would be occupied and therefore scheduled TB should wait to the next free TTI to be sent, resulting on an impact into delay statistics (see Fig. 1). Modeling this effect is not trivial. On the other hand, TBs are considered to be lost independently, which is also a simplification. In fact, as explained before, a correlation and a bursty behavior exists, which is not modeled by the authors.

Similarly, the analytical expressions proposed in [10] regarding SR-ARQ delay in the UTRAN, for the two reception buffer policies, are deduced on a greedy source situation, where packets are sent continuously, and later validated with simulation. However, the greedy source assumption is not always realistic. Furthermore, related mathematical expressions are based on iterative computation, so that obtained results are quite complex in order to be implemented in real mobile devices.

Considering the inability of different theoretical proposals for modeling delay in the UTRAN we decided to obtain a model of UTRAN SR-ARQ delay by analizing delay distributions out of simulation results.

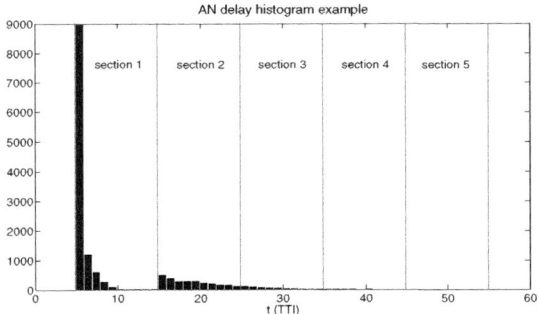

Fig. 2. Retransmission sections in an d_{AN} histogram

The basic analysis of the AM RLC mode and obtained simulation results for AN delay show a RTT_u-spaced periodic pattern due to successive retransmissions. Within each retransmission section most of the frames arrive at the first slot (TTI) available but there appear also delayed frames in successives slots due to the so-called "side effect" shown in Fig. 1. Then, it is clear that the solution must be defined in retransmission sections due to SR-ARQ working mode; where the first section represents the retransmission-free frames. When analysing simulation data, a RTT_u of 100ms is considered, so that the distance between retransmission periods will be 10 TTIs (a TTI duration of 10ms is used). At very high BLERs or bitrate situations, it is possible to find samples from previous section (so $i-1$ retransmissions) that appear in the following one (i^{th}

retransmission section) because they have been delayed more than a whole retransmission interval. This situation appears whenever the retransmission buffer
is so full that no free slot is available for the originally scheduled ones for a
whole RTT_u period. However, the contribution of those samples is considered
negligible, so that this effect is not taken into account and the retransmission
sections are considered as a single function for the whole retransmission period
(note the regular function in (1)). Besides, there are no samples between the
origin ($t = 0$) and the first sample in the first section. This delay is due to fixed
transmission delay and packetization and reception buffer effects, and it is not
taken into account for fitting. For convenience we shifted all delay samples to
the origin, considering this the arrival of the first packet without losses. Fig. 2
shows the per-retransmission section behavior along up to 50 TTIs.

In order to obtain both the shape and empirical parameters for every network
state and application setting considered an extensive set of simulations have
been carried out. The main features about the simulation scenario are collected
in Table 1.

Table 1. Simulation parameters for AN modeling

Element	Value
Loss model	Modified Gilbert Elliot's in [7]
Reception policy	Not-in-order delivery
Discard Timer (ms)	500
Bearer configuration	(384kbps DL, 64kbps UL)
TTI (ms)	10
Transport format (TB/TTI)	12
Transmission channel type	Dedicated Channel (DCH)
Simulation tool	OPNET Modeler

For PMF fitting, nonlinear Least Squares method [11] is used, which minimizes the summed square of residuals. We have modeled the PMF of the delay
in the AN, $D_{AN}[n] = D_{AN}(t)|_{t=n \cdot TTI}$, by an approximation with delayed ex-
ponentials, so that the resulting analytic expression of the PMF of the delay in
the UTRAN is shown in equation (1).

$$\sum_{i=1}^{N} a_i \cdot e^{b_i \cdot (t-(i-1) \cdot RTT_u)} \cdot \sqcap_{RTT_u} \left(t - i \cdot \frac{RTT_u}{2} \right) \qquad (1)$$

where:

- $\sqcap_{RTT_u} \left(t - i \cdot \frac{RTT_u}{2} \right)$ is the rectangular function
  - width RTT_u
  - centered in $i \cdot \frac{RTT_u}{2}$
- i identifies the retransmission section.
- N is the maximum number of retransmission sections.
- a_i and b_i parameters are the result of the exponential fitting done with each
  section samples.

To validate the fitting (i.e. to check the correlation between the normalized histogram and the estimated delay PMF for each configuration) the R-Square parameter has been analyzed. Chosen exponential functions for all the retransmission sections provide good results (see Table 2).

Table 2. R-square values for packet size=30bytes, bitrate=12.2kbps, ABL=1.75 and different BLERs

BLER (%)	R-square (section 1)	R-square (section 2)	R-square (section 3)	R-square (section 4)	R-square (section 5
1	0.999	0.837	-	-	-
5	0.999	0.717	0.726	-	-
10	0.988	0.854	0.953	-	-
20	0.994	0.926	0.974	0.701	0.633
30	0.983	0.797	0.919	0.848	0.567

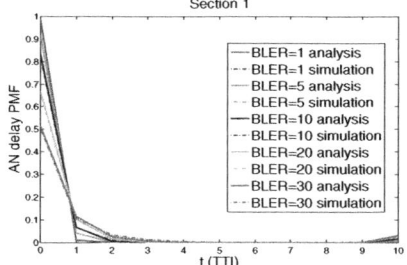
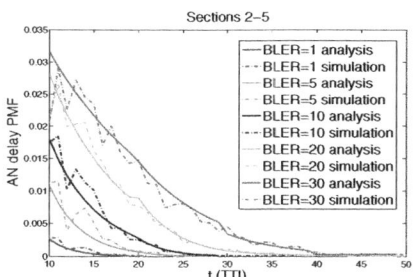

Fig. 3. Comparison between empirical and approximated AN delay PMF for packet size=30byte, bitrate=12.2kbps, ABL=1.75 and different BLERs

As shown in Fig. 3, the impact of BLER on the delay is clearly reflected, as when this is higher, retransmissions are more probable resulting in lower values for the histogram at the beginning of the first section (free of retransmissions). This behavior is also reflected on the values of a_i and b_i coefficients. Besides, the analysis of Table 3 shows that the fitting mechanism provides all the coefficients (including $i = 5$) at high BLERs only, since there are no delay samples for low BLERs in latter sections.

After carrying out the simulations with all possible combination of interesting input values (BLER, ABL, packet size and bitrate -see Fig. 4-) we have built a set of tables of a_i and b_i coefficients. Later, by using the expression (1) and a_i and b_i associated to the specific network state under study we can obtain the model for $D_{AN}[n]$ that will be finally used for the estimation of e2e delay. These lookup tables and the expression for $D_{AN}[n]$ provide researchers with a simple but effective method for forecasting the behavior of a UMTS network for certain RLC configuration sets and network states.

Table 3. a_i and b_i coefficients at the i-th retransmission section for different BLER (packet size=30bytes, ABL=1.75, bitrate=12.2kbps)

BLER(%)	a_1	a_2	a_3	a_4	a_5	b_1	b_2	b_3	b_4	b_5
1	0.98	2.5E-03	—	—	—	-458.97	-38.53	—	—	—
5	0.89	10.5E-03	6.0E-04	—	—	-312.54	-28.30	-35.88	—	—
10	0.79	0.017	2.6E-03	—	—	-253.49	-18.49	-33.12	—	—
20	0.62	0.026	8.2E-03	7.0E-04	2.0E-04	-189.40	-12.58	-23.75	-15.97	−20.36
30	0.46	0.029	0.013	3.0E-03	2.0E-04	-151.04	-7.61	-11.54	-14.84	−2.01

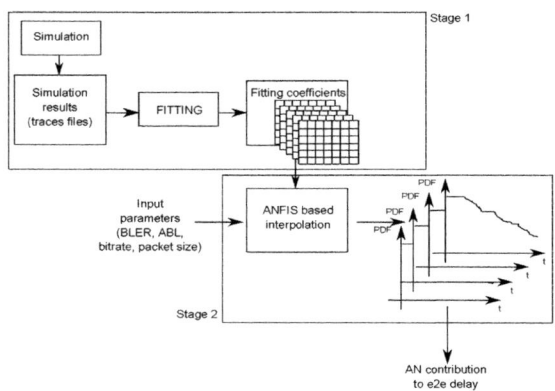

Fig. 4. Proposed process for modeling AN delay

Although a limited set of discrete input values for the AN modeling will be enough for most services, since they allow a limited range of values for bitrate or packet size, other multimedia services may require a full estimation of e2e delay for continuous ranges of service and network states. For these special cases, we propose a second stage consisting of an ANFIS (Adaptive Neuro-Fuzzy Inference System) [12] [13]) based interpolation mechanism to obtain the estimation for all posible combinations of inputs. Note here that, regardless the granularity needed, the first stage has to be carried out just once. After the ANFIS system is deployed and trained, it will provide the estimation of the delay distribution for any possible input parameters (BLER, ABL, bitrate and packet size) with no need of additional modeling. We have developed an ANFIS–based learning model to predict a_i and b_i values for different combinations of BLER, ABL, packet size and bitrate. In our system, after training the ANFIS network with simulation–derived dataset, we have obtained a nearly "infinite" interpolated shape for different a_i and b_i parameters of the exponentials in (1).

In Fig. 5 we can see the comparison between the original, empirical data-based approach and the ANFIS extended one for different combinations of BLER and packet sizes.

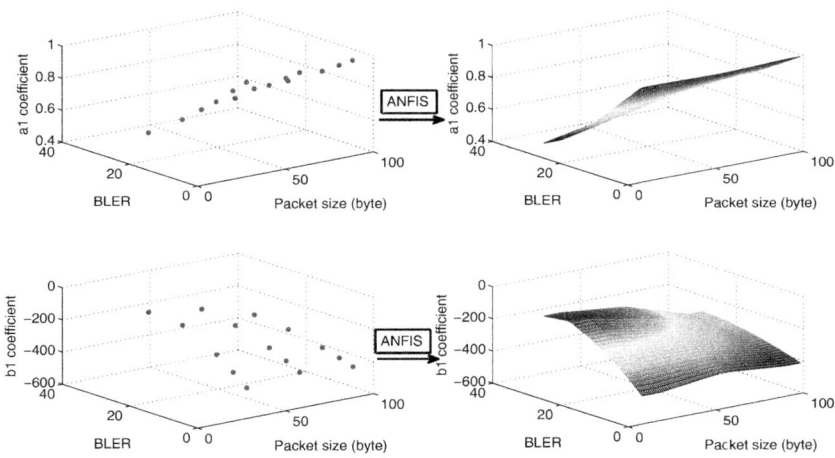

Fig. 5. Extension of a_1 and b_1 using ANFIS

As a conclusion, in comparison with previous studies about the SR-ARQ in UMTS networks, we have obtained a much simpler expression of the PMF but still capable of providing a good approximation of UTRAN behavior for different BLERs (and different packet size, bitrates and lower level parameters). Nevertheless, in order to know when the delay exceeds a threshold the CCDF is needed. Formally,

$$CCDF\{d_{AN}\}(t) = 1 - \sum_{k=0}^{\lfloor \frac{t}{TTI} \rfloor} D_{AN}[k] \qquad (2)$$

As the $AN\ PMF$ is defined on the interval $\left[0, N \cdot \frac{RTT_u}{TTI} - 1\right]$ the value of k is limited, so that k_{max} is:

$$k_{max} = N \cdot \frac{RTT_u}{TTI} - 1 \qquad (3)$$

As was expected, since PMF approximation was good, obtained CCDFs match those obtained with simulation (see Fig. 6).

Moreover, recall that the fixed delay (d_f) introduced by propagation, serialization and forwarding processes both in the wireless and wired segments is not included in the previous expression. As obtained from simulation, this delay is about 60ms. Therefore, the e2e delay CCDF is computed as follows:

$$CCDF\{d_{e2e}\}(t) = \begin{cases} 1 & t < d_f \\ CCDF\{d_{AN}\}(t - d_f) & t >= d_f \end{cases} \qquad (4)$$

This expression will be used in next section to infer QoE in VoIP.

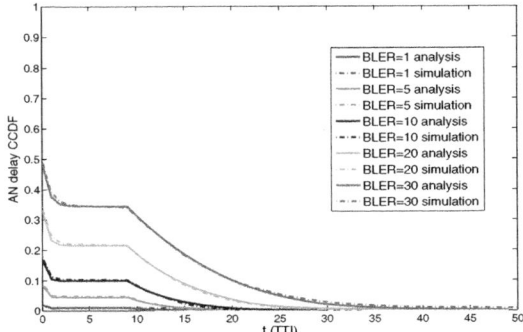

Fig. 6. Comparison between modeled and simulated AN delay CCDF for packet size=30bytes, bitrate=12.2kbps, ABL=1.75 and different BLERs

4   Forecasting VoIP QoE

Finally, we will show the capability of the proposed system for forecasting the evolution of the VoIP QoE under different network states. AMR-based VoIP services [14] will be analyzed, comparing the obtained analytical results with the simulated ones.

The predominant method for assessing the QoE of VoIP services is the E-model, defined by the ITU-T in [15] and applied to AMR-based VoIP services over 3G UMTS connections in [16] resulting in expression (5).

$$R = R_o - I_s - I_d - I_{e-eff} + A \tag{5}$$

This model allows computing the value of a rating factor, R, which provides an evaluation of the communication impairment, and relating this score to the corresponding MOS (Mean Opinion Score) score. The additional impairment due to the network performance is caused by two main factors: I_d is defined as the impairment due to the total mouth–to–ear delay, while I_e is the impairment factor due to the combined effect of the codec and packet losses.

For a specific e2e delay distribution, the values experienced by I_d and I_e factors will depend on the application buffer size. At high buffer sizes, user experiences a higher delay in the communication, and thus I_d increases. Yet, all those voice samples that reach the destination endpoint with a variable delay inferior to the buffer value are successfully presented to the end user, resulting on lower I_e impairments. On the contrary, lower buffer sizes will result on lower I_d values but likely increased I_e impairment, depending on the e2e delay statistics. Therefore, the proposed e2e delay estimation method is worth to estimate the expected QoE at different network states, as well as to infer the most suitable buffer policy for each case.

Table 4 represents the estimated MOS scores for a concrete VoIP configuration (AMR–12.2 and 1 voice frame per IP packet), using a dejittering buffer of

200ms, for different combinations of network states (represented by the experienced BLER value at the radio segment). In order to validate the proposed QoE estimating approach, two MOS values (MOS_{VoIP}) are provided for each network state. The first one illustrates the outcome of the proposed model: the e2e delay CCDF is computed with (4) and the specific application–level delay and loss ratio is obtained for the established buffer size. The second value is the average MOS as directly computed from simulation traces. As can be observed, the proposed system provides good estimations of the expected values of the QoE for this kind of VoIP services.

Table 4. Forecasted and simulated MOS values for VoIP AMR calls (MOS_{VoIP})

BLER (%)	MOS_{VoIP} Model	MOS_{VoIP} Simul.
1	3.84	3.78
5	3.50	3.70
10	2.91	3.01
20	2.07	2.14
30	1.54	1.55

The expressions presented in this paper can be further used for optimizing the provisioning of multimedia content in convergent networks. Once identified the current network state for a VoIP user, it is easy to check the enhancement in terms of QoE of possible cross–layer adaptations that will lead to a different state.

Additionally, the method proves to be useful for real–time buffer dimensioning also. As illustrated in Fig. 7, once obtained the e2e delay CCDF for the specific case, estimating the associated application–level losses for different buffer size values is straightforward. Later, obtained new application–level delay and loss ratio values can be taken as inputs for the MOS estimation expressions, in order to infer the most suitable configuration for the current state. According to Fig. 7 we can conclude that, for that particular network state, a buffer of 100ms is better in terms of QoE than a buffer of 150ms since the CCDF is almost identical (so that application level losses -and I_e- are the same) while the constant delay due to buffering (included in I_d) is higher on the second case.

Therefore, reception buffer dimensioning plays an important role in order to obtain the best service configuration for a certain network state, and our analytical results provide finding the most appropriate buffer length for each case. The expected MOS map for different network and service states is summarize in Fig. 8. Network states are defined with BLER values 1, 5, 10, 20 and 30, while service state, s, is defined as:

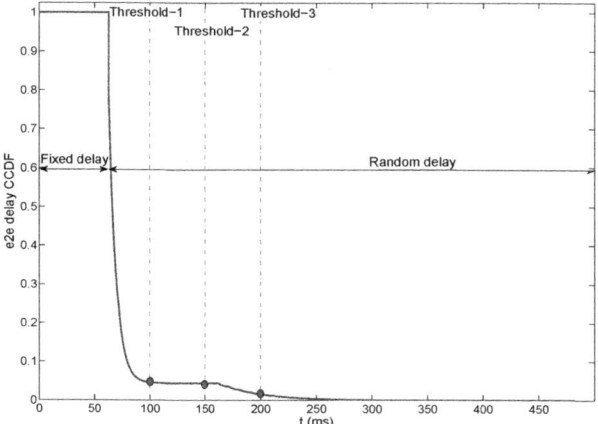

Fig. 7. CCDF of delay and associated impact into MOS_{VoIP} of different delay thresholds

$$s = (i - 1) \cdot 123 + (j - 1) \cdot 41 + k \qquad (6)$$

where:

- AMR mode: i = (1 - 3) for 12.20, 10.20, 7.95 kbps.
- Packetization: j = (1 - 3) for 1, 2, 3 frames/packets.
- Buffer size: k = (1 - 41) for 100, 110, 120, ..., 500 ms of buffer.

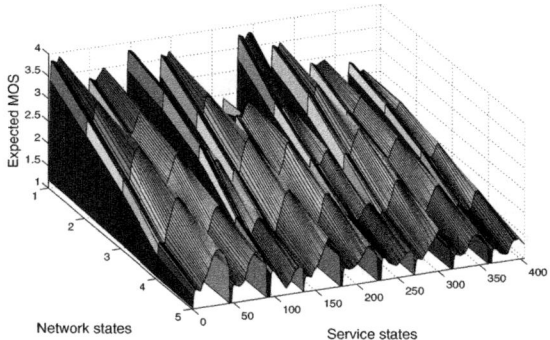

Fig. 8. Expected MOS for different network and service states

The VoIP configuration will be represented as a 3-tuple of parameters (i, j, k). For all the network states defined, we can find the most suitable VoIP configuration for each considered network case. Table 5 summarizes the results obtained in the whole process. As was expected, the best service configuration for all network states is the AMR–12.20 codec mode with one voice frame per packet. However, if the distance in the MOS scale between two possible VoIP configurations is low, it could be preferable the configuration with a lower bitrate requirement; this is, with lower AMR mode or higher packetization scheme. This is useful from the standpoint of the optimization of resource and battery usage. On the other hand, different optimum dejittering buffer sizes are obtained for each network state, which clearly reflects the importance of using an adaptative buffer.

Table 5. Best choice of service level parameters

Network state (BLER %)	Optimum service state	Expected MOS
1	(1,1,180)	3.90
5	(1,1,210)	3.51
10	(1,1,270)	3.21
20	(1,1,320)	2.72
30	(1,1,370)	2.23

5   Conclusions

In this paper, we propose a model for estimating e2e delay aimed at estimating QoE in multimedia services over 3G networks. This way, application level losses due to excessive delay have been considered.

The main contribution of this paper is that a simple yet realistic model for e2e delay statistics has been described. A new retransmission–based empirical model has been proposed for the analysis of the delay statistics in the UTRAN. This model overcomes the limitations found in the state of the art, and allows estimating the random delay contributions at different service and network conditions. We have obtained both a fitted closed–form expression, for coarse grained network-service states, and ANFIS based estimation for finer granularity.

Finally, e2e delay characteristics have been mapped to user experienced QoE for the specific case of AMR VoIP service using the E–model. Estimated MOS_{VoIP} values fit simulation results and make our system suitable to be included in decision makers of multimedia cross–layer adaptation systems. This way, by forecasting MOS_{VoIP} for the network state that would result after every possible adaptation, we can choose the best suitable combination of service level paramaters and network states. For the same network state, the model allows us to calculate the length of the buffer associated the most convenient tradeoff solution between losses and delay.

References

1. Bacioccola, A., Cicconetti, C., Stea, G.: User-level performance evaluation of VoIP using ns-2. In: Proc. of the 2nd International Conference on Performance Evaluation Methodologies and Tools. ICST (2007)
2. Schwartz, M.: Telecommunication networks: protocols, modeling and analysis. Addison-Wesley Longman Publishing Co., Inc., Boston (1986)

3. 3rd Generation Partnership Project.: Ts 25.322: rlc protocol specification (2009), http://www.3gpp.org/ftp/Specs/html-info/25322.htm
4. 3rd Generation Partnership Project.: Ts. 25.331: radio resource control (rrc), protocol specification (2009), http://www.3gpp.org/ftp/Specs/html-info/25331.htm
5. Gilbert, E.: Capacity of a burst-noise channel. Bell Systems Technical Journal 39(5), 1253–1265 (1960)
6. Elliott, E.: Estimates of error rates for codes on burst-noise channels. Bell Syst. Tech. J. 42(9), 1977–1997 (1963)
7. Karner, W., Nemethova, O., Svoboda, P., Rupp, M.: Link Error Analysis and Modeling for Video Streaming Cross-Layer Design in Mobile Communication Networks. ETRI Journal 29(5) (2007)
8. B. V. Ericsson Telecommunicatie. EURANE - enhanced UMTS radio access network extensions for ns-2, http://eurane.ti-wmc.nl/eurane/
9. Necker, M.: A simple model for the IP packet service time in UMTS networks. In: International Teletraffic Congress, ITC (2005)
10. Rossi, M., Zorzi, M.: Analysis and heuristics for the characterization of selective repeat ARQ delay statistics over wireless channels. IEEE Transactions on Vehicular Technology 52(5), 1365–1377 (2003)
11. Madsen, K., Nielsen, H.B., Tingleff, O.: Methods for Non-Linear Least Squares Problems. Informatics and Mathematical Modelling, Technical University of Denmark (1999)
12. Jang, J.: ANFIS: adaptive-network-based fuzzy inference system. IEEE Transactions on Systems, Man, and Cybernetics 23(3), 665–685 (1993)
13. Khan, A., Sun, L., Ifeachor, E.: An ANFIS-based hybrid video quality prediction model for video streaming over wireless networks. In: The Second International Conference on Next Generation Mobile Applications, Services and Technologies, NGMAST 2008, pp. 357–362. IEEE (2009)
14. 3rd Generation Partnership Project.: Ts 26.975: performance characterization of the adaptive multi-rate (amr) speech codec (2008), http://www.3gpp.org/ftp/Specs/html-info/26975.htm
15. ITU-T, Recommendation g.107: the e-model, a computational model for use in transmission planning (2003)
16. Sun, L., Ifeachor, E.: Voice quality prediction models and their application in VoIP networks. IEEE Transactions on Multimedia 8(4), 809–820 (2006)

Route Recovery Algorithm for QoS-Aware Routing in MANETs

Wilder Castellanos, Pau Arce, Patricia Acelas, and Juan C. Guerri

Multimedia Communication Group, ITEAM Institute,
Universitat Politècnica de València
{wilcashe,patacdel,paarvi}@iteam.upv.es, jcguerri@dcom.upv.es

Abstract. The communications inside MANETs usually show frequent disruptions due to changes in topology and the condition of having a shared physical channel. It is necessary to implement a mechanism to maintain connectivity and ensure Quality of Service (QoS). In this paper we introduce a route recovery mechanism for a QoS routing protocol called AQA-AODV (Adaptive QoS-Aware AODV) which is an extension of the AODV protocol. Our proposal provides a mechanism to detect the link failures in a route and reestablish the connections taking into account the conditions of QoS that have been established during the route discovery phase. The simulation results reveal performance improvements in terms of packet delay, number of link failures and connection setup latency while the end-to-end throughput is not affected compared with the throughput achieved by other protocols like AODV.

Keywords: Wireless ad hoc networks, quality of service-aware routing, route recovery, link failure.

1 Introduction

A mobile ad hoc network (*MANET*) is a group of autonomous wireless devices organized themselves dynamically in a mesh topology. The key feature of this type of networking is the nonexistence of any permanent infrastructure. Due to these infrastructure-less and self-organized characteristics, *MANET* encounters different problems from infrastructure-based wired network, such as bandwidth-constrained, variable capacity links and energy-constrained operation. Moreover, routes may include multiple hops because communications need to use intermediate nodes as routers in order to communicate with nodes that are out of its transmission range. This mobility of nodes causes frequent link failures and high error rates, so it makes difficult to maintain the desired QoS in the network. Additionally, due to the fact that the wireless channel is shared among neighbor hosts and that network topology can change as hosts move, the transmission of time-sensitive data (e.g. video packets) is made more difficult. Especially in applications that generate a huge data volume that is delay-sensitive and bursty, since losses of some important data segments (such as synchronization data) may seriously disrupt a long sequence of frames [1].

J. Del Ser et al. (Eds.): MOBILIGHT 2011, LNICST 81, pp. 81–93, 2012.
© Institute for Computer Sciences, Social Informatics and Telecommunications Engineering 2012

The main issue is how to efficiently transmit a large volume of delay-sensitive data when many packets are dropped due to the fact that network resources are limited and time-varying. We propose in [2] a QoS-aware routing protocol (*AQA-AODV, Adaptive QoS-Aware AODV*) which is a modified and enhanced version of the Ad hoc On-demand Distance Vector (*AODV*) [3] that allows the source to adapt the transmission rate. More precisely, we have introduced into the original AODV protocol, an adaptive feedback scheme and two mechanisms: one for the estimation of the available bandwidth in the node and the other for the prediction of the consumed bandwidth for a route of multihops. In addition, a QoS extension is added to the *AODV* control packets and the routing table. The result is a QoS path finding mechanism that can provide feedback to the application about the current network state in order to allow the application to appropriately adjust the transmission rate.

To support QoS routing on MANETs, new protocols, like *AQA-AODV*, need to use an efficient route maintenance mechanism. In this paper, we propose a route recovery mechanism for *AQA-AODV*, which not only has to re-establish the connections but it also has to take into account the conditions of QoS that have been established during the route discovery phase.

In order to test the performance of our route recovery model, we have implemented the proposed solution in the ns-2 simulator [4]. Results indicate that the packet delay, link failures and the connection setup latency decrease significantly while the overall end-to-end throughput is not impacted.

This paper is organized as follows. Section 2 briefly reviews some related works. Section 3 describes the main components of *AQA-AODV* protocol. Section 4 presents the performance evaluation of our route recovery mechanism and Section 5 offers some conclusions.

2   Related Works

Several approaches have been proposed based on AODV routing protocol. In [5], Pan *et al.* suggest the approach with two routing protocols called AODV - Local Repair TTL (*AODV-LRT*) and AODV - Local Repair Quota (*AODV-LRQ*). In these approaches are decreased the breadth and depth, of route repair requests. Decreasing the breadth of the repair mechanisms means to limit the maximum number of hops that *RREQ* packets have to pass, assuming that the size of the network topology and transmission range of every node are known. Decreasing the depth of the repair mechanisms means to limit the number of times a node is allowed to forward the route repair request. In [6], Youn et al. propose a new local repair scheme using promiscuous mode which is mainly composed of two parts: adaptive promiscuous mode and quick local repair scheme. Adaptive promiscuous mode repeats the switching processes between promiscuous mode and non-promiscuous mode. The proposed scheme adopts promiscuous mode such that each node keeps monitoring the overheard packets from which the routing information about the route path in adjacent nodes can be obtained. This action can cause excessive energy consumption and reduce network efficiency.

The solutions proposed by Pan *et al.* [5] and Youn *et al.*[6] aim to change the *AODV* protocol to make it more efficient in relation to route maintenance, but do not take into account the conditions of QoS, since these were designed for a routing protocol without QoS support, like *AODV*. Other proposed studies can be consulted in reference [7] which offers a survey of AODV-based approaches.

Sarma *et al.* [8] proposed two route maintenance mechanisms for *AODV* with QoS support. One is based on a special local route repair by limiting route recovery flooding to one hop neighbors only. Other one is route recovery by the destination node itself. However these mechanisms fail when try to re-establish connection to destination with the QoS conditions that had been negotiated during the initial route discovery phase. Other QoS routing protocols for *MANETs* with route recovery mechanisms are described in references [9], [10] and [11]. They are based on AODV with QoS extensions using a model of admission control according to *QAODV* internet draft [12]. However, these solutions do not integrate an adaptive feedback scheme by which the source node can easily adapt its transmission rate according to the state of the route.

3   AQA-AODV: QoS Routing Protocol with Adaptive Feedback Scheme for MANETs

The AQA-AODV (*Adaptive QoS-Aware Ad-hoc On-demand Distance Vector*) protocol, is a QoS routing protocol based on *AODV*, designed with the following modifications:

1. New fields in the packets used in the route discovery phase (*RREQ*, Route Request and *RREP*, Route Reply) to the bandwidth requirements and a "session ID", used to identify each QoS flow that is established.
2. An intermediate node receiving *RREQ/RREP* packets with QoS extension must examine whether it can satisfy the QoS requirements or not in order to rebroadcast/forward the packet to the next hop
3. Algorithms used for the estimation of the available bandwidth that allow nodes along the path to know their available resources (in terms of bandwidth).
4. An adaptive feedback scheme by which the source node can easily adapt its transmission rate according to the state of the route.

3.1   Route Discovery in AQA-AODV

If a source node requires a route to a destination node with specific bandwidth requirements, it broadcasts a *RREQ* packet with the QoS extension (*QRREQ*) to its neighbor nodes. When a node receives a *QRREQ* packet, a reverse route entry is created with the *session ID*, and the *QRREQ* packet is rebroadcasted as in *AODV*. This process continues until the *QRREQ* packet reaches the destination node. In *AODV*, when a destination node or an intermediate node has a "fresh enough" route to the destination, it sends a route reply message to the source [3]. However, only the destination will be able to send the route reply packet (*QRREP*) in *AQA-AODV*. This will ensure that all nodes in the selected route satisfy the bandwidth constraints. When the destination node receives a *QRREQ* packet, if it is a new request, a reverse route entry for the new session is created. Before sending the *QRREP* to the source, local available bandwidth is checked. However, it is not enough to affirm that the route can offer the required bandwidth indicated in the *QRREQ*. The reason is the mutual interference between packets of the same flow, also called "Intraflow contention" [13]. Therefore, one final check is necessary in the destination node. To estimate the intraflow contention, we use the relation between the number of hops and the end-toend throughput. Since the destination node is the last host, it can determine its distance from the source (by the number of hops in *QRREQ*). This information will allow the node to estimate the bandwidth along a path taken into account the contention between packets of the same flow. Figure 1a shows the host's working procedure after receiving a *QRREQ*. Finally, the *QRREP* will be transmitted to the source with a modified header that includes the minimum value between required bandwidth for the source and the maximum bandwidth that all hosts along the route could support taken into account the intraflow contention. Once an intermediate node receives the *QRREP* packet, it compares its available bandwidth with the bandwidth indicated in the *QRREP*. If its local available bandwidth is lower, it updates the min-bandwidth field in *QRREP*, using its available bandwidth. Otherwise, the node forwards the *QRREP*. This procedure will ensure that the source knows the minimum bandwidth along the path which will be the maximum rate that it may transmit. The procedure is shown in Figure 1b. Figure 2 illustrates the overall operation of the key phases of *AQA-AODV*.

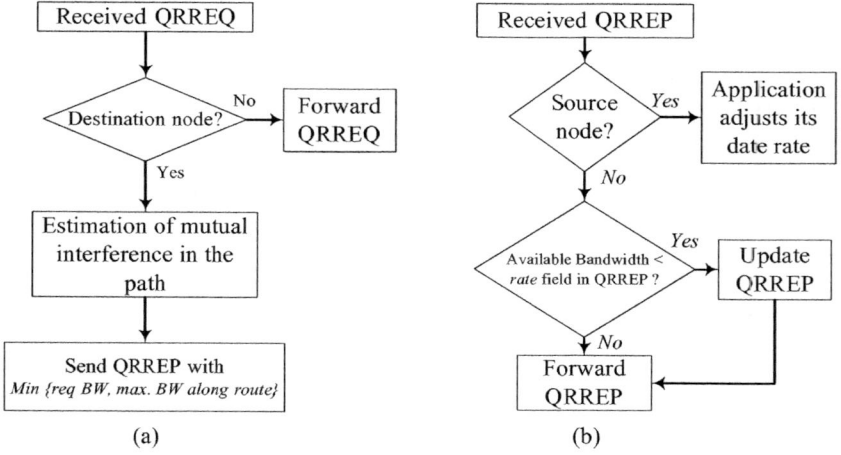

Fig. 1. Procedure in nodes after receiving a QRREQ (a) and QRREP (b)

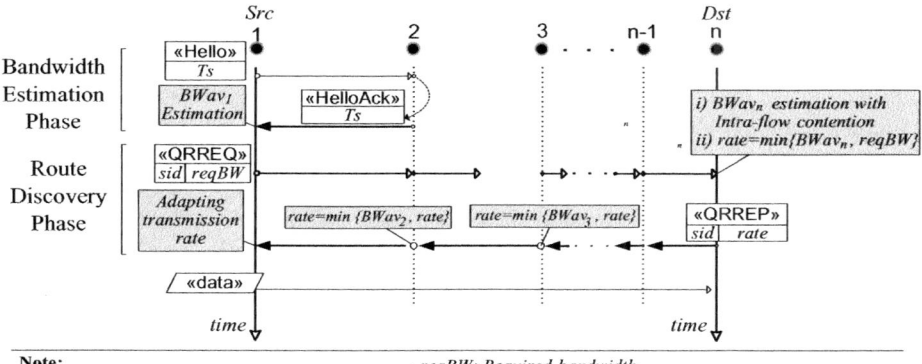

Fig. 2. Overview of AQA-AODV

3.2   Route Recovery Mechanisms for AQA-AODV

Due to changes in topology because of the mobility of the nodes and the condition of having a shared physical channel, the communications inside *MANETs* usually show frequent disruptions. For this reason, it is necessary to implement a route recovery mechanism. This mechanism not only has to re-establish the connections but also take into account the conditions of QoS that have been established during the route discovery phase.

The implemented route discovery mechanism in *AQA-AODV* detects the connection losses in a route when a host doesn't receive a *Hello* message from a neighbor during an interval of time. The *Hello* messages may not be received for three main reasons:

> Case 1. There is total connectivity but due to congestion some of the *Hello* messages are lost.
> Case 2. The neighbor node is no longer available because it is out of transmission range and the node should look for a new path to the destination.
> Case 3. The node is no longer available in the ad hoc network.

Our route recovery mechanism implemented in *AQA-AODV*, perfectly works in any of the two previous cases in which connection recovery is possible. To explain the functionality of the proposed route recovery mechanism in detail, we show two examples for case 1 and case 2. The example showed in Figure 3 consists of a network with four nodes in which each node is inside the transmission range of its one hop neighbors and inside the interference range of its two hop neighbors. Node 1(source node) broadcast a *QRREQ* message to obtain a route to node 4 with a transmission rate of 1 Mbps. When the destination node checks that the maximum transmission rate available is 0.5 Mbps, it sends the information to the source using a *QRREP* message. During the backward process of the *QR-REP*, each node checks its own available bandwidth and compares it to the value included in the *QRREP* message. This process has been described in previous sections and we will refer to it as standard procedure of route discovery. Moreover, each node adds a register in the session cache list associated to a session identifier (sid) and an expiration time (*Expiration Time*) with the aim of erasing the old registers (see Figure 3a). Once the route from node 1 to node 4 is defined, data packets are sent through the network. Each time a node gets a data packet related to that session, it updates the expiration time of the registers, avoiding the elimination of the register and keeping the session alive. When some of the hello messages sent by node 4 are lost due to congestion, node 3 detects a link failure and it sends an error message (*RERR*) to the source, including the affected session identifier (*esid*) (See Figure 3b).

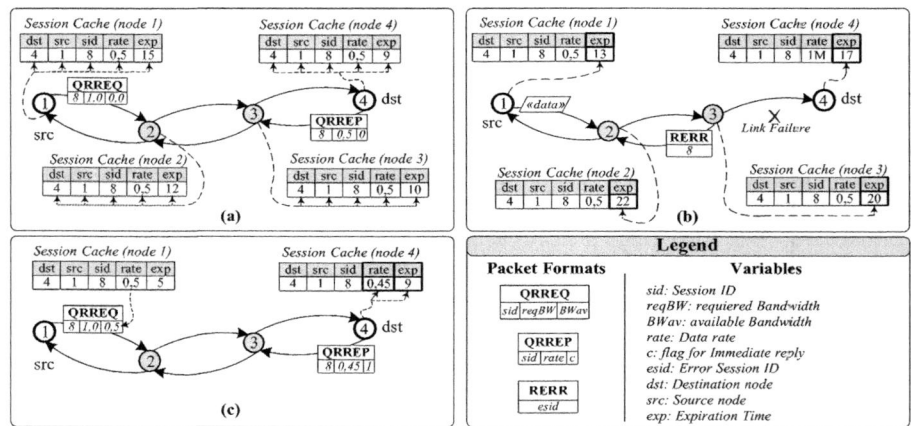

Fig. 3. Example of route recovery mechanism. Case 1.

When node 1 receives the *RERR* message, it queries its session cache list using the session identifier received in the *RERR* message (esid, Error Session ID). Therefore the source sends a *QRREQ* message which includes the required bandwidth, the actual data rate and the session identifier (see Figure 3c).

When the destination node receives the *QRREQ* message it checks if it has a register with the same sid as the one sent by the source in the *QRREQ*. If it does have one, the destination creates a *QRREP* message with the same session identifier, the supported maximum data rate – this rate may be different from the original one – and an immediate reply flag (*c = 1 immediate reply, c = 0 standard reply*). The immediate reply flag warms the intermediate nodes not to execute the standard procedure to verify the available bandwidth but send the *QRREP* message directly to the next hop back to the source. As a conclusion, the route recovery mechanism tries to re-establishconnection to destination with the QoS conditions that had been negotiated during the initial route discovery phase.

In Figure 4, we have the same conditions as in the previous example. However, a new node (node 5) has been added and it does not take part in the present route between nodes 1 and 4. In Figure 4a, we can see the information of the previous session established in the ad hoc network, using the procedure mentioned before. We also suppose that node 4 is moving in the opposite direction of node 3 and, at that moment, there will be a link failure.

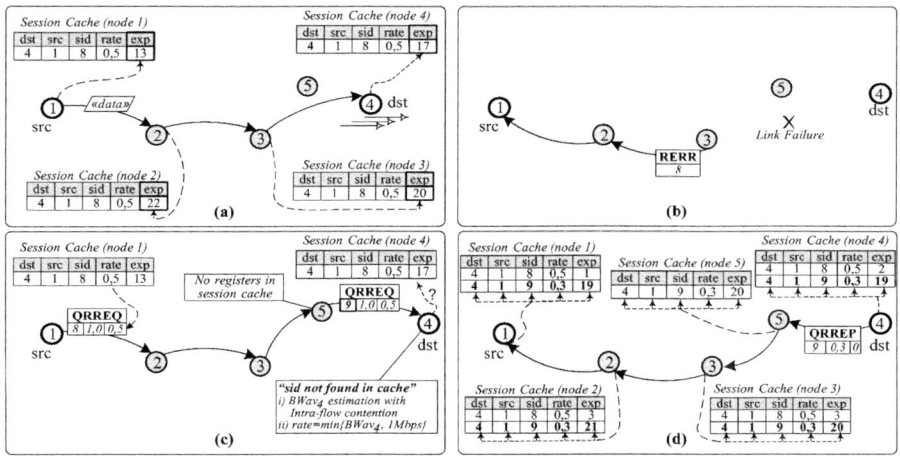

Fig. 4. Example of route recovery mechanism. Case 2.

The link failure between nodes 3 and 4 will be detected in a similar way as it was shown in Figure 3 using a *RERR* message sent to the source (Figure 4b). Nevertheless, now it is possible to achieve the destination node (node 4) through node 5, which is inside the transmission range of nodes 3 and 4. When node 1 receives the error message, it sends a *QRREQ* message and node 5, after processing the message, without finding a register associated to a session identifier (*sid*), proceeds to generate a new *sid* (Figure 4c). This makes the difference with the example shown in Figure 3. For this reason, node 4 does not take into consideration the information of the previous session and it analyses the route request in the standard way. Therefore, it calculates the available bandwidth again and compares it with the bandwidth requested by the source (1 Mbps). The response to the route request is sent to the source through the intermediate nodes using the standard way. These nodes create a new register in session cache and check if they have enough bandwidth to transmit the traffic (Figure 4c). Once the source receives the *QRREP* message, it adapts its transmission rate according to the available bandwidth calculated in the route recovery mechanism, and it starts sending packets to the destination. Registers in the session cache in each node are erased when the time-out expires.

4   Performance Evaluation

The performance of the proposed route recovery algorithm was evaluated using Network Simulator (*NS-2*). This simulator implements the IEEE802.11 protocol for the MAC layer, working in the Distributed Coordination Function (DCF) mode with a channel data rate of 2 Mbps. The radio propagation model is Two Ray Ground and queue type is Drop Tail with maximum length of 50. The transmission range and interference range are 250 m and 550 m respectively. The performance of our route recovery mechanism was evaluated by comparing it with conventional AODV protocol, using two simulation scenarios: the first scenario consists of a static linear topology with variable length where the parameters was evaluated as a function of the chain length and the second scenario consists of 30 mobile nodes in a rectangular field, 1000m x 1000m, and the mobility model uses the random waypoint model.

4.1   Simulations Results

Scenario 1: Static linear topology with variable length. The first scenario consists of a chain of nodes where the performance was evaluated as a function of the chain length. In this scenario, the performance of *AQA-AODV* is tested as function of the number of hops on the path. Node 1 is the source of data traffic and the last node in the chain is the traffic sink. Initially the source required a transmission rate of 0,9 Mbps which be maintained constant when *AODV* is used, but can be changed by the source when *AQA-AODV* is used in the network due to the adaptive feedback scheme.

As seen in Figure 5, using *AQA-AODV*, the network congestion is significantly reduced. Therefore, the time used for waiting in the packet queue and contending for the channel decreases. In other words, our adaptive feedback scheme and our route recovery mechanism allow getting an important decrease in packet loss (Figure 5a) and delay (Figure 5b) without any bandwidth sacrifice.

Figure 5c shows the number of link failures and the Connection Setup Latency (*CSL*). CSL is the latency incurred in establishing new connection from source to destination after the previous connection is lost (which includes route break detection time and recovery time). In Figure 5c, we notice that, when the chain has 3 or more nodes, the transmission rate (0.9Mbps) is not supported efficiently and the number of link failures drastically increases for *AODV*, which is about 25% higher than link failures for *AQA-AODV*. CSL (*Connection Setup Latency*) in case of *AQA-AODV* is the lowest and varies from 0.05s (for a chain of 4 nodes) to 0.5s (for a chain of 15 nodes) due to reduced recovery time in comparison with *AODV*.

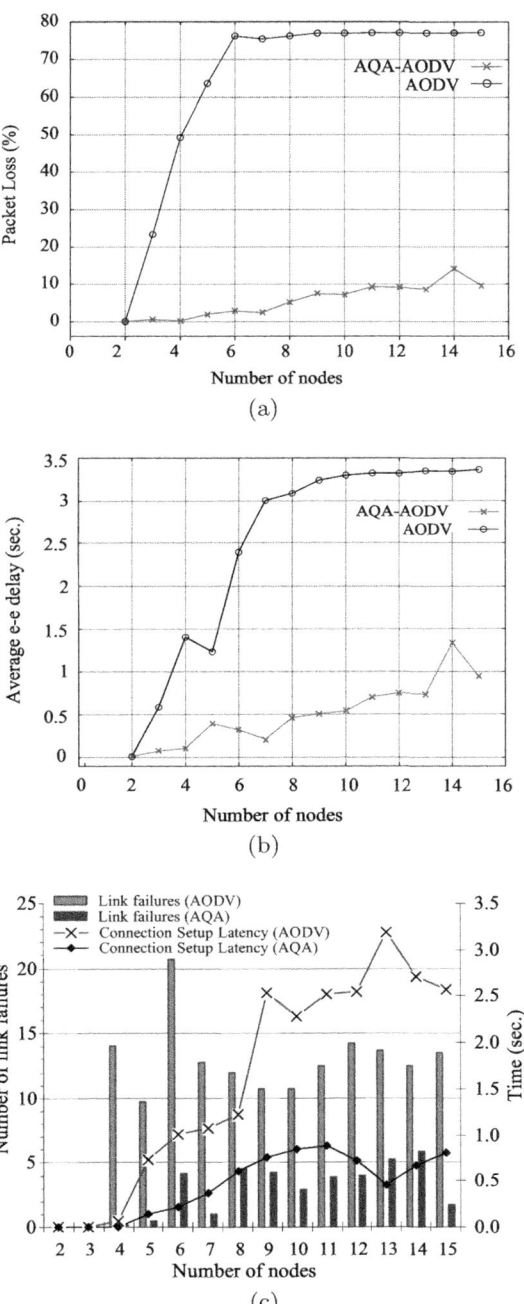

Fig. 5. Packet Loss (a), Average end-to-end delay (b) and Number of link failures and CSL, Connection Setup Latency (c) with variable chain length

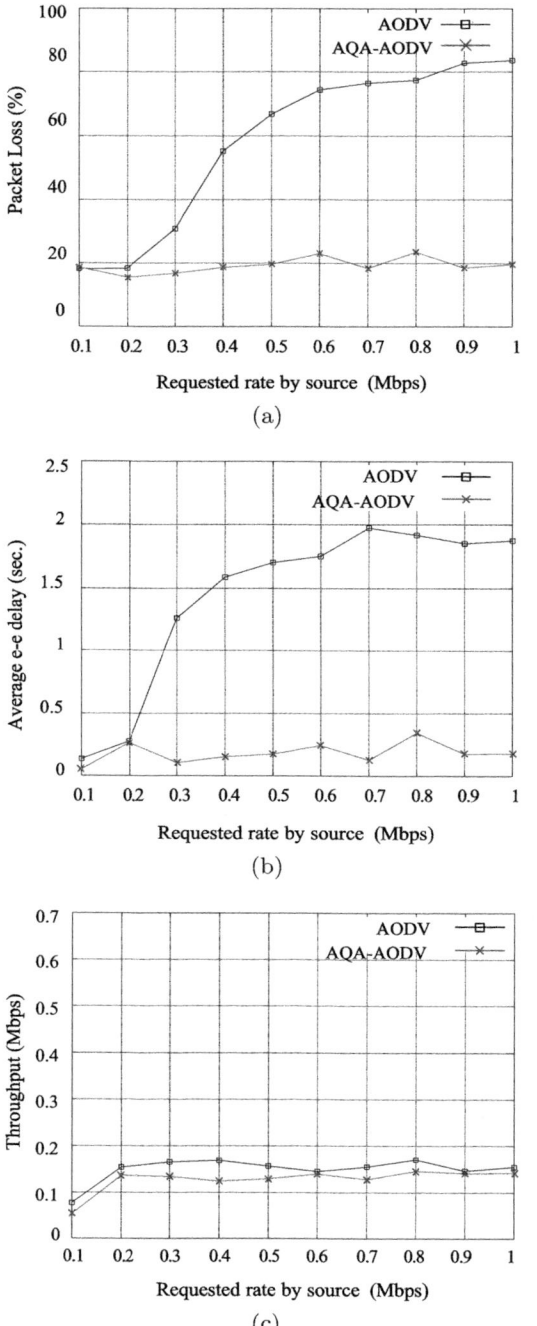

Fig. 6. Packet Loss (c), Average end-to-end delay (b) and Throughput (c) with variable requested rate

Scenario 2: Mobile Topology. The second scenario consists of 30 nodes move in a 1000m x 1000m area according to the random waypoint model with pause time set to 20sec. The nodes move toward a random destination using a speed between 0 – 3 m/s. A random source-destination pair sends packets using a request rate between 0,1 and 1,0 Mbps. All traffic flows are Constant Bit Rate (*CBR*) streams over UDP with a packet size of 1000 bytes. Figure 6 shows the results of our simulations in which the packet loss, average end to end delay and throughput are plotted versus the requested rate by source node. In terms of packet loss (Figure 6a), *AQA-AODV* shows great improvement over *AODV*, which achieves very high packet losses for some requested rates. For example, the packet loss is between 19% and 83% using *AODV*, whereas using *AQA-AODV* the packet loss remains lower than 24%.

Figure 6b shows that the average end to end delay of *AQA-AODV* is always below 0,4s, whereas, the end to end delay of *AODV* increases badly when the transmission rate increases from 200 kbps to 1000 kbps. With *AODV*, the maximum average end to end delay reaches 1,9s at 700 kbps, about 16 times higher than using *AQA-AODV*. As seen in Figure 6c the total network throughput achieved with *AQA-AODV* is very close to throughput achieved using *AODV*. We would expect the *AQA-AODV* protocol's performance will degrade in scenarios with high mobility because the nodes will need a specific time for exchanging information about the network status. We observe in figure 7 that in a scenario with mobile nodes the frequencies of route break increase in the network compared with previous static scenario where the link failures were caused by congestion in the nodes. Each time a route breaks due to node mobility, there is some latency in new connection setup (which includes route break detection time, route discovery time and recovery time) and packet gets lost during

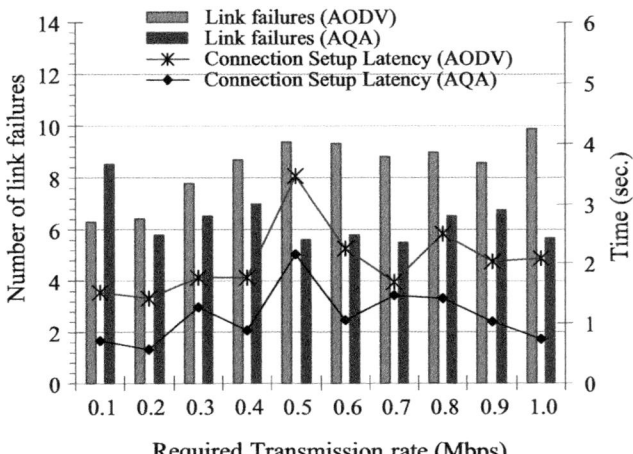

Fig. 7. Number of link failures and CSL (Connection Setup Latency) with variable requested rate

connection setup period which could explain the growth of the packet loss. Figure 7 shows that the *CSL* of *AQA-AODV* is always lower than *CSL* of *AODV* (about 50% of average lower than *CSL* for *AODV*).

5   Conclusions

The proposed route recovery algorithm for QoS routing protocol (*AQA-AODV*) can contribute to the diminishing of the latency incurred in establishing new connection from source to destination after the previous connection is lost. Our approach incorporates new fields in the route request and route reply packets, an extension of the routing table and the implementation of a session cache table where registers of the active sessions are stored. Simulations show that our proposed mechanism is perfectly integrated into adaptive feedback scheme of *AQA-AODV*. This reduces significantly the dropping rate, the end-to-end delay and the connection setup latency, without impacting the overall end-to-end throughput. In the future, we plan to examine how to implement a hybrid algorithm using source and local repair, which would decrease the connection setup latency and improve the performance in mobile environments.

Our main goal is to implement a framework where the video source exploits the feedback information from the underlying protocol (*AQA-AODV*) to tune a parameter on the source coding in order to adapt the traffic rate to the path. Moreover, this framework must include the route recovery mechanism that would allow nodes to repair link failures with previous QoS conditions.

References

1. Wu, D., Hou, T., Zhu, W., Lee, H.-J., Chiang, T., Zhang, Y.-K., Chao, H.J.: On End-to- End Architecture for Transporting MPEG-4 Video Over the Internet. IEEE Transactions on Circuits and Systems for Video Technology 10(6), 923–941 (2000)
2. Castellanos, W., Acelas, P., Arce, P., Guerri, J.: Evaluation of a QoS-Aware Protocol with Adaptive Feedback Scheme for Mobile Ad Hoc Networks. In: 6th International ICST Conference on Heterogeneous Networking for Quality, Reliability, Security and Robustness, QShine 2009, Las Palmas, Gran Canaria (2009)
3. Perkins, C., Royer, E.M., Das, S.: Ad hoc on-demand distance vector (AODV) routing. IETF RFC 3561 (2003)
4. The Network Simulator NS-2, http://www.isi.edu/nsnam/ns/
5. Pan, M., Chuang, S., Wang, S.: Local repair mechanisms for on-demand routing in mobile ad hoc networks. In: Proceedings of Pacific Rim International Symposium on Dependable Computing (2005)
6. Youn, J., Lee, J., Sung, D., Kang, C.: Quick local repair scheme using adaptive promiscuous mode in mobile ad hoc networks. Journal of Networks 1(1), 1–11 (2006)
7. Pereira, N., de Moraes, R.M.: A comparative analysis of AODV route recovery mechanisms in wireless Ad Hoc networks. In: IEEE LATINCOM 2009 Conference on Communications, pp. 1–6 (2009)

8. Sarma, N., Nandi, S., Tripathi, R.: Enhancing Route Recovery for QAODV Routing in Mobile Ad Hoc Network. In: The International Symposium on Parallel Architectures, Algorithms, and Networks (2008), doi:10.1109/I-SPAN
9. De Renesse, R., Friderikos, V., Aghvami, H.: Towards Providing Adaptive Quality of Service in Mobile Ad-Hoc Networks. In: IEEE VTC, Melbourne (2006)
10. Zhang, Y., Gulliver, T.: Quality of service for ad hoc on-demand distance vector routing. In: Proceedings of WiMob 2005, vol. 3, pp. 192–196 (2005)
11. Kumar, V., Ilanchezhiapandian, G.: Enhancement of AODV Routing protocol to provide QOS for MANET. In: Proceedings of the National Conference on Emerging Trends in Computing Science NCETCS (2010)
12. Perkins, C., Royer, E.M.: Quality of service for ad hoc on-demand distance vector routing. IETF Draft (2004),
   http://www.psg.com/~charliep/txt/aodvid/qos.txt
13. Sanzgiri, K., Chakeres, I., Belding-Royer, E.: Determining intra-flow contention along multihop paths in wireless networks. In: Proc. Broadnets 2004 Wireless Netw. Symp., pp. 611–620 (2004)

Cross–Layer Adaptation of H.264/AVC over 3G UMTS Mobile Video Services

Jose Oscar Fajardo, Ianire Taboada, Fidel Liberal, and Armando Ferro

University of the Basque Country, Spain
joseoscar.fajardo@ehu.es
http://det.bi.ehu.es/NQAS

Abstract. This paper deals with the analysis of user perceived visual quality for mobile multimedia services. H.264/AVC is selected for low resolution video encoding, and 3G UMTS is considered for mobile data services. From subjective tests results, the combined impact of different service– and network–level parameters is inferred. As a result, different cross–layer adaptation alternatives are proposed to maximize the perceived quality level under different service conditions.

Keywords: Mobile multimedia services, H.264/AVC, 3G UMTS, adaptation.

1   Introduction

Multimedia applications have become increasingly popular over Internet. Furthermore, the wide–spreading of enhanced mobile Internet access and the continuous evolution of multimedia encoding techniques allow provisioning video services over mobile data connections at acceptable quality levels. For instance, the mobile version of the YouTube video sharing platform currently offers H.264/AVC–encoded video clips, which can be accessed from a mobile handset via RTSP with a standard video client. Typically, two versions of the video clips are available. On one hand, the normal version is based on the low spatial resolution (SR) of QCIF (176x144) and low encoding bitrate (about 80kbps). On the other one, the High Quality (HQ) versions offer a higher SR at QVGA –or square pixel SIF– (320x240) at higher encoding bitrates (about 250kbps).

The choice of the most suitable version depends on different parameters. The screen resolution of the mobile device imposes a first requirement. Nowadays, typical screen resolutions are QCIF, QVGA and VGA. As well, the performance variability associated to mobile data services must be considered. Taking into account a typical 3G Universal Mobile Telecommunications System (UMTS) service, the 384kbps bearer supports the requirements of HQ versions under perfect reception conditions. However, the variable radio conditions may introduce degradations in the transmission, and the normal version could be preferred.

As widely studied, an adaptive approach could provide the best quality level all over the service time by introducing real–time modifications in the service

J. Del Ser et al. (Eds.): MOBILIGHT 2011, LNICST 81, pp. 94–108, 2012.

provision configuration. This paper deals with this topic, focusing on the provision of H.264/AVC–based video services over 3G UMTS data connections. Specifically, the work presented here tries to offer a consolidated performance study by taking into account the combined effects of the specific characteristics of the UMTS Terrestrial Radio Access Network (UTRAN) on one hand, and the specific characteristics of the H.264 encoding on the other one. Additionally, both service– and network–level adaptations are proposed in a cross–layer approach. In order to perform the most suitable adaptation actions, the decision making process shall be driven by the expected visual quality level as perceived by end users. Extensive subjective tests have been performed in order to model the combined impact of the service and network conditions into the visual quality.

1.1 Background

The analysis of user perceived visual quality has been subject of many studies in recent years. With regard to H.264/AVC at mobile resolutions, several studies present their subjective tests results for service contexts similar to the considered in this paper.

From the subjective tests results presented in [1], the H.264/AVC codec can be assumed as the best performing codec for QCIF mobile video services. As well, the most suitable audio–video (A/V) bitrate ratio is inferred as a function of the content type (CT). In [2] authors study the combined effects of the source bitrate (SBR) and frame rate (FR) for mobile resolutions. The main outcome is a regression–based expression which relates the Mean Opinion Score (MOS) to the SBR and FR values for different CT. The FR is considered in the range of 5 fps to 15 fps, while the target SBR varies from 24kbps to 105kbps. In this case, all the subjective test sequences are presented at QCIF resolution screen, which justifies the low values considered for SBR. This range of SBR values is considered insufficient for the aims of this paper, where higher SR values are considered. As well, in these studies only the encoding parameters are considered, assuming ideal transmission conditions.

Concerning higher SR, e.g. results in [3] present quality evaluations of CIF (352x288) video sequences in terms of blocking, blurring and flickering artifacts. The SBR is considered from 100kbps to 300kbps for this SR at 25fps. However, only the encoding effects are considered for the quality analysis as well.

Both H.264/AVC CIF video sequences and transmission effects are considered in the analyses of perceived video quality presented in [4] and [5]. In both cases, a 2–state Markov model is used to implement the bursty packet losses. However, in both cases the loss model is implemented at IP level. In our case, the bursty error pattern is implemented at UMTS Radio Link Control (RLC) level following the results shown in [6], where the error pattern is obtained from live UMTS network traces. This feature entails a better emulation of the combined effects of the service–level settings and the experienced UMTS performance conditions.

As a result, although all the reviewed studies are close to the case study, none of them covers all the objectives proposed in this paper and hence new subjective tests have been performed.

1.2   Scope and Objectives

The main objective of the study is to in–depth study the combined impact of different service– and network–level parameters into the experienced visual quality from a consolidated standpoint, in order to analyze the most suitable cross–layer adaptations that maximize the Quality of Experience (QoE). In this sense, the main contributions of this paper are twofold:

- A thorough analysis of the visual quality which could be expected in a mobile multimedia service, as currently being provided in real–world services.
- A dynamic cross–layer adaptation mechanism based on three configurable parameters, aimed at maximizing the QoE under variable UMTS service conditions.

One of the main novelties of the paper is the mobile multimedia service awareness. All the subjective tests have been performed resembling actual service conditions, including mobile–oriented media encoding and presentation to end users. As well, the used UMTS reception patterns are based on real–world measurements under different mobility scenarios.

The remainder of this paper is structured as follows. Section 2 presents the methodology followed for performing the subjective tests. Section 3 focuses on the analysis of perceived visual quality, in function of the encoding settings. From subjective tests results, the evolution of expected MOS with the SBR is inferred for each CT and SR. This way, the most suitable SR can be identified for the achievable SBR and per CT. Likewise, Section 4 focuses on the analysis of perceived visual quality in function of the UMTS performance. From subjective tests results, the evolution of expected MOS with the experienced UMTS conditions is inferred for each CT. This way, the most suitable SBR can be identified for the experienced Block Error Rate (BLER) and per CT. Section 5 states the considered adaptation capabilities and discusses the expected improvement in terms of QoE. Based on the subjective testing results, a cross–layer adaptation mechanism for mobile video streaming services is proposed and evaluated. Finally, Section 6 gathers the main conclusions to this paper.

2   Subjective Testing Methodology

The experimental design for the subjective video tests is mainly based on ITU recommendations and tutorials [7] [8] [9]. Different aspects are taken into account for planning the subjective tests for this kind of multimedia applications.

Concerning the viewing conditions, video sequences are presented to end users resembling a mobile UMTS service in order to enhance the accuracy of results [10]. All video sequences are displayed in a mobile handset and users are asked to hold it in their hands. The device used in the tests is a Nokia N95-8Gb, which provides a 320x240 screen resolution. The tests are carried out with RealPlayer for s60, which uses image re–scaling for presenting QCIF sequences at full screen.

The video sequences are a priori generated including both encoding and transmission effects, and they are stored in the mobile device for its presentation to subjects. So, an appropriate device and displaying format is used to achieve the proposed objectives in a fixed environment conditions for all the subjects. In order to capture the combined effects of the specific encoding and transmission techniques, long duration video sequences (about 2 minutes) have been used instead of the typical short (about 10s) reference video sequences.

Taking as reference the proposed test structure in [7] [8] and [9], before starting with test sessions, written instructions were shown to subjects and a training phase was done in which some videos are presented and evaluated, without taking into account these results. Then this is followed by several test sessions. In each test session different types of test scenes are shown. These are presented in random order and some implicit replications are included to check coherence [7]. Due to fatigue issues, break periods between sessions are introduced [9]. For carrying out the different experiments proposed in this study, each subject participates in the experiments for three different days.

In order to reproduce viewing conditions that are as close as possible to real–world contexts, Single Stimulus (SS) tests are used and the audio track is also included in the multimedia stream. The evaluations are based on Absolute Category Rating (ACR). Thus, after each test sequence presentation the subjects are asked to evaluate the quality of the presented sequence in the MOS scale of 1 to 5.

3   Service–Level Parameters

Three different CT have been considered: low–motion (LM), medium–motion (MM), and high–motion (HM) video sequences. As seen in [11], there are different alternative approaches for content classification based on the spatial and temporal activity of the whole video sequences. Besides, other studies (e.g. [12]) propose the use of a single content motion estimation metric in a dynamic way all over the video sequence. In this paper, the content categorization has been performed following the parameters proposed in [13], which associates an scene complexity (C) and level of motion (M) value to each video sequence based on the average bits per frame and the average quantization parameter (QP) for I and P frames respectively. These metrics allow implementing an easy estimation of the content complexity directly from the results of the encoding process. All the video sequences have been encoded with the H.264/AVC Joint Model Reference Software. The considered content and associated complexity metrics are described as follows:

- LM sequence is made up of news video clip with two people, including change of planes and a commercial in middle of the sequence (C=0.62; M=0.96).
- MM sequence is a typical TV series scene, featuring different people and including change of planes (C=0.49; M=0.42).
- HM sequence is taken from a basketball top 10 best plays video sequence, with changes of planes from field to face planes (C=0.46; M=0.28).

Instead of analyzing the visual quality as a function of the combined setting of FR and SBR, in this study only the SBR is modified. The FR values are set up

Table 1. Encoding settings for subjective tests

CT	SBR (kbps)	SR	FR (fps)
LM	{80, 130, 200}	320x240	10
	{48, 88, 128}	176x144	10
MM	{80, 130, 200}	320x240	12.5
	{48, 88, 128}	176x144	12.5
HM	{80, 200, 256}	320x240	15
	{48, 88, 128}	176x144	15

in function of the content dynamics: 10, 12.5 and 15 fps are established for LM, MM and HM respectively. Table 1 summarizes the considered values.

The H.264/AVC encoder is set up to its Baseline Profile at level 1.2. The frame structure is IPPP, with a Group of Pictures (GOP) size of 10 seconds. Thus, each 10 seconds the video sequence is refreshed with a new I frame.

3.1   Encoding Quality

First, subjective tests were performed for evaluating the impact of SBR into the encoding quality of QVGA sequences. A total of 20 people evaluated each sequence, for a total of 180 tests. The obtained results are presented in Fig. 1, which gathers the box plots of the subjective testing results for the whole set of considered CT–SR conditions. For each combination of CT and SBR considered, the median, Q1 and Q3 percentiles and minimum/maximum values are illustrated. As well, those values which are considered outliers in the data sample are individually plotted. Outlier values are determined by the Chauvenet's Criterion and they are not taken into account in the computation of the MOS. Table 2 shows the average quality values derived from the subjetive tests for QVGA resolutions.

Table 2. Subjective tests results for QVGA

SBR (kbps)	LM	MM	HM
80	3.49	2.90	1.98
130	4.26	3.92	–
200	4.53	4.49	3.73
256	–	–	4.26

Similarly, new tests were performed for the QCIF versions of the same videos. In order to resemble actual service conditions, QCIF video sequences are presented at full screen in the mobile handset with RealPlayer, so image scaling is required. A total of 20 people evaluated each sequence, for a total of 180 tests. In this case only the averaged MOS values are provided in Table 3.

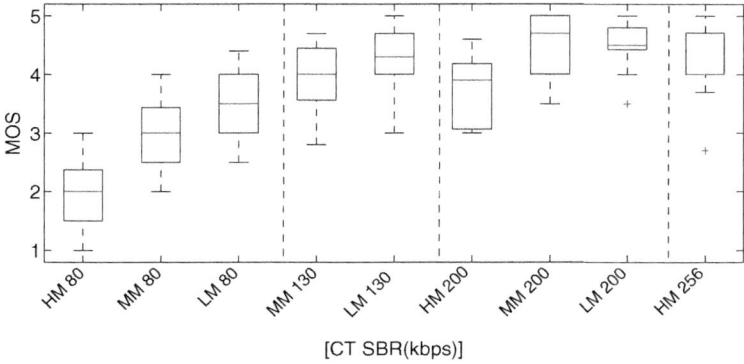

Fig. 1. Boxplot of subjective testing results for encoding of QVGA sequences

Table 3. Subjective tests results for QCIF

SBR (kbps)	LM	MM	HM
48	3.49	2.75	1.83
88	3.52	3.46	2.65
128	3.66	3.60	2.91

The QCIF versions, being half the size than the QVGA sequences, require less SBR to achieve an acceptable visual quality. However, the increase of SBR does not entail a proportional increase in the perceived quality for different SR values. As described in [5], the video encoding process may reach the visual quality threshold and above this threshold a higher SBR is not captured by the human visual system.

3.2 Impact on Service–Level Adaptation

In order to evaluate the evolution of both alternatives, we applied fitting techniques to the subjective results and obtained an approximation of the relationship between MOS and SBR values for both SR. The fitting function is given in equation 1 as proposed in [5].

$$MOS = a \cdot log(SBR) + b \qquad (1)$$

Parameters a and b are related to the activity level of the content and spatial resolution, and the obtained values are illustrated in Table 4.

By using these results, Fig. 2 presents the expected evolution of the subjective visual quality in terms of MOS for each pair of considered CT–SR values. For each CT, the two resulting curves cross at a specific point (SBR_{th}) and two

Table 4. Experimental coefficients for the fitting function

	LM QVGA	MM QVGA	HM QVGA	LM QCIF	MM QCIF	HM QCIF
a	1.141	1.739	2.274	0.159	0.608	1.124
b	-1.442	-4.665	-7.954	2.858	0.556	-2.48

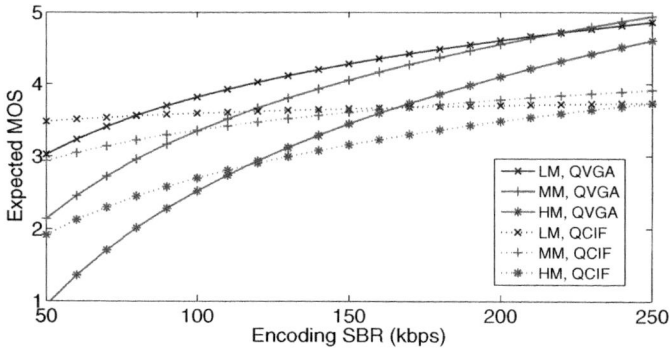

Fig. 2. Evolution of expected MOS vs. SBR per CT and SR

differentiated regions are identified. For SBR values lower than SBR_{th}, it is preferable to switch to a lower SR in order to control the impact of the limited SBR. The SBR_{th} value is higher for more dynamic sequences. For the three CT considered in this study, we find that this threshold is around 80, 100 and 115 kbps for LM, MM and HM respectively.

4   UMTS Access Network

The scope of this paper regards to the provision of mobile video streaming services over a typical wide–area 3G UMTS data service, as defined in 3GPP TR 25.993 [14] for the $Interactive or Background/UL : 64 DL : 384 kbps/PSRAB$. This kind of bearer service provides a maximum downlink bitrate (DLBR) of 384kbps. A detailed description of the considered service provision and the impact of the UTRAN is given in [15].

From the total bitrate, the final amount available for the video encoding process is reduced by several factors. First of all, the RTP/UDP/IP packetization introduces an overhead in the data transmission. Second, we must consider the effect of the audio stream within the multimedia transmission. The analysis of the audio quality and the integral quality (as shown e.g. in [16]) is out of the scope of this paper. Yet, the impact of its transmission is simulated by adding a 64 kbps stream to the considered video stream. Finally, part of the DLBR is used by the RLC AM functions for the local recovery of lost MAC PDUs. In this sense, the performance of the video streaming service highly depends on the radio error pattern.

As cited in Sect. 1.1, one of the novelties of this paper is that the UMTS error model is implemented at RLC level from real–world measurement results, instead of using a typical 2–state Markov model for simulating the IP–level loss events. For the simulation of different network conditions, the implemented error model is a 2–state Markov model with variable Block Error Rate (BLER) values.

Two characteristics are adopted from the results presented in [6]:

– For mobile users, the radio errors can be grouped at Transmission Time Interval (TTI) level.
– The Mean Burst Length (MBL) of erroneous TTIs can be approximated to 1.75.

The error model, as well as the simulation methodology, is further detailed in [17]. The RLC–level error model, in combination with the application–level settings, determines the performance of the service. The RLC errors may derive to additional delays if the RLC is able to recover the lost PDU, or to video frame losses otherwise.

Taking into account the relevance of the different frame types, a content–aware scheduling is implemented in a similar way to the concepts proposed in [18] for 3G UMTS and in [19] for HSDPA. In this case, the priority of different RLC retransmissions is modified in order to implement an enhanced protection for I frames. This way, we prevent severe degradations in the initial picture of each GOP and its propagation all over the 10 seconds period.

4.1   3G UMTS Transmission Quality

Considering the mentioned characteristics for 3G UMTS mobile multimedia services, different combinations of service and network conditions are simulated. All the simulations are run with OPNET Modeler, where both the specific UMTS error pattern and a H.264/AVC RTP trace injector have been implemented. QVGA video traces corresponding to the three CT have been used in the simulations with several SBR values. 80, 130 and 200 kbps have been considered for LM and MM video sequences, while an additional 256 kbps version has been used for HM traces. All the traces have been transmitted several times from a video server to the mobile endpoint, traversing the UMTS network segment. The downlink BLER value in the radio part is set up from 1% to 30% at 5% steps. At 30% of BLER, all the sequences experience high degradations except for the LM 80kbps versions.

From the whole set of results obtained, a mapping between different BLER values and experienced IP Loss Ratio (IPLR) patterns is established. For those points with negligible IPLR values (under 0.1%), the service performance is considered accurate. Similarly, high IPLR values (over 5%) indicate unaffordable service conditions.

The rest of intermediate points are considered for subjective evaluation of the visual quality perceived by users. Table 5 presents all the service– and network–level conditions that have been included in this set of subjective tests. The application–level performance (in terms of IPLR) is not only determined by the experienced BLER values, but also by the different traffic patterns associated to the different CT. Thus, different conditions require subjective evaluation in function of the CT. For the aims of this study, a total of 114 subjective tests were performed by 20 people.

Table 5. SBR and BLER settings for subjective tests

CT	SBR (kbps)	BLER(%)
LM	130	15, 20, 25
	200	5, 10
MM	130	20, 25
	200	5, 10
HM	256	5

Fig. 3 illustrates the results obtained from subjective tests for the considered encoding/transmission conditions. As can be observed, especially two conditions (namely LM–130–20 and MM–200–10) show a high variability in the quality scores provided by users. In those cases, the associated MOS are 2.15 and 2.58 respectively.

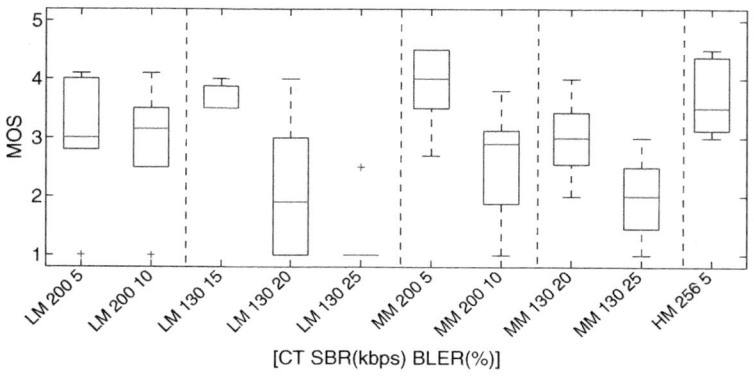

Fig. 3. Boxplot of subjective testing results for UMTS transmission of QVGA sequences

4.2   Impact on Cross–Layer Service Adaptation

The extensive results obtained both from the analysis of simulation results and from subjective tests allow us depicting a mapping between the application and transmission conditions to the expected video quality in terms of MOS, as shown in Fig. 4 for each considered CT.

From the behavior illustrated in Fig. 4, each combination of CT, SBR and BLER determines the expected visual quality value. Thus, for a specific BLER value, the video streaming session can be set up to the new SBR value that maximizes the expected MOS. For the aims of this paper, this adaptation is considered as a standalone decision making process.

The BLER value is not modified (e.g. by modifying the power control functions of the link layer) so the impact of the adaptation of a video stream on the performance experienced by other users in the same cell is limited in this case.

If power or rate control mechanisms are considered in multi–user environments, where several users are contending for the access to limited cell resources at the same time, the optimization problem can be studied as shown in [20].

Another alternative in the standalone management of mobile services is the capability of modifying the Radio Bearer settings in function of the experienced conditions. If for the experienced BLER condition none of the highest considered

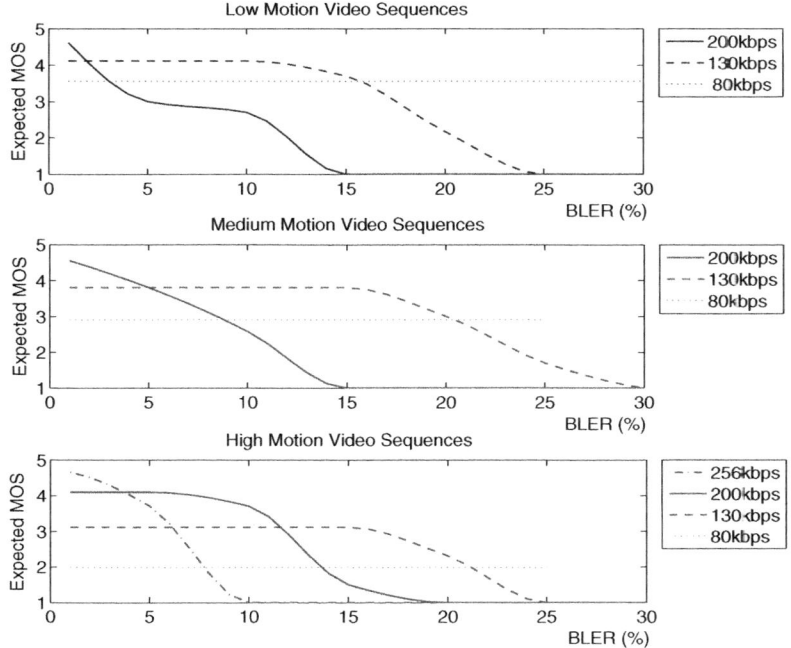

Fig. 4. Evolution of expected MOS vs. BLER per CT and SBR for QVGA sequences

SBR values is suitable, the UMTS Radio Bearer can be switched to a DLBR value of 128 kbps, which exhibits a better resilience to noise and interference at the same transmission power levels. Thus, low BLER values can be expected in order to guarantee no further impairments than the encoding process itself.

At the same time, the multimedia streams are switched to a lower SBR in order to cope with the new DLBR requirements. As a result, this approach entails a combined encoding/network service adaptation. As in the previous case, the impact on other users is limited and each user can be adapted by itself.

5   Evaluation of Cross–Layer Adaptations

Based on the results obtained from the subjective tests, different adaptation approaches can be evaluated. Four alternative approaches are considered, each option including a new adaptation capability from zero to three configurable parameters.

- *No Adaptation (NA)*. In this case, no adaptations actions are considered. Thus, the achieved MOS values are those corresponding to the initial service conditions. In order to offer accurate quality levels, for LM and MM sequences the initial SBR is set up to 200kbps, while HM sequences are configured to 256kps.
- *Network–Aware Bitrate Adaptation (NABA)*. This alternative considers the adaptation of SBR to the value that maximizes the expected MOS for the specific CT and experienced BLER, driven by the estimation curves shown in Fig. 4. For the purposes of this paper, the allowed SBR values are 256, 200, 130 and 80 kbps.
- *Network–Aware Bitrate and Spatial Adaptation (NABSA)*. This approach takes into account the two configurable service–level parameters considered in this paper: SBR and SR. Thus, based on the resulting SBR adaptation, the SR is also switched from QVGA to QCIF if the optimal SBR is lower than the SBR_{th} for the specific CT as illustrated in Sect. 3.2.
- *Network–Aware Cross–Layer Adaptation (NACLA)*. In this case, a cross–layer adaptation is adopted by taking into account the three parameters considered. Besides the application–level adaptation (combined SBR/SR), the DLBR can be decreased to overcome severe degradation conditions in the UMTS data connection.

Fig. 5 shows the obtained results for each dynamic adaptation approach, taking into account the different CT considered.

For LM sequences, only two curves are differentiated. If no dynamic adaptation is applied, the mobile video service gets completely degraded around the 15% of BLER. However, the service can be kept at suitable quality levels (MOS=3.5)

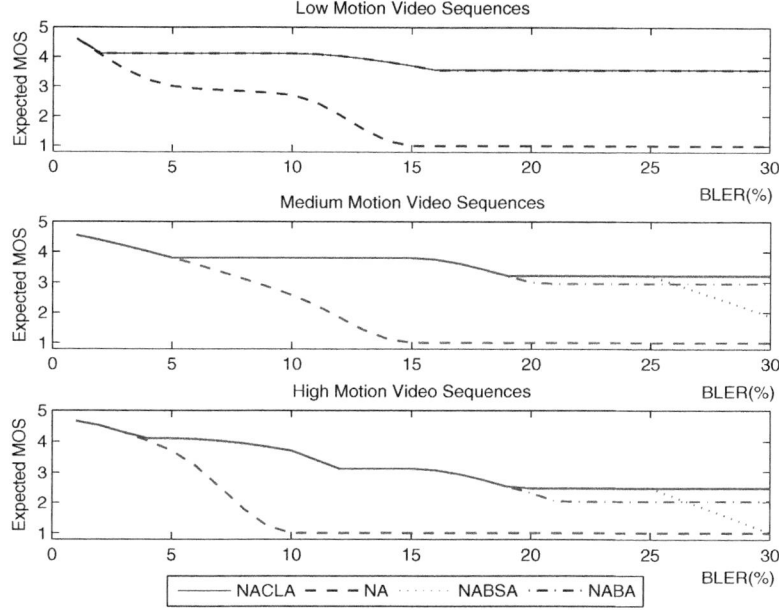

Fig. 5. Quality achieved by the different adaptation approaches

just with bitrate adaptation under the analysed UMTS conditions. The LM sequences exhibit no degradations at BLER=30%, and both SR versions provide similar quality levels at SBR=80kbps. Hence, the three dynamic adaptation approaches offer analogous behaviours at the whole range of the BLER conditions studied. Thus, just SBR adaptation provides the maximum achievable quality under different conditions in this case.

As can be observed, this is not case for MM and HM sequences. On one hand, the SBR_{th} is located above 80kbps in both cases, and thus the SR should be switched to QCIF when SBR is set up to 80kbps. On the other hand, the QCIF versions exhibit frame losses when the experienced BLER is above 25%. As a result, the three dynamic adaptation approaches provide different quality levels under severe UMTS degradations and different type of adaptations are required in order to maximize the QoE.

From the analysis of results, a dynamic network–aware cross–layer adaptation mechanism can be proposed, as illustrated in Fig. 6. Following the depicted logic, a mobile endpoint could be capable of launching the required adaptation procedures to keep the mobile video service in the maximum achievable quality level along the service time.

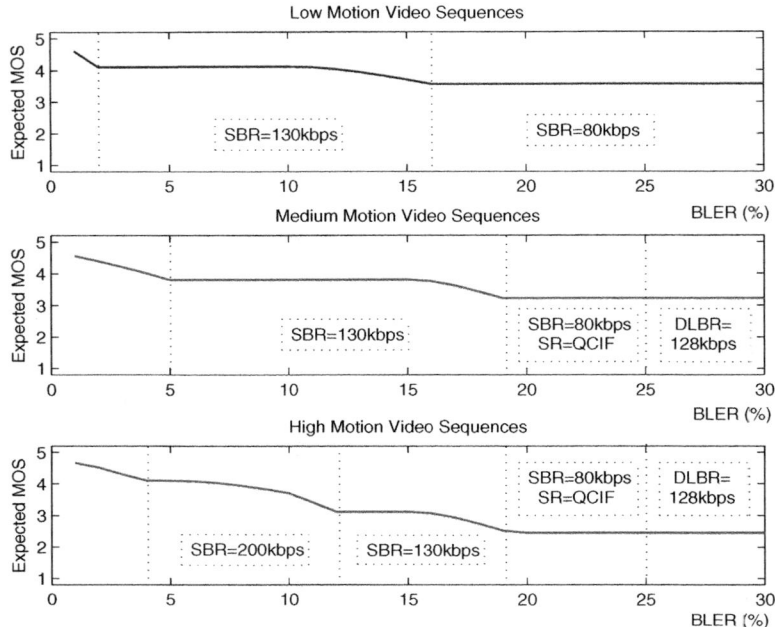

Fig. 6. Dynamic configuration and maximum achievable quality

6 Conclusions

This paper deals with possible dynamic adaptations for H.264/AVC based video services over 3G UMTS mobile connections. In order to get an optimal configuration, different cross–layer adaptations are proposed and evaluated. Thus, both service– and network–level adaptations are considered to cooperate in a coordinated way in order to maximize the QoE.

Since the proposed adaptations are driven by the expected QoE, specific subjective tests have been performed to cope with the specific service context. Thus, both specific H.264/AVC parameters and specific error models for commercial UMTS networks have been used. The outcomes of the subjective testing phase are later considered as inputs for the dynamic service adaptation logic.

Different adaptation approaches have been evaluated, considering different number of variable parameters. As a main result of this paper, we can state the enhancements of the three–parameter adaptation approach, based on the combination of encoding bitrate, spatial resolution and UMTS bearer bitrate. As described in Section 5, the considered adaptations do not have an impact on other users, so it can be implemented in a per–user basis.

In order to implement the adaptation logic at the mobile endpoint, the mobile device requires access to low–level parameters. Currently, several Android–based commercial mobile devices are capable of providing BLER statistics in real–time. As a work in progress, we have developed the software to capture these statistics from the chipset in order to make them available for the applications.

References

1. Jumisko–Pyykkö, S., Häkkinen, J.: Evaluation of subjective video quality of mobile devices. In: 13th Annual ACM International Conference on Multimedia – MULTIMEDIA, pp. 535–538. ACM, New York (2005)
2. Ries, M., Nemethova, O., Rupp, M.: Video Quality Estimation for Mobile H.264/AVC Video Streaming. Journal of Communications 3(1), 41–50 (2008)
3. Ho, H.H., Wolff, T., Salatino, M., Foley, J.M., Mitra, S.K., Yamada, T., Harasaki, H.: An investigation on the subjective quality of H. 264 compressed/decompressed videos. In: Third International Workshop on Video Processing and Quality Metrics for Consumer Electronics – VPQM (2007)
4. De Simone, F., Naccari, M., Tagliasacchi, M., Dufaux, F.D., Tubaro, S., Ebrahimi, T.E.: Subjective assessment of H.264/AVC video sequences transmitted over a noisy channel. In: International Workshop on Quality of Multimedia Experience – QoMEX, pp. 1–4. IEEE (2009)
5. Koumaras, H., Lin, C.-H., Shieh, C.-K., Kourtis, A.: A Framework for End–to–End Video Quality Prediction of MPEG Video. Journal of Visual Communication and Image Representation 21(2), 139–154 (2010)
6. Karner, W., Nemethova, O., Svoboda, P., Rupp, M.: Link Error Analysis and Modeling for Video Streaming Cross–Layer Design in Mobile Communication Networks. ETRI Journal 29(5), 569–595 (2007)
7. ITU-R BT 500-10: Methodology for the subjective assessment of the quality of the television pictures (2000)
8. ITU-T Rec. P 910: Subjective video quality assessment methods for multimedia applications (1999)
9. ITU-T Rec. J.148: Requirements for an objective perceptual multimedia quality model (2003)
10. Jumisko–Pyykkö, S., Hannuksela, M.M.: Does Context Matter in Quality Evaluation of Mobile Television? In: 10th International Conference on Human Computer Interaction with Mobile Devices and Services, pp. 63–72. ACM, New York (2008)
11. Korhonen, J., You, J.: Improving objective video quality assessment with content analysis. In: Fifth International Workshop on Video Processing and Quality Metrics for Consumer Electronics – VPQM, pp. 1–6 (2010)
12. Unanue, I., Del Ser, J., Sanchez, P., Casasempere, J.: H. 264/SVC rate-resiliency tradeoff in faulty communications through 802.16 e railway networks. In: International Confwerence on Ultra Modern Telecommunications & Workshops – ICUMT, pp. 1–6. IEEE (2009)
13. Hu, J., Wildfeuer, H.: Use of content complexity factors in video over IP quality monitoring. In: International Workshop on Quality of Multimedia Experience – QoMEX, pp. 1–4. IEEE (2009)
14. 3GPP TR 25.993: Typical examples of Radio Access Bearers (RABs) and Radio Bearers (RBs) supported by Universal Terrestrial Radio Access (UTRA) (2008)
15. Fajardo, J.O., Liberal, F., Bilbao, N.: Impact of the video slice size on the visual quality for H.264 over 3G UMTS services. In: Sixth International Conference on Broadband Communications, Networks, and Systems – BROADNETS, pp. 1–8. IEEE (2009)

16. Winkler, S., Faller, C.: Perceived audiovisual quality of low-bitrate multimedia content. IEEE Transactions on Multimedia 8(5), 973–980 (2006)
17. Khan, A., Sun, L., Ifeachor, E., Fajardo, J.O., Liberal, F.: Video Quality Prediction Model for H.264 Video over UMTS Networks and their Application in Mobile Video Streaming. In: 2010 IEEE International Conference on Communications – ICC, pp. 1–5. IEEE (2010)
18. Benayoune, S., Achir, N., Boussetta, K., Chen, K.: Content–aware ARQ for H.264 Streaming in UTRAN. In: IEEE Wireless Communication and Networking Conference – WCNC, pp. 1956–1961. IEEE (2008)
19. Superiori, L., Wrulich, M., Svoboda, P., Rupp, M.: Cross-Layer Optimization of Video Services over HSDPA Networks. In: Granelli, F., Skianis, C., Chatzimisios, P., Xiao, Y., Redana, S. (eds.) MOBILIGHT 2009. LNICST, vol. 13, pp. 135–146. Springer, Heidelberg (2009)
20. Khan, S., Thakolsri, S., Steinbach, E., Kellerer, W.: QoE–based Cross-layer Optimization for Multiuser Wireless Systems. In: 18th ITC Specialist Seminar on Quality of Experience, pp. 63–72. Blekinge Institute of Technology, Karlskrona (2008)

Concurrent and Distributed Projection through Local Interference for Wireless Sensor Networks

Xabier Insausti[1], Pedro M. Crespo[1], Baltasar Beferull[2], and Javier Del Ser[3]

[1] CEIT and TECNUN (University of Navarra). 20018 Donostia-San Sebastián, Spain
{xinsausti,pcrespo}@ceit.es
[2] University of Valencia. 46980, Paterna (Valencia), Spain
baltasar.beferull@uv.es
[3] TECNALIA Research and Innovation. 48170 Bilbao, Spain
javier.delser@tecnalia.com

Abstract. In this paper we use a gossip algorithm to obtain the projection of the observed signal into a subspace of lower dimension. Gossip algorithms allow distributed, fast and efficient computations on a Wireless Sensor Network and they can be properly modified to evaluate the sought projection. By combining computation coding with gossip algorithms we proposed a novel strategy that leads to important saving on convergence time as well as exponentially decreasing energy consumption, as the size of the network increases.

Keywords: Wireless Sensor Networks, Computational Codes, Signal Subspace Projection, Neighborhood Gossip.

1 Introduction

The fast spreading of wireless sensor networks has recently encouraged researchers to design and develop fast and efficient algorithms for such networks. The most common approach for computation and information exchange in wireless sensor networks has been done by using a class of decentralized algorithms known as *Randomized Gossip Algorithms* [1]. Wireless sensor networks are characterized for having very particular properties, of which we can highlight the limited computation power and energy resources. Besides, such networks usually do not have a centralized entity that synchronizes communication and therefore the knowledge that nodes have about the topology of the entire network is very limited.

In order to be more precise, consider a network composed of N sensors distributed randomly (uniformly) within the unit area circle. Sensors are assigned a limited transmission power P_T per source symbol. Therefore, they can communicate reliably with a certain number of neighbors within their coverage area, which will be the set of sensors located to a distance less than d (that depends on P_T). Let $\mathcal{N}_d(i) \subset \{1, \ldots, N\}$ denote the local neighborhood of node i, i.e., the set of nodes within distance d of node i.

J. Del Ser et al. (Eds.): MOBILIGHT 2011, LNICST 81, pp. 109–119, 2012.
© Institute for Computer Sciences, Social Informatics and Telecommunications Engineering 2012

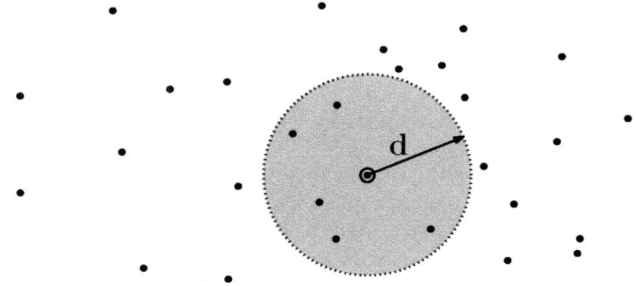

Fig. 1. Sensor network with N nodes. The gray circle is the local neighborhood of the active node, which is the set of sensors within distance d of the active node.

Randomized gossip algorithms have been mainly used for computing the average in an arbitrarily connected network of nodes as in figure 1 in a decentralized fashion. These algorithms are proven to be fast, efficient and to have a low computational cost and their operating basics are as follows: In the tth time slot, a sensor awakens at random and randomly chooses another sensor within its neighborhood, setting their values equal to the average of their current values. As $t \to \infty$ the entire network will converge to the desired average. For a proof and for further details refer to the work of Boyd et al. [1].

In a recent work, Nazer et al. [2][3] have used a new coding technique, known as *computation coding* [4], which together with a modification of a randomized gossip algorithm, leads to both time and energy savings in computing the average. Nazer's algorithm, called *neighborhood gossip*, is based on the following modifications of Boyd's algorithm: A node wakes up randomly in the tth gossip round and it requires all of the nodes within its neighborhood to transmit their values using a computation code that is designed so that the central node receives only the average of the values. The central node uses the received information and its own value to compute the average of the entire neighborhood, broadcasting the updated value to all nodes in its neighborhood. That way, the entire neighborhood gets the average in a single gossip round.

In this work, we are interested in using the random gossip technique to solve a much more general problem than averaging. We are looking for the projection of the observation vector into a subspace. Barbarossa was the first to propose a decentralized technique to perform the projection minimizing convergence time in [5]. We will use computation coding in order to achieve both time and energy savings with respect to Barbarossa's technique.

Computing the projection of the observed signal into a subspace of a known dimension is very useful for reducing the measurement noise. In general, the observation of a single sensor may be unreliable due to noise or malfunctioning. But, if the environment we are monitoring is smooth, the projection into a subspace will improve the reliability of the observation thanks to the interaction among nodes. Mathematically speaking, being smooth means that the signal exhibits spatial correlation and therefore belongs to a subspace of dimension smaller than the number of nodes.

Consider for instance that the network composed of N sensors is measuring a signal that can be expressed in a Fourier basis of dimension r, $r < N$. Then, the projection of the observed signal (which lies in a space of dimension N) into a subspace of dimension r will lead to very important noise reduction. The goal of the current work is to compute the projection achieving both time and energy savings.

The remainder of the paper is organized as follows: Section 2 precisely defines the framework of the current problem. In section 3 a valid strategy for performing neighborhood gossip is concluded based on the general gossip algorithm proposed by Boyd et al. Section 4 introduces Computation Coding and states the conditions for the computation code to exist and to work properly in our framework. Section 5 compares the gain of the algorithm when it uses computation coding and finally, Section 6 concludes the paper.

2   Problem Statement

Given the random wireless sensor network introduced in section 1, we consider that sensors do not have any geographic information available, although they know the sensors within their local neighborhood. Sensors have an initial observation $\mathbf{z} \triangleq (z_1, \ldots, z_N)^T$, which is a version of the useful signal vector $\boldsymbol{\xi} \in \mathbb{R}^N$ corrupted by some additive noise vector $\mathbf{v} \in \mathbb{R}^N$, that is,

$$\mathbf{z} = \boldsymbol{\xi} + \mathbf{v} \tag{1}$$

where, z_i denotes the observation taken by sensor $i \in \{1, \ldots, N\}$. Furthermore, it will be assumed that the useful signal lies in some subspace of dimension $r < N$. That is, $\boldsymbol{\xi} = \mathbf{U}\mathbf{s}$ for some $\mathbf{s}$, where $\mathbf{U}$ is an $N \times r$ matrix representing a basis of the r-dimensional useful signal subspace.

The least square estimator of the $\boldsymbol{\xi}$ is $\widehat{\boldsymbol{\xi}}$, which is given by the orthogonal projection of the observed vector onto the subspace spanned by the columns of $\mathbf{U}$ as shown in Figure 2. Without any loss of generality, we consider that the columns of $\mathbf{U}$ are orthonormal and therefore, the projector can be written as

$$\widehat{\boldsymbol{\xi}} = (\widehat{\xi}_1, \ldots, \widehat{\xi}_N)^\top = \mathbf{U}\mathbf{U}^T\mathbf{z} \tag{2}$$

Now the goal is to compute $\widehat{\xi}_i$ at each node of the network $i \in \{1, \ldots, N\}$ in a decentralized fashion by performing neighborhood gossip using computation codes, so that both time and energy savings can be achieved. For that purpose, we seek to obtain the estimate $\widehat{\boldsymbol{\xi}}$ iteratively by using the following recursive dynamics

$$\widehat{\boldsymbol{\xi}}[t+1] = \mathbf{W}\widehat{\boldsymbol{\xi}}[t], \quad t = 0, 1, \ldots \quad \mathbf{W} \in \mathbb{R}^{N \times N} \tag{3}$$

where $\widehat{\boldsymbol{\xi}}[t]$ denotes the value of the estimate at iteration t, $\widehat{\boldsymbol{\xi}}[0] = \mathbf{z}$, and the matrix $\mathbf{W}$ rules the transmissions between sensors at every iteration or time

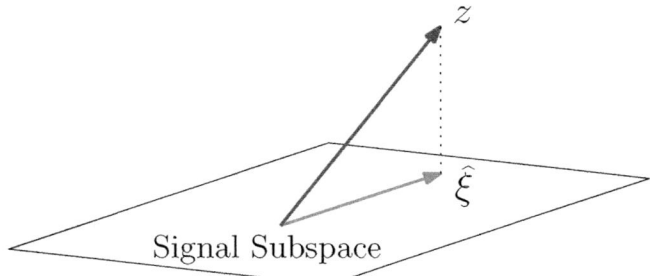

Fig. 2. Projection of the observed vector **z** into the target signal subspace. The goal is to have the network compute $\widehat{\boldsymbol{\xi}}$.

instant t. There are multiple choices for **W** but we are particularly interested in a sparse one, such that (3) can be performed using neighborhood gossip techniques. Therefore, we set that $w_{ij} = [\mathbf{W}]_{ij} \neq 0$ only if $j \in \mathcal{N}_d(i)$ or $j = i$. In words, the only nonzero elements of **W** correspond to pairs of sensors that are within a distance d. We will therefore say that **W** satisfies the topology constraint.

It is shown in [5] that, if the transmission power P_T is high enough (and equal for all the sensors), a symmetric and sparse matrix **W** exists, which satisfies the topology constraint, minimizes the *convergence time* and makes the network converge to $\widehat{\boldsymbol{\xi}}$, i.e.,

$$\lim_{t \to \infty} \widehat{\boldsymbol{\xi}}[t] = \lim_{t \to \infty} \mathbf{W}^t \mathbf{z} = \mathbf{U}\mathbf{U}^T \mathbf{z} = \widehat{\boldsymbol{\xi}} \tag{4}$$

and therefore,

$$\lim_{t \to \infty} \mathbf{W}^t = \mathbf{U}\mathbf{U}^T \tag{5}$$

Necessary and sufficient conditions for (5) are given next [5]:

Proposition 1.- *Given the dynamical system in (3) and the projection matrix* $\boldsymbol{UU}^T$, *the vector* $\widehat{\boldsymbol{\xi}} = \boldsymbol{UU}^T z$ *is globally asymptotically stable for any fixed* $z \in \mathbb{R}^N$, *if and only if the following conditions are satisfied:*

$$\boldsymbol{WUU}^T = \boldsymbol{UU}^T$$
$$\boldsymbol{UU}^T \boldsymbol{W} = \boldsymbol{UU}^T$$
$$\rho(\boldsymbol{W} - \boldsymbol{UU}^T) < 1 \tag{6}$$

where $\rho(\cdot)$ *denotes the spectral radius operator. Under these three conditions, the error vector* $e[t] \triangleq \widehat{\boldsymbol{\xi}}[t] - \widehat{\boldsymbol{\xi}}$ *satisfies the following dynamics:*

$$e[t + 1] = (\boldsymbol{W} - \boldsymbol{UU}^T)e[t], \quad t = 0, 1, \ldots \tag{7}$$

Intuitively, the previous proposition can be understood as follows:

- There should be a communication path between any two nodes with at most $N - 1$ hops.
- The number of neighbors of each node should not be smaller than the dimension of the signal subspace r for generic subspaces.

In one word, the radius of the neighborhoods d (and so the transmission power P_T) must be large enough to satisfy the convergence conditions.

Provided that the iteration in (3) minimizes the *convergence time* of the network, the goal now is to reduce the energy and time of each local iteration. This can be done using computation coding, which efficiently converts the wireless channel into a set of reliable equations between users [4].

In order to describe the algorithm required for the distributed projection we will rewrite (3) to clearly define the set of equations that should be transmitted within the local neighborhoods:

$$\widehat{\xi}_i[t+1] = \sum_{j=1}^{N} w_{ij}\widehat{\xi}_j[t], \quad i = 1, 2, \ldots, N \quad t = 0, 1, \ldots \tag{8}$$

Due to the sparsity of matrix $\mathbf{W}$ (given by the topology of the network and the transmission power) most of its terms will be zero. Actually, the only non-zero terms will be those referring to the local neighborhood and to the sensor itself, and therefore, the previous expression reduces to

$$\widehat{\xi}_i[t+1] = \sum_{j \in \mathcal{N}_d(i) \cup \{i\}} w_{ij}\widehat{\xi}_j[t], \quad i = 1, 2, \ldots, N \tag{9}$$

Notice that in order to perform the distributed projection, the ith sensor only needs to know the ith column of $\mathbf{W}$, which in turn means that sensor i only needs to keep $|\mathcal{N}_d(i)| + 1$ values.

We now consider a wireless channel with finite bandwidth so that we can model it as a discrete-time channel. Assuming transmissions occur within the neighborhood, the received signal at time instant $t \in \mathbb{N}$ by node i is

$$y_i[t] = \sum_{j \in \mathcal{N}_d(i)} h_{ij}[t] x_j[t] + n_i[t],$$

$$h_{ij}[t] = d_{ij}^{-\alpha/2} e^{j\theta_{ij}[t]} \tag{10}$$

where d_{ij} is the distance between nodes i and j, $\alpha \in \mathbb{R}_+$ is the power path loss coefficient, $\theta_{ij}[t]$ are the phases, and $x_j[t]$ is the signal transmitted by sensor j at time t. The channel noise samples $\{n_i[t]\}$ are realizations of i.i.d circular symmetric Gaussian random variables with variance σ^2.

As already mentioned, the main goal of our technique is to reduce power consumption. To that end, we propose the use of computation codes for reliable communication among nodes, strategy that will lead to important power gains.

Notice that in order to apply coding, one has to assume that each sensor has an i.i.d sequence of observation (vector) instead of just a single observation, as it has been assumed so far. The proposed strategy can be performed by using a general gossip algorithm as in [1] and it follows on a similar strategy as the one used in [2] and [3] for including computation codes[1].

3   Neighborhood Gossip Algorithm

The most important aspect when computing (9) in a distributed fashion is to try to maintain synchronized the number of iterations at the nodes, so that convergence is guaranteed using as few gossip rounds as possible. In other words, all the estimated quantities $\widehat{\xi}_j$, $j \in \mathcal{N}_d(i)$, available at the nodes inside $\mathcal{N}_d(i)$ should have been updated the same number of times t (i.e., all should be at iteration t), before computing the next iteration $t+1$ of $\widehat{\xi}_i$, at node i. This requires a substantial change in the general averaging gossip algorithm proposed by Boyd et al. in [1], and also in the neighborhood gossip algorithm with computation codes introduced by Nazer et al. in [2] and [3]. The reason is that the goal of these algorithms is to obtain a single parameter estimation at all the nodes of the network, mainly the average of the observations, and the convergence to the average does not require to keep the same number of updates for the nodes inside a neighborhood. Therefore, none of those algorithms can be directly applied in the current context.

Let us look at the gossip algorithm proposed in [2]. A node i wakes up randomly and requires all the nodes within its neighborhood $\mathcal{N}_d(i)$ to transmit their observations by using a computation code so that the decoded value at node i is the average of the observations of all the nodes belonging to $\mathcal{N}_d(i)$. Then, node i broadcasts the new computed average to all the nodes in $\mathcal{N}_d(i)$ so that these nodes get in a fast way the new average value in a single iteration or gossip round. This process is repeated until a good estimate of the average is obtained. If one tries to apply this gossip algorithm into our problem, it will fail for two reasons. First, the final broadcasting stage is clearly inapplicable in our problem since we are not looking for a single parameter estimator (average) at all the nodes, but rather for a set of N parameter estimators, one for each node of the network. Second, the system dynamics given in (3) would be violated (i.e., synchronization of iterations at the nodes would fail) whenever a node woke up more times than others within a neighborhood. To better understand this problem let us consider two sensors, s_1 and s_2 within a distance d. Assume that s_1 starts the tth gossip iteration: node s_1 wakes up at random and requires s_2 and the rest of its neighborhood to send their observations. They will be sending their observation that comes from gossip iteration $t - 1$. After receiving the messages, s_1 will update its current observation. Now consider that s_2 wakes up and requires their neighbors to send their observations too. All neighbors of node s_2 will be transmitting their last estimation except node s_1, which will be required

---

[1] However, here the goal is to compute the projection, whereas in [2] [3], the goal was to compute the average of the sensors.

not to send its latest update but the previous one (observation referring to iteration $t-1$). This fact requires that nodes keep track of previous observations for each neighbor and then Boyd's general gossip algorithm can be applied to our problem.

According to [5], the *convergence time* $\tau_0(\mathbf{W})$ of our strategy, defined as the number of gossip rounds required for the error (7) to decrease by a factor $1/e$ for the worst possible initial vector, given that all nodes have performed the same amount of gossip rounds, is given by

$$\tau_0(\mathbf{W}) \triangleq \frac{1}{\ln\left(\frac{1}{\rho(\mathbf{W}-\mathbf{U}\mathbf{U}^T)}\right)}. \tag{11}$$

The problem of finding the matrix $\mathbf{W}$ that minimizes the *convergence time* given a particular network topology can be converted into a convex problem [5] and solved by using classical *Semidefinite Programming* (SDP) tools [6].

4 Computation Coding

The most energy consuming stage is when all nodes in the local neighborhood transmit their corresponding observations to the central node. The key here is to realize that the central node does not need to know the observation of each neighbor. Rather, it only needs to know the weighted sum of the observations. Since the weighting process is already done in the source nodes, the central node only requires the sum of the incoming messages. This can be done very efficiently by using a code construction recently developed in [4] and known as computation coding.

First, we will assume that sensors know the channels in their local neighborhood from themselves to the central node. For node i, this is equivalent to knowing the channel coefficients $(d_{ij}, \theta_{ij}[t])$ in (10) for every $j \in \mathcal{N}_d(i)$. Exploiting this knowledge leads to a simplification of the multiple-access channel and it can be considered to be the following simple multiple-access channel:

$$y_i[t] = \sum_{j \in \mathcal{N}_d(i)} x_j[t] + n_i[t] \tag{12}$$

Now that the channel behaves as a simple adder, we know that we can efficiently and reliably compute sums by using computation coding to improve the performance of the transmissions as in [3]. All nodes in the neighborhood of i encode and send their values using identical linear codebooks. The transmitted codewords will be added on the channel and node i will receive the sum of the codewords. Since the codebook is linear, the sum of the codewords is also a codeword and it is actually the codeword corresponding to the desired sum. Node i has to simply add the received message to its observation weighted by the coefficient w_{ii} to get its desired new observation.

This said, we are ready to state the following theorem from [3], which establishes the required condition for the computation code to exist.

Proposition 2.- *Choose $\epsilon > 0$. Assume each node in a local neighborhood has a length L bounded real-valued weighted observation vector, that is, $\|s_{ij}[t]\|^2 \leq LP_T$ $\forall i, j, t$. For L large enough, there exists a coding scheme such that the receiving node i can make an estimate $\widehat{\xi}_i[t_i+1]$ of the sum $\xi_i[t_i+1]$ as in (9) that satisfies[2]:*

$$\mathcal{P}\left(\|\xi_i[t_i+1] - \widehat{\xi}_i[t_i+1]\|^2 \geq \frac{P_T}{L}2^{-2B}\right) < \epsilon \quad \forall i, t_i \tag{13}$$

so long as:

$$T\log\left(\frac{1}{|\mathcal{N}_d(i)|} + \frac{P_T}{(\max_{j\in\mathcal{N}_d(i)} d_{ij})^{-\alpha/2}\sigma^2}\right) > B \tag{14}$$

for some choice of T channel uses per observation symbol and precision B bits.

See [4] for a proof.

The computation code is based on using the same lattice code by all sensors, which is chosen to simultaneously be a good channel and a good source code. The transmitters within a neighborhood quantize their weighted observation vector to the lattice and they transmit such quantization simultaneously. The receiver decodes the sum and makes an estimate of it. Next, the transmitters send their quantization errors using the same lattice code. This continues until the total number of channel uses T is exhausted and the desired precision B is reached.

5   Performance Comparisons

We now compare the performance of the proposed strategy when it uses the best possible separation scheme or when it uses a computation code. We will conclude that we can achieve an exponentially increasing power gain by using computation codes as the density of sensors in the network increases. To that end, we will make use of two theorems from [4], which compute the achievable distortion for sending a Gaussian sum over a Gaussian MAC. Notice that distortion here is again measured by the mean-squared error criterion,

$$D = \max_{i, t_i}\left(\frac{1}{L}\|\xi_i[t_i+1] - \widehat{\xi}_i[t_i+1]\|^2\right). \tag{15}$$

Proposition 3.- *For T channel uses per observation symbol, $T \in \mathbb{Z}_+$, the following distortion is achievable for sending a Gaussian sum (of M transmitters) over a Gaussian MAC with noise variance σ^2 so long as $P_T > \frac{M-1}{M}\sigma^2$:*

$$D = M\sigma_S^2\left(\frac{\sigma^2}{\sigma^2 + MP_T}\right)\left(\frac{M\sigma^2}{\sigma^2 + MP_T}\right)^{T-1}, \tag{16}$$

where σ_S^2 is the variance of the Gaussian sources.

On the other hand, we have an interesting result for separation-based schemes.

---

[2] Notice that $\xi_i[t_i+1]$ refers to the actual sum of the observations and $\widehat{\xi}_i[t_i+1]$ refers to the value that node i gets after T channel uses.

Proposition 4.- *The best achievable distortion for a separation-based scheme for sending a Gaussian sum (of M transmitters) over a Gaussian MAC with noise variance σ^2 is given by:*

$$D = M\sigma_S^2 \left(\frac{\sigma^2}{\sigma^2 + MP_T} \right)^{\frac{T}{M}}, \tag{17}$$

where σ_S^2 is the variance of the Gaussian sources and now the number of channel uses per observation symbol T might not be an integer. See [4] for the proofs.

Since computation coding is optimal for symmetric linear MACs [4], the distortion given in proposition 3 is achievable using computation codes. Although we cannot guarantee that the messages of the sensors follow a Gaussian distribution, we will consider it as the worst case result. Therefore, based on the previous propositions, we can compute the number of channel uses per source symbol that computation codes and the best separation-based scheme need to achieve the given distortion D. Consider that node i wakes up and requires all the nodes within its neighborhood $\mathcal{N}_d(i)$ to transmit their observations. The number of channel uses per symbol is then given by:

$$T_{COMP}(i) = \frac{\log\left(\frac{\sigma_S^2}{D}\right)}{\log\left(\frac{1}{|\mathcal{N}_d(i)|} + \frac{P_T}{\sigma^2}\right)}. \tag{18}$$

$$T_{SEP}(i) = \frac{|\mathcal{N}_d(i)|\log\left(\frac{D}{|\mathcal{N}_d(i)|\sigma_S^2}\right)}{\log\left(\frac{\sigma^2}{\sigma^2 + |\mathcal{N}_d(i)|P_T}\right)} \tag{19}$$

If we consider that all channel uses require the same energy and since we know that both schemes will converge in the same number of gossip rounds, the energy ratio will be given by the difference in the number of channel uses per source symbol.

Nevertheless, both quantities depend on the number of transmitters, i.e. the neighborhood of the current active node $\mathcal{N}_d(i)$, and that quantity is in general a random variable which we will model next.

Consider a circular region of unitary area (radius $1/\sqrt{\pi}$) and distribute N sensors uniformly on it, so that we have N sensors per unit square. The distribution of the distances of the nodes to the center of the circle R will be:

$$f_R(r) = 2\pi r \qquad 0 < r < \frac{1}{\sqrt{\pi}} \tag{20}$$

Now assume without loss of generality that our target node i is on the center of the circle and has a coverage area given by $d < 1/\sqrt{\pi}$. The probability of a sensor falling within the coverage area of sensor i is then given by integrating (20) between 0 and d, which yields $P_d = \pi d^2$. Therefore, the number of neighbors M of node i is given by the set of nodes that fall within its coverage area, which is distributed as

$$P_M(m) = \binom{N}{m} \left(\pi d^2\right)^m \left(1 - \pi d^2\right)^{N-m} \tag{21}$$

We only have to finally compute the average of the channel uses per symbol, which are given by:

$$\overline{T_{COMP}} = \sum_{m=1}^{N} \binom{N}{m} \frac{\log\left(\frac{\sigma_S^2}{D}\right)}{\log\left(\frac{1}{m}+\frac{P_T}{\sigma^2}\right)} \left(\pi d^2\right)^m \left(1 - \pi d^2\right)^{N-m} \qquad (22)$$

$$\overline{T_{SEP}} = \sum_{m=1}^{N} \binom{N}{m} \frac{m\log\left(\frac{D}{m\sigma_S^2}\right)}{\log\left(\frac{\sigma^2}{\sigma^2+mP_T}\right)} \left(\pi d^2\right)^m \left(1 - \pi d^2\right)^{N-m} \qquad (23)$$

And therefore the average gain is given by

$$g = \frac{\overline{T_{COMP}}}{\overline{T_{SEP}}} = \frac{\sum_{m=1}^{N} \binom{N}{m} \frac{\log\left(\frac{\sigma_S^2}{D}\right)}{\log\left(\frac{1}{m}+\frac{P_T}{\sigma^2}\right)} \left(\pi d^2\right)^m \left(1 - \pi d^2\right)^{N-m}}{\sum_{m=1}^{N} \binom{N}{m} \frac{m\log\left(\frac{D}{m\sigma_S^2}\right)}{\log\left(\frac{\sigma^2}{\sigma^2+mP_T}\right)} \left(\pi d^2\right)^m \left(1 - \pi d^2\right)^{N-m}} \qquad (24)$$

so long as $P_T \geq \sigma^2$ and $d < 1/\sqrt{\pi}$.

Figure 3 shows the shape of the gain when the density of nodes increases. In general terms, the gain is very little dependent on the distortion but it grows exponentially as the density of sensors increases.

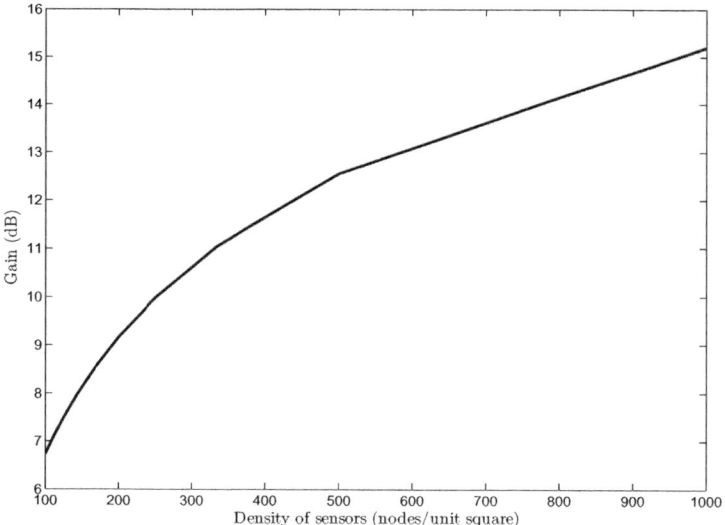

Fig. 3. Energy gain of the projection algorithm with computation codes with respect to the best possible separation-based scheme. $d = 0.2$, $\sigma^2 = 1$, $\sigma_S^2 = 1$, $P_T = 3$, $D = 10^{-5}$.

6   Conclusion

The main goal of the current paper has been to modify a general gossip algorithm by introducing computation coding in order to achieve both time and energy savings in a projection problem. The results that lead to the goal are summarized in the following list:

- The problem of finding the projection of the observed signal into a subspace can be solved in a distributed fashion by solving an SDP problem.
- A general gossip algorithm has been modified to allow the use of computation coding.
- Combining computation coding with a gossip algorithm to perform the projection leads to exponentially increasing power savings when the number of nodes increases.

Acknowledgments. This work was supported in part by the Spanish Ministry of Science and Innovation through the projects COSIMA (TEC2010-19545-C04-02) and COMONSENS (CSD2008-00010), as well as by the Basque Government through the Ph.D. grant 2009/3723.

References

1. Boyd, S., Ghost, A., Prabhakar, B., Shah, D.: Randomized Gossip Algorithms. IEEE Transactions on Information Theory 52, 2508–2530 (2006)
2. Nazer, B., Dimakis, A.G., Gastpar, M.: Local Interference Can Accelerate Gossip Algorithms. In: 46th Annual Allerton Conference, Monticello, IL (2008)
3. Nazer, B., Dimakis, A.G., Gastpar, M.: Neighborhood Gossip: Concurrent Averaging through Local Interference. In: IEEE International Conference on Acoustics, Speech and Signal Processing, Taipei, Taiwan (2009)
4. Nazer, B., Dimakis, A.G., Gastpar, M.: Computation over Multiple-Access Channels. IEEE Transactions on Information Theory 53, 3498–3516 (2007)
5. Barbarossa, S., Scutari, G., Battisti, T.: Distributed Signal Subspace Projection Algorithms with Maximum Convergence Rate for Sensor Networks with Topological Constraints. In: IEEE International Conference on Acoustics, Speech and Signal Processing, Taipei, Taiwan (2009)
6. Boyd, S., Vandenberghe, L.: Convex Optimization. Cambridge University Press (2004)

SA-RI-MAC: Sender-Assisted Receiver-Initiated Asynchronous Duty Cycle MAC Protocol for Dynamic Traffic Loads in Wireless Sensor Networks

Shagufta Henna

Computer Science Department, University of Leicester,
University Road, Leicester, United Kingdom
sh334@le.ac.uk

Abstract. Duty cycling is an efficient mechanism to conserve energy in wireless sensor networks. Several existing duty cycling techniques have been proposed to conserve energy, but they are not able to handle the contention under dynamic traffic loads. In this paper we propose a protocol called Sender-Assisted-Receiver-Initiated MAC (SA-RI-MAC) which solves this problem without sacrificing the energy efficiency. SA-RI-MAC employs the receiver initiated transmissions mechanism of RI-MAC with a sender assisted approach to handle the contention at the receiver. Senders tend to cooperate with each other to resolve the contention dynamically based on the contention level at the receiver. A further improvement is achieved by prioritizing the sender transmissions which has been starved long for the channel occupancy. Our simulation results in ns-2 show that SA-RI-MAC achieves significant improvement in conserving energy over RI-MAC. It can handle traffic contention much more efficiently than RI-MAC; thus improving end to end delivery ratio with a reduction in the latency. Under light traffic load, the performance of SA-RI-MAC is comparable with RI-MAC in terms of end to end delivery ratio, latency and energy efficiency.

Keywords: MAC, Duty Cycling, energy efficient wireless sensor networks, asynchronous duty cyling, dynamic traffic loads.

1 Introduction

One of the major limitations considered in wireless sensor networks is scarcity of energy. In order to conserve energy, power efficient protocols are desirable. These protocols tries to mitigate energy consumption by devising different clever mechanisms at different layers of the protocol stack. Among these methods, mechanisms deployed at the Medium Access Layer (MAC) are more power efficient due to its direct access to the wireless medium.

Generally, a wireless Radio has four power levels depending on its state: idle, sleeping, receiving and transmitting. During the active state a node is able to

J. Del Ser et al. (Eds.): MOBILIGHT 2011, LNICST 81, pp. 120–135, 2012.

transmit and receive data but in sleep state it completely turns its radio off. Idle listening is one of the main reasons of energy consumption as it requires the same amount of energy as to transmit and receive. This consumption can be saved by turning the radio of a sensor off as frequently off as possible. Duty cycling is an efficient mechanism to handle the problem of idle listening [1, 2]. In duty cycling, wireless nodes periodically turn their radios on and off to reduce the idle listening time.

Different approaches to duty cycling MAC can be categorized as synchronous and asynchronous. Synchronous approaches include RMAC [3], T-MAC [4], DW-MAC [5] and S-MAC [6]. In these approaches, neighbouring nodes synchronize their active and sleep schedules by using some synchronizing protocol. These approaches greatly reduce idle listening but are complex and need extra overhead to synchronize different neighbours with different sleep and active schedules. On the other hand, asynchronous approaches such as WiseMAC [7], X-MAC [8], B-MAC [9] and RI-MAC [10] allow nodes to have their own sleep and active schedules independent to any neighbouring nodes. Asynchronous schemes work efficiently for light traffic loads but become less efficient in terms of latency, energy consumption and delivery ratio under high traffic loads.

In some applications of wireless sensor networks such as convergecast [11] and correlated-event workload traffic [12] where sensors are used for event monitoring, communication demand may suddenly increase in a burst. For example, in the event of fire several sensors report this event to some common sink. If contention created by such events is not handled well, the data sent to the sink may experience longer delays or may be lost. Under such dynamic traffic loads, MAC layer protocols should be able to handle the contention at the sink.

In this work, we present a sender-assisted asynchronous duty cycling MAC protocol, called Sender-Assisted-Receiver-Initiated MAC (SA-RI-MAC). SA-RI-MAC attempts to resolve the contention among the senders with a common intended receiver and helps them to find a rendezvous time to communicate with the receiver. SA-RI-MAC differs from RI-MAC and previous asynchronous duty cycling protocols the way different contended senders resolve the contention at the receiver by cooperating with each other. In SA-RI-MAC, a sender waits for an explicit beacon from the receiver to initiate the transmissions. An explicit beacon containing the value of channel access failure is exchanged among the neighbours which have a common intended receiver. This value of channel access failure is used to resolve the contention among the senders for the medium access. Another improvement is achieved by prioritizing the transmissions of the senders which have been starved longer for the channel occupancy.

We believe this the first attempt which combines the idea of receiver initiated transmissions with the sender assisted contention resolution. This sender assisted coordination adaptively increases the channel utilization which improves the packet delivery ratio, and power efficiency under dynamic traffic loads. We have implemented SA-RI-MAC in ns-2 [15] simulator for evaluation in different network scenarios under dynamic traffic loads.

The rest of the paper is organized as follows. Section 2 discusses related work. Section 3 discusses the contention resolution mechanism in RI-MAC and its weaknesses. Section 4 presents the detailed SA-RI-MAC design. Section 5 reports the performance evaluation of SA-RI-MAC using ns-2 simulation. Finally in section 6, we present our conclusions.

2   Related Work

In wireless sensor networks where energy is a scarce resource, transmissions between sender and receiver can be classified as sender or receiver initiated. The idea of Receiver initiated transmissions in a MAC protocol has been recently introduced in [10]. We make the first attempt to combine the idea of receiver initiated transmissions with the sender assisted contention resolution in ad hoc wireless sensor networks.

Receiver initiated collisions avoidance schemes for general wireless networks have been proposed in [16]. In these approaches collision avoidance is more important than energy efficiency. However under high traffic loads when the degree of contention rises, these approaches lack any coordination among the senders to resolve the contention. Low power probing (LPP) is an asynchronous receiver initiated transmission mechanism used in Koala systems [17]. In koala systems, downloads of the bulk data are initiated by the gateway nodes which allows other nodes to sleep most of the time to conserve energy. In LPP, each node broadcasts a preamble periodically. Other nodes which receive the preamble sends an acknowledgement. After receiving an acknowledgement, a node stays awake and starts acknowledging the probes of other nodes. However LPP approach triggers the false wake ups and sleeps affecting the throughput and energy efficiency.

B-MAC [1] and X-MAC [8] are asynchronous duty cycling MAC protocols in which transmissions are initiated by the senders. Prior to transmissions a sender sends a wake up signal to the receiver by using a long preamble. The length of the preamble is longer than the sleep interval of a node to ensure that the node will wake up at least once during this duration. B-MAC is optimized under light traffic loads for energy consumption. However, an increase in the traffic load may keep a node awake unnecessarily spending a significant amount of time in the active state even if the packets are destined for other nodes. X-MAC solves this problem by sending the preamble as a series of short preambles prior to any transmission and waits for an acknowledgement generated by the receiver which reduces the channel occupancy significantly. X-MAC preamble contains the target address which allows the irrelevant nodes to go to sleep immediately to conserve energy and allows the intended receiver to send an acknowledgement to the sender to stop probing the channel. After receiving the first DATA transmission, a receiver in X-MAC stays awake for a duration equal to maximum back-off window size. This time interval termed as dwell time is used by the sender to send any queued packets.

RI-MAC [10] uses the concept of receiver initiated transmissions. In R-MAC, it is the receiver which initiates the transmissions by sending beacons at regular intervals. Sender wakes up asynchronously at regular intervals to receive an invitation for transmission from the receiver. In response to an invitation from the receiver, sender sends a DATA frame to acknowledge the reception of the beacon. In RI-MAC collisions are handled by receiver dynamically. On detecting the contention, receiver sends an explicit beacon with an increased value of Contention window to the senders to reduce the contention at the receiver. In RI-MAC medium access among senders is controlled by the receiver, however such contention resolution is not very power efficient and reliable under dynamic traffic loads. Senders back off according to the back off value specified by the receiver, however under dynamic traffic loads an increased value of back offs affect the energy efficiency and delivery ratio significantly.

Previous synchronous and asynchronous duty cycling approaches such as X-MAC, B-MAC and RI-MAC achieves greater energy efficiency under light traffic loads. However SA-RI-MAC differs from these approaches by dynamically triggering the coordination among the senders to handle the contention under high traffic loads. Other asynchronous duty cycling approaches give no preference to nodes to transmit which have been starved longer for channel occupancy. SA-RI-MAC on the other hand, prioritizes the transmissions from the starved senders after contention resolution.

3 Contention Resolution Mechanism in RI-MAC

In RI-MAC, a receiver coordinates the DATA transmissions from the contending senders by exchanging an explicit beacon. In the beacon, receiver specifies back off window size(BW) which senders should use to contend the channel. The size of BW is controlled by the receiver depending on the number of collisions. Receiver can know about an incoming packet with the help of Start Frame Delimiter (SFD). Clear channel assessment (CCA) is used to detect any channel activity. If CCA detects any channel activity, receiver assumes a collision and generates another beacon with an increased value of BW. Depending on the network conditions, receiver in RI-MAC adjusts the value of BW by using binary exponential back off (BEB). Receiver turns its radio off if it keeps detecting continuous collisions after consecutive beacon transmissions or if the value of BW exceeds maximum back off window size. Due to high contention, a sender could potentially miss the beacon with the exact value of BW. Sender increments the retry count, if no beacon has been received from the intended receiver for duration equal to three times of the sleep interval. Further, retry count is incremented if no acknowledgement has been received from the receiver for a DATA transmission within the maximum back off window time. If the value of retry count exceeds a predefined value of retry limit, sender cancels the further transmission of the DATA frame.

Contention resolution mechanism used by RI-MAC does not involve the contended senders to resolve the contention at the receiver. RI-MAC tries to handle the contention with the help of a BW value specified by the receiver. We show that by using only BW value specified by the receiver is not sufficiently enough to cope efficiently with the collisions at the receiver. This increased BW value increases back off at the contended senders which decreases throughput and increases latency. Further, increased back offs at the contended senders does not help to conserve any energy even when the sender is not able to access the medium. If the contending senders are not well coordinated , they may deviate from the BW value specified by the receiver and may decide to transmit without any back off. This deviation from the BW value specified by the receiver only can degrade the throughput [18].

Figure 1 shows continuous collisions at the receiver caused by simultaneous transmissions from the contending senders in RI-MAC. RI-MAC tries to handle these collisions by adjusting the size of the BW. However under high traffic loads, the probability of loss of the beacon with the exact BW value by the contending senders increases which makes their transmissions more likely to collide at the receiver.

4   SA-RI-MAC Design Overview

Sender in SA-RI-MAC tracks the number of times it has failed to access the channel when trying to transmit a packet to the receiver. A counter CHANNEL_ACCESS_FAILURE is maintained to record the failure to access the channel. This counter is updated every time retry limit exceeds the maximum retry limit threshold. Contending senders exchange an explicit beacon with each other containing the value of CHANNEL_ACCESS_FAILURE counter at regular intervals. Prior to transmission, sender estimates the contention level at the receiver by using the BW value specified by the receiver. If the value of BW specified by the receiver exceeds the maximum contention window size, sender considers it as an indication of high contention. However, under high traffic loads, possibility to drop the beacon containing the BW value increases. In this case, maximum value of CHANNEL_ACCESS_FAILURE among all the contending senders is compared with the CHANNEL_ACCESS_FAILURE_THRESHOLD. If CHANNEL_ACCESS_FAILURE exceeds CHANNEL_ACCESS_FAILURE_THRESHOLD; this indicates the high contention at the receiver.

If the contention at the receiver is significantly high, the sender node will evaluate if any of its neighbours has the value of CHANNEL_ACCESS_FAILURE higher than its own value. If there is such a contending neighbour which starved longer , it turns its radio off immediately to conserve energy and to minimize further contention at the receiver and wakes up asynchronously. More importantly, this sender assisted contention resolution increases fairness among the senders and gives priority to starved senders. This design choice is more energy efficient to resolve the contention at the receiver.

Figure 2 shows how SA-RI-MAC avoids collisions at the receiver. It shows that contending senders S1 and S2 coordinate with each other under high traffic loads by exchanging their recent CHANNEL_ACCESS_FAILURE values. S1 has CHANNEL_ACCESS_FAILURE value greater than S2; therefore S1 starts transmissions to the receiver R. In order to avoid collisions at the receiver, S2 turns its radio off to conserve energy.

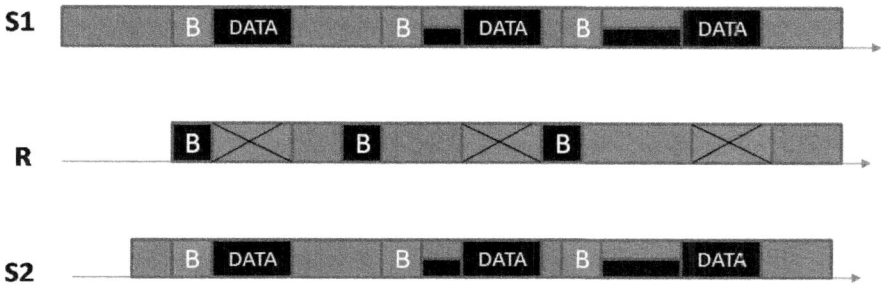

Fig. 1. RI-MAC: DATA frame transmissions from contending senders. Simultaneous transmissions from the contending senders can cause continuous collisions at the Receiver.

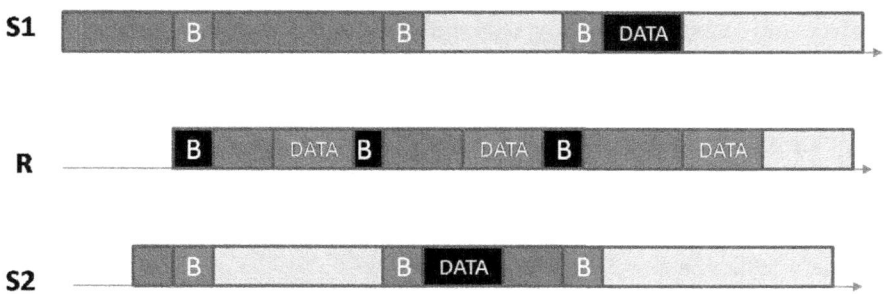

Fig. 2. SA-RI-MAC: DATA frame transmissions from contending senders. Transmissions from the contending senders are well coordinated to avoid continuous collisions at the Receiver.

4.1 Beacon Frame in SA-RI-MAC

When a receiver wakes up it sends a base beacon containing the value of source field. Base beacon can have two optional fields, destination field and BW size. If destination field is set in the base beacon, it means beacon frame is an acknowledgement to the sender with the destination field and other senders can treat this it as a request to send data.

BW value is specified by the receiver according to the contention level at the receiver. In RI-MAC, this BW value is used by the contending senders to back off before transmissions to reduce the chance of collision at the receiver. However, in SA-RI-MAC this value is used as an indication of the contention at the receiver and triggers the coordination among the contending senders prior to further transmissions. It is a better design choice instead of continuous back offs which are not very any energy efficient and does not improve delivery ratio significantly. Further these back offs increase the latency of the transmissions significantly.

After receiving a beacon from the receiver, a sender always makes a Clear Channel Assessment(CCA) before transmission in order to avoid collisions at the receiver. CCA must indicate the medium idle for at least SIFS plus maximum propagation delay time. If no activity is detected during this time, receiver R turns its radio off.

4.2   Collisions in SA-RI-MAC

By coordinating the senders to transmit data at the receiver, SA-RI-MAC reduces collisions significantly at the receiver and thus cuts down unnecessary retransmissions. As data transmissions among contending senders are explicitly controlled and coordinated based on the contention level at the receiver, contending senders know when not to send the data and thus can turn their radio off to conserve energy. In RI-MAC, if the back off value reaches the maximum back off window and receiver keeps detecting collisions, it turns its radio off. On the other hand, SA-RI-MAC tries to reduce the continuous collisions at the receiver and thus prevents any back offs forced by the collisions which affects latency and delivery ratio significantly.

5   Performance Evaluation

We evaluated SA-RI-MAC in ns-2 simulators and compared its performance with the RI-MAC. We simulated SA-RI-MAC under different network scenarios with dynamic traffic loads.

5.1   Simulation Evaluation

we used two-ray ground reflection radio propagation model for all the scenarios. Different simulation parameters used are shown in Table 1. These parameters are similar to CC2420 radio [18] used in MICAz motes. CCA check is performed by sampling RSSI delay as reported by Ye et al [14]. This check is performed every 20ms longer than the interval between two short preambles. The transmission and sensing range are modelled according to 914 MHz lucent WaveLAN radio, as similar ranges have been observed in some sensor nodes [20]. In both RI-MAC and SA-RI-MAC, BW value is adjusted based on BEB which takes value of 0,31,63,127 and 255. We used Initial value back off window size of 32 and congestion window of size 8 which are default values used in UPMA package distributed as TinyOS [21]. Dwell time for both RI-MAC and SA-RI-MAC is dynamically adjusted based on the BW specified by the receiver plus propagation delay and SIFS. Initial wake up for both the protocols is randomized and a value of 1 second is used for the sleep interval.

We compared the performance of SA-RI-MAC with RI-MAC in random networks, clique networks and a 49 node (7×7) grid network.

Table 1. Simulation parameters for Radio

Tx range	250 m
Slot time	320 us
SIFS	192 us
Bandwidth	250 Kbps
CCA check Delay	128 us
Carrier Sensing Range	550 m
Duty Cycle	1 %
CHANNEL_ACCESS_FAILURE_THRESHOLD	5
Tx Power	31.2 mW
Rx Power	22.2 mW
idle Power	22.2 mW
Contention Window (CW)	32 ms

Clique Networks. We compare the performance of SA-RI-MAC and RI-MAC in clique networks. In a clique network all the nodes are within the transmission range of each other. We varied the number of flows in the network to vary the traffic load in the network. We allow flows to share the same destination to cause contention.

For clique network, number of nodes in the network are twice the number of flows. Each source node generates packets 10 seconds after the start of the simulation. The interval between two packets is uniformly distributed between 0.5 and 1.5 seconds. Next wakeup time for each node in the network is randomly chosen between 0 and 10 seconds. A packet is not considered delivered if it is in the queue. Each simulation runs for 100 seconds. We have taken an average on three random clique network scenarios.

Figure 3 shows the delivery ratio of SA-RI-MAC for a clique network with an increase in the number of contending flows. Both RI-MAC and SA-RI-MAC achieves delivery ratio close to 100 when the number of flows are fewer than 15. However, as the number of flows exceed 15, the delivery ratio of RI-MAC drops significantly due to an increase in the contention level at the receiver. The delivery ratio of SA-RI-MAC does not drop significantly as an immediate coordination will be triggered among the contending senders to resolve the contention at the receiver.

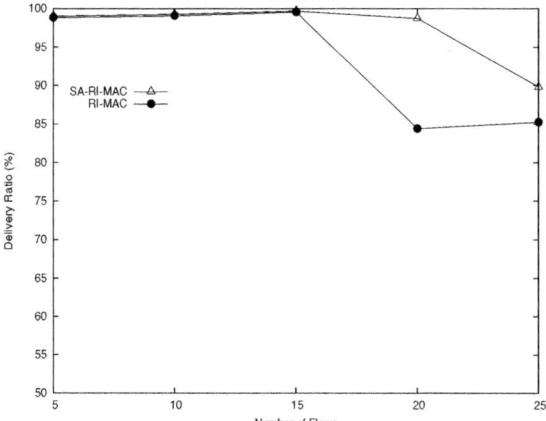

Fig. 3. Delivery Ratio vs. number of Flows

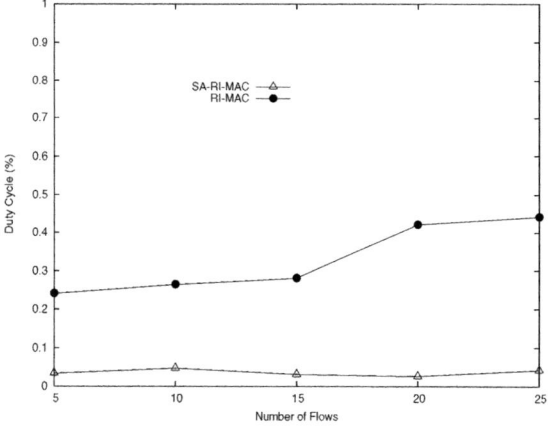

Fig. 4. Duty Cycle vs. Number of Flows

The overall duty cycle of the nodes is shown in figure 4. In addition to a gradual drop in the delivery ratio, SA-RI-MAC conserves much more energy than RI-MAC. It can be observed from the figure that for all contending flows the energy consumption of SA-RI-MAC is less than RI-MAC. For all flows, SA-RI-MAC saves more than 75% energy compared to RI-MAC. SA-RI-MAC conserves much more energy during high traffic loads by triggering coordination among the senders giving them a chance to conserve energy by turning their radio off compared to back off mechanism used by RI-MAC.

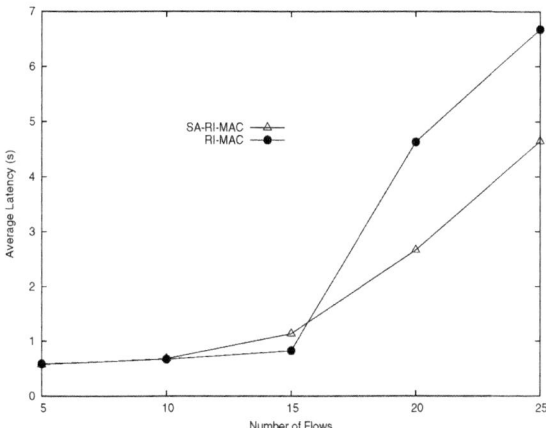

Fig. 5. Avergae Latency vs. Number of Flows

In addition to high duty cycle, RI-MAC also has higher latency compared to SA-RI-MAC as shown in figure 5. This increase in latency is due to an increased value of back off by the receiver to handle the contention. However, SA-RI-MAC avoid collisions at the receiver by prioritizing a starved node among the contending senders to transmit during high traffic loads which reduces unnecessary back offs and helps to conserve the energy.

Grid Network under Correlated Event Workload. We compare the performance of SA-RI-MAC with RI-MAC in a grid network with 49 nodes. Maximum distance between two neighbouring nodes is 200 meters. Target sink node to receive event notifications is at the centre of the grid. We used a Random Correlated-Event (RCE) model to generate traffic in the grid network [13]. RCE is based on the correlated-event workload which simulates spatially-correlated events in a sensor networks. This model simulates a synchronized triggered traffic load in the network which is a common case for tracking and detection applications. In RCE, an event is generated on some randomly selected location (x,y) in the network. A node in the network can sense and report an event if it is in the radius R centred at (x,y). We generated a new event once every 200 seconds. Each node with in radius R senses the event and reports it to the sink. In a 7×7 grid network, path traversed by each packet varies from 1 to 6 hops and on average 3.05 hops. We perform each simulation for 3 random runs for a series of 48 events triggered from random locations. Unicast packets are transmitted by the nodes within the radius R to notify the sink. Each simulation run lasts for 10,000 seconds.

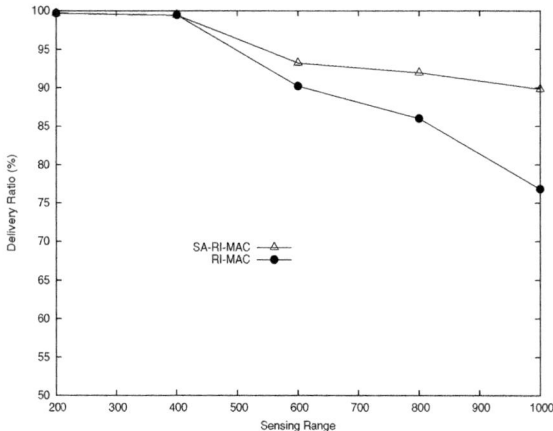

Fig. 6. Delivery Ratio vs. number of Flows

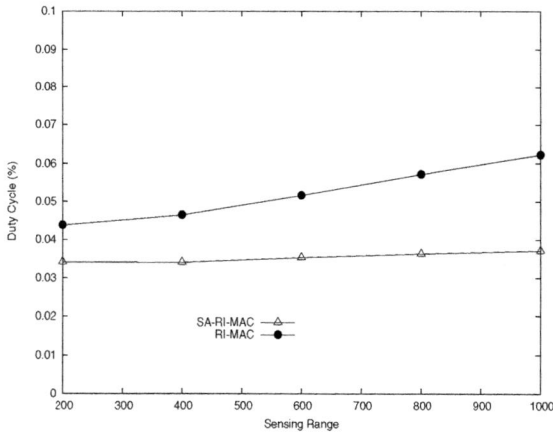

Fig. 7. Duty Cycle vs. Number of Flows

Performance comparison of RI-MAC and SA-RI-MAC is shown in figure 6. Figure 6 shows the packet delivery ratio. When the traffic load in the network is not very high RI-MAC and SA-RI-MAC maintains packet delivery ratio upto 100%. However, with an increase in sensing range high contention is caused for medium access so the performance of RI-MAC and SA-RI-MAC drops. However, SA-RI-MAC is augmented with sender assisted coordination which maintains its delivery ratio higher than RI-MAC in high traffic loads. SA-RI-MAC as shown in figure 7 in addition to achieving the better delivery ratio than RI-MAC also

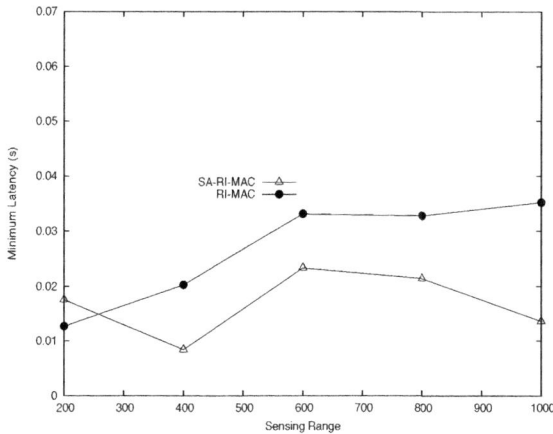

Fig. 8. Minimum Latency vs. Number of Flows

achieves lower duty cycles. In RI-MAC, contention is handled at the receiver end only which increases the value of back off for senders. This unnecessary back off does not conserve any energy. On the other hand, in SA-RI-MAC, sender coordination allows contended sender to turn off their radio to conserve energy and reduce contention at the receiver. For all sensing ranges, the duty cycle for SA-RI-MAC is significantly lower than RI-MAC. For example, for a sensing range of 1000 m, SA-RI-MAC duty cycle is better than RI-MAC duty cycle at 200m. Figure 8 show the minimum end to end latency for packets reported to a sink for RCE model as the sensing range increases in the grid network. It is apparent from the figure that, in SA-RI-MAC an event notification is received earlier than the RI-MAC. This event reporting is faster than RI-MAC for all the sensing ranges. Sender coordinated contention resolution conserve energy and at the same time allows contended senders to deliver packets without collisions at the receiver. For the RCE model, how quickly an even has been notified to a sink is more important than average and maximum latency of all the packets received at the receiver.

Random Networks. We compare the performance of SA-RI-MAC and RI-MAC in 3 random networks with 40 nodes randomly located in 1000m × 1000m simulation area. Flows are generated between a random source and a randomly selected sink node. The interval between two consecutive packets is 1 second. Each simulation run lasts for 100 seconds. Figure 9 shows the delivery ratio achieved by SA-RI-MAC and RI-MAC. For random network scenario, with flows between random source and destination pairs, SA-RI-MAC outperforms RI-MAC. SA-RI-MAC shows a substantial improvement over RI-MAC in terms of delivery ratio as the traffic load in the network increases. SA-RI-MAC

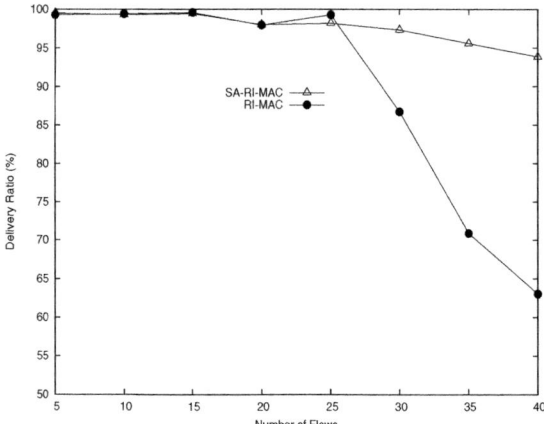

Fig. 9. Delivery Ratio vs. number of Flows

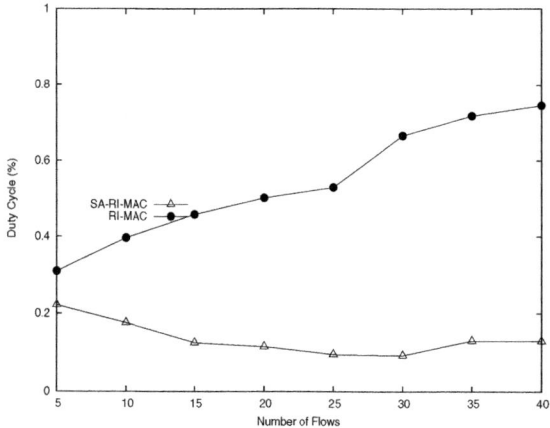

Fig. 10. Duty Cycle vs. Number of Flows

maintains delivery ratio above 90% for all the traffic loads. SA-RI-MAC conserves much more energy than RI-MAC by turning off the radio of a contended sender under high traffic loads as shown in figure 10. Figure 11 shows that for light traffic loads, RI-MAC has lower latency than SA-RI-MAC. However, as the contention in the network increases, RI-MAC triggers increased back offs at the senders which causes an increase in the latency. On the other hand, SA-RI-MAC triggers sender assisted coordination among the contended senders to avoid collisions at the receiver which reduces its latency significantly.

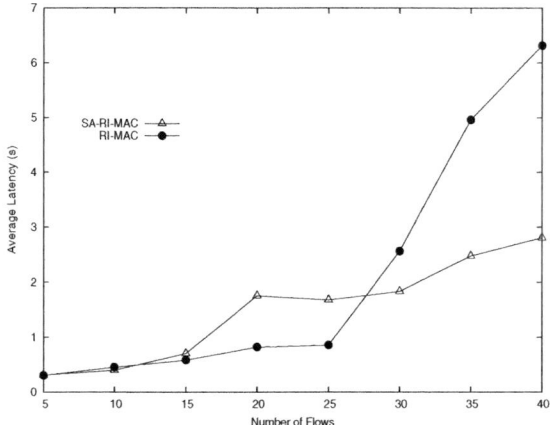

Fig. 11. Avergae Latency vs. Number of Flows

6 Conclusion

In this paper, we have presented a sender assisted receiver initiated asynchronous duty cycling MAC protocol for wireless sensor networks. SA-RI-MAC adaptively resolve the contention at the senders as traffic load increases, allowing SA-RI-MAC to achieve higher delivery ratio, lower delivery latency and less energy consumption under dynamic traffic loads. To achieve this, SA-RI-MAC turns off the radio of the contending senders to minimize the collisions at the intended receiver while still decoupling sender and receiver clocks. SA-RI-MAC significantly improves fairness among the contending senders by prioritizing the transmissions from the most starved senders.

We compared SA-RI-MAC with RI-MAC through extensive simulations. We found through evaluation that SA-RI-MAC significantly outperforms RI-MAC, with higher delivery ratio, lower delivery latency and higher power efficiency under high traffic loads. For example, under high traffic loads in clique networks, SA-RI-MAC conserve more than 75% energy than RI-MAC. In addition, SA-RI-MAC improves delivery ratio and latency under all scenarios in our simulations.

References

1. Polastre, J., Hill, J., Culler, D.: Versatile Low Power Media Access for Wireless Sensor Networks. In: Proceedings of the Second International Conference on Embedded Networked Sensor Systems (SenSys 2004), pp. 95–107 (November 2004)
2. Ye, W., Heidemann, J.S., Estrin, D.: An Energy-Efficient MAC Protocol for Wireless Sensor Networks. In: Proceedings of the 21st Annual Joint Conference of the IEEE Computer and Communications Societies (INFOCOM 2002), pp. 1567–1576 (June 2002)

3. Du, S., Saha, A.K., Johnson, D.B.: RMAC: A Routing-Enhanced Duty-Cycle MAC Protocol for Wireless Sensor Networks. In: Proceedings of the 26th Annual IEEE Conference on Computer Communications (INFOCOM 2007), pp. 1478–1486 (May 2007)
4. van Dam, T., Langendoen, K.: An Adaptive Energy-Efficient MAC Protocol for Wireless Sensor Networks. In: Proceedings of the First International Conference on Embedded Networked Sensor Systems (SenSys 2003), pp. 171–180 (November 2003)
5. Sun, Y., Du, S., Gurewitz, O., Johnson, D.B.: DW-MAC: A Low Latency, Energy Efficient Demand-Wakeup MAC Protocol for Wireless Sensor Networks. In: Proceedings of the Ninth ACM International Symposium on Mobile Ad Hoc Networking and Computing (MobiHoc 2008), pp. 53–62 (May 2008)
6. Ye, W., Heidemann, J.S., Estrin, D.: An Energy-Efficient MAC Protocol for Wireless Sensor Networks. In: Proceedings of the 21st Annual Joint Conference of the IEEE Computer and Communications Societies (INFOCOM 2002), pp. 1567–1576 (June 2002)
7. El-Hoiydi, A., Decotignie, J.-D.: WiseMAC: An Ultra Low Power MAC Protocol for Multi-hop Wireless Sensor Networks. In: Nikoletseas, S.E., Rolim, J.D.P. (eds.) ALGOSENSORS 2004. LNCS, vol. 3121, pp. 18–31. Springer, Heidelberg (2004)
8. Buettner, M., Yee, G.V., Anderson, E.: X-MAC: A Short Preamble MAC Protocol for Duty-Cycled Wireless Sensor Networks. In: Proceedings of the 4th International Conference on Embedded Networked Sensor Systems, pp. 307–320 (2006)
9. Polastre, J., Hill, J., Culler, D.: Versatile Low Power Media Access for Wireless Sensor Networks. In: Proceedings of the Second International Conference on Embedded Networked Sensor Systems (SenSys 2004), pp. 95–107 (November 2004)
10. Sun, Y., Gurewitz, O., Johnson, D.B.: RI-MAC: A Receiver Initiated Asynchronous Duty Cycle MAC Protocol for Dynamic Traffic Loads in Wireless Sensor Networks. In: Proceedings of the Second International Conference on Embedded Networked Sensor Systems (SenSys 2008), pp. 1–14 (November 2008)
11. Zhang, H., Arora, A., Choi, Y.-R., Gouda, M.G.: Reliable Bursty Convergecast in Wireless Sensor Networks. In: Proceedings of the Sixth ACM International Symposium on Mobile Ad Hoc Networking and Computing (MobiHoc 2005), pp. 266–276 (May 2005)
12. Hull, B., Jamieson, K., Balakrishnan, H.: Mitigating Congestion in Wireless Sensor Networks. In: Proceedings of the Second International Conference on Embedded Networked Sensor Systems (SenSys 2004), pp. 134–147 (November 2004)
13. Sun, Y., Du, S., Gurewitz, O., Johnson, D.B.: DW-MAC: A Low Latency, Energy Efficient Demand-Wakeup MAC Protocol for Wireless Sensor Networks. In: Proceedings of the Ninth ACM International Symposium on Mobile Ad Hoc Networking and Computing (MobiHoc 2008), pp. 53–62 (May 2008)
14. Ye, W., Silva, F., Heidemann, J.: Ultra-Low: Duty Cycle MAC with Scheduled Channel Polling. In: Proceedings of the Second International Conference on Embedded Networked Sensor Systems (SenSys 2006), pp. 321–334 (October 2006)
15. UCB/LBNL/VINT Network Simulator-ns-2, http://www.isi.edu/nsnam/ns
16. Garcia-Luna-Aceves, J.J., Tzamaloukas, A.: Reversing in Collision Avoidance Handshake in wireless networks. In: Proceedings of 5th Annual ACM/IEEE International Conference on Mobile Computing and Networking, pp. 120–131 (1999)
17. Musaloiu-E., R., Liang, C.-J.M., Terzis, A.: Koala: Ultra-Low Power Data Retrieval in Wireless Sensor Networks. In: Proceedings of the 7th International Conference on Information Processing in Sensor Networks, pp. 421–432 (2008)

18. CC2420 Datasheet, `http://www.ti.com`
19. Kyasanur, P., Vaidya, N.: Detection and Handling of MAC Layer Misbehavior in Wireless Networks. In: Proceedings of the International Conference on Dependable Systems and Networks (DSN 2003), pp. 173–182 (June 2003)
20. Anastasi, G., Falchi, A., Passarella, A., Conti, M., Gregori, E.: Performance Measurements of Motes Sensor Networks. In: Proceedings of the 7th ACM International Symposium on Modeling, Analysis and Simulation of Wireless and Mobile Systems (MSWiM 2004), pp. 174–181 (October 2004)
21. UPMA Package: Unified Power Management Architecture for Wireless Sensor Networks,
    `http://tinyos.cvs.sourceforge.net/tinyos/tinyos-2.x-contrib/`
    `wustl/upma/`

Bluetooth Indoor Positioning System Using Fingerprinting

Christian Frost, Casper Svenning Jensen, Kasper Søe Luckow,
Bent Thomsen, and René Hansen

Department of Computer Science, Aalborg University
Selma Lagerlöfs Vej 300, DK-9220 Aalborg Øst, Denmark
{chrfrost,semadk,luckow,bt,rhansen}@cs.aau.dk

Abstract. Indoor Positioning has been an active research area in the last
decade, but so far, commercial Indoor Positioning Systems (IPSs) have
been sparse. The main obstacle towards widely available IPSs has been
the lack of appropriate, low cost technologies, that enable indoor positioning. While Wi-Fi infrastructures are ubiquitous, consumer-oriented Wi-
Fi enabled mobile phones have been missing. Conversely, while Bluetooth
technology is present in the vast majority of consumer mobile phones, Bluetooth infrastructures have been missing. Bluetooth infrastructures have
typically been installed as part of complete hardware-/software IPSs that
often incur a substantial hardware cost. Furthermore, Bluetooth has low
power consumption compared to Wi-Fi devices, which promotes longer
battery life-time on mobile phones. In this paper, we present a Bluetooth
IPS based entirely on commodity-grade products. The positioning accuracy is evaluated by using the so-called location fingerprinting technique
which is well-known from Wi-Fi positioning literature. The results show
that 2 meters median accuracy is achievable - a result that compares
favourably to results for Wi-Fi based systems.

Keywords: indoor positioning, bluetooth, fingerprinting, radio map.

1 Introduction

Since the turn of the century, Location-Based Services (LBSs) have been hailed
as one of the next "killer apps". To provide such services, an underlying positioning system is required. One such system is the Global Positioning System
(GPS), which has uncovered the potential of LBSs as witnessed by the prolifera-
tion of LBSs that work outdoors. Unfortunately, GPS radio waves are unable to
penetrate most building structures, leaving large areas of indoor positioning potential untapped. Thus, an Indoor Positioning System (IPS) is needed to enable
LBSs in indoor environments.

Wi-Fi is an appealing technological alternative in GPS-less environments due
to the ubiquity of Wi-Fi infrastructures and has thus been the subject of much
recent research [1,2,3,4,5,6,7,8]. However, consumer-oriented Wi-Fi enabled mobile phones currently constitute only 18% of the total mobile phone market [9,10].

J. Del Ser et al. (Eds.): MOBILIGHT 2011, LNICST 81, pp. 136–150, 2012.
© Institute for Computer Sciences, Social Informatics and Telecommunications Engineering 2012

Although, the market share is rising (e.g. Nokia provides Wi-Fi capabilities in 35% of their new models in the UK [11]), the vast majority of mobile phones have Bluetooth capabilities. It is estimated that 75% of all existing phones and 95% of all new phones on the Australian market have Bluetooth [12]. Thus, it can be expected, that for some time to come, Bluetooth will remain more common, especially since manufacturers only provide Wi-Fi capabilities in their high-end mobile phones. Although Bluetooth has been an integrated part of mobile phones for years, Bluetooth infrastructures have been rare. Typically, Bluetooth infrastructures have been set up as part of a complete hardware-/software Indoor Positioning System (IPS), e.g., as developed by companies like BLIP Systems [13]. The cost of a complete hardware-/software IPS may be too prohibitive for many, especially smaller companies, but recently, a cheaper alternative has emerged. It is now possible to set up a Bluetooth infrastructure at a very modest cost simply by extending modern commodity-grade Wi-Fi access points with Bluetooth dongles [14,15]. This extension also provides the possibility of making a combined Wi-Fi/Bluetooth IPS based on the same infrastructure.

Another advantage of using Bluetooth is that it has a low power consumption. Specifically, it only uses 81-120mW compared to Wi-Fi which uses 890-1600mW under load [16]. Thus, Bluetooth is ideal for mobile phones where a desirable property is to increase battery life-time.

Positioning in an indoor environment is a non-trivial task. As signals propagate through an indoor environment, they are reflected, scattered, and subject to multipath fading. Moreover, signals are affected by transient effects such as changes in humidity level and human bodies absorbing the signals.

The most widely used technique for accurate indoor positioning is called *location fingerprinting* and is a technique that counters the adverse effects from environmental factors by relying on empirically measured signal strengths. The term *location fingerprinting* refers to the fact that the Received Signal Strength (RSS) values from nearby access points form spatio-temporally, approximately unique, RSS vectors.

The location fingerprinting technique is divided into two phases. In the *offline phase*, before an IPS becomes operational, a location fingerprint is created by empirically measuring the RSS values at a particular location for a period of time. At the end of the measurement period, the $\langle location, RSSvector \rangle$ pair is saved as an entry in a *radio map*. This measurement process is typically conducted for locations spaced 2-3 meters apart throughout the indoor environment. In the operational *online phase*, a location estimate is obtained by comparing the RSS values recorded by an end-user's device with the entries saved in the radio map. The location of the closest matching RSS vector is returned as location estimate.

Accuracy and precision are two common performance metrics of a positioning system whose definitions are described below and used in the following:

Accuracy. describes the extent to which the estimated location deviates from the actual location. That is, accuracy denotes the euclidian distance between an estimated location and the actual location.

Precision. refers to the percentage of measurements retaining a particular accuracy.

In this paper, we examine the positioning accuracy that can be obtained by a cheap dongle-based Bluetooth IPS using the location fingerprinting technique. Additionally, we investigate how different Bluetooth power-classes influence the performance, and evaluate the performance/cost tradeoff. Finally, we investigate the effect of different ways of constructing a radio map with respect to user orientation. Traditionally, a fingerprint is constructed by measuring RSS values in four orthogonal orientations and then averaging them [1,2]. However, this may mask important RSS differences dependent on orientation as noted in [14,15]. Thus we have compared three different radio map construction methods:

- Average Method
- Four Directions Method
- Single Direction Method

The *average method* refers to the traditional approach of constructing a fingerprint. The *four directions method* keeps all four fingerprints without averaging them. Finally, the *single direction method* is identical to the *four directions method* in the offline phase. However, when matching the fingerprints in the online phase, only a subset of fingerprints, corresponding to the user's current orientation, is searched. The intuition is that accuracy may be improved by disregarding fingerprints from other directions that may potentially yield erroneous results. The vision is that the system is able to deduce orientation of the user either by harnessing a compass, which is available in many smartphones today, or by using the history of previously estimated locations to deduce a path, which may provide a heuristic of user orientation.

The remainder of this paper is organised as follows: Section 2 describes related work on IPSs, Section 3 presents the challenges of using Bluetooth as an IPS technology, Section 4 describes the methodology of our research, followed by Section 5 where the results are evaluated, and finally, Section 6 concludes the paper.

2   Related Research

As mentioned, signals are vulnerable and influenced by several environmental factors such as reflection, changing humidity levels, presence of people, and multipath fading. The interplay of influencing factors means that traditional trilateration and signal propagation techniques are not equipped to deliver accurate position estimates [3,4,17].

Wi-Fi is a widely adopted as technology for indoor positioning. As Wi-Fi and Bluetooth operate on the same frequency band, they are both vulnerable to the same impacting environmental factors. One of the primary motivations for using Wi-Fi is that it is ubiquitous; many facilities such as educational institutions and companies include a Wi-Fi infrastructure for extending the Intranet for the mobile users. Hence, the deployment cost of a Wi-Fi based IPS is low.

The RADAR project [17], carried out at Microsoft Research, pioneered the field of IPSs by proposing the *location fingerprinting* technique to enable accurate indoor positioning in the face of a noisy Wi-Fi channel. The RADAR project used a deterministic Nearest Neighbour algorithm to match online fingerprints against the offline fingerprints saved in the radio map resulting in a median accuracy of 2.94 meters. Subsequent research efforts have tried to improve the obtainable positioning accuracy by different classification algorithms, including Support Vector Machines [5,6], Neural Networks [7,8], and Bayesian Inference [18,19]. Moreover, motion models have been applied to counter adverse effects caused by increased signal strength fluctuations when users are moving [1,20] (We refer to survey papers such as   [21,22] for an additional overview of existing algorithmic approaches). However, the obtainable accuracy can not be attributed exclusively to the use of a particular location determination algorithm, as it is also affected by factors such as the number and placement of access points, the number of samples used in the measurement process, the density of the radio map, the sensitivity of the antennas, orientation, and the environment [1,3,23]. In fact, the obtainable accuracy is inherently limited by the very nature of the Radio Frequency (RF) signals: Elnahrawy et al. [24] performed an extensive study that compared a wide range of different positioning algorithms. The study concluded that ten feet accuracy represents a feasible lower bound due to inherent limitations of differentiating RSS values at closer distances. Therefore, at roughly the ten feet mark, algorithms are only able to improve the precision, i.e., the percentage or confidence with which a given accuracy is obtained. However, ten feet accuracy is still more than enough to support a wide range of IPSs.

Commercial Bluetooth IPSs have typically used trilateration or discrete positioning to deliver context-aware information to end-users in certain information zones [13,25]. Lower granularity position estimates are acceptable for pushing content to end-users whereas navigation scenarios impose higher demands on positioning accuracy.

3   Bluetooth Challenges

Using Bluetooth for indoor positioning introduces some challenges which we have not observed elsewhere in the Wi-Fi literature. The following describes the challenges we have encountered in our work and our solutions to these.

3.1   Received Signal Strength Indicator (RSSI) Values

The Bluetooth specification [26] dictates that a signal strength can be read in terms of an RSSI value, which, as the name implies, is a metric indicating the strength of the signal. However, the problem is that the Bluetooth specification does not prescribe a standardised mapping between the RSS values measured in dBm and the RSSI values. This means that individual Bluetooth vendors are responsible for implementing their own mapping. Typically, an interval of RSS values are mapped to one RSSI value, hence the distribution granularity is relatively coarse. However, the granularity may be sufficient if the RSSI values are distributed such that small changes in distance yield distinguishable RSSI values. To ensure that the Bluetooth implementation on the mobile phone exhibits this behaviour, we have made preliminary experiments where RSSI values are measured at different distances using a Class 2 Bluetooth device [14]. As shown in Figure 1, the RSSI values are distinguishable at the different distances and the Bluetooth implementation on the mobile phone is thus applicable for indoor positioning.

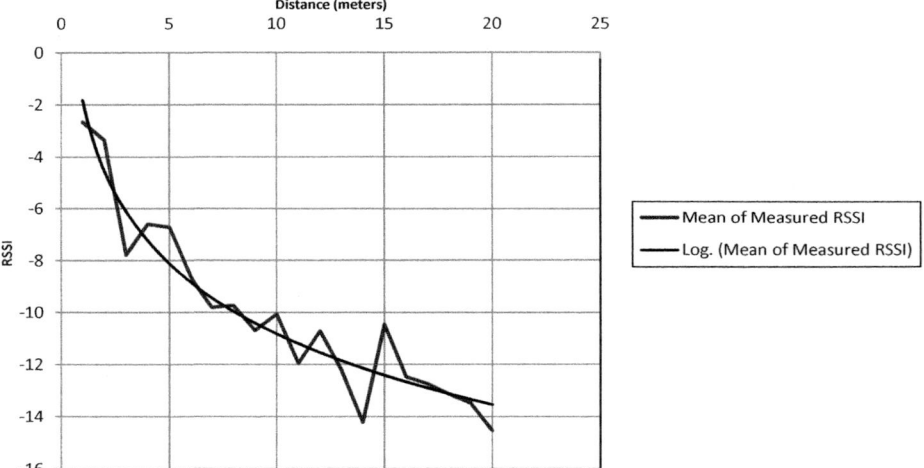

Fig. 1. Relationship between RSSI values as a function of distance

Furthermore, it should be noted that they follow the radio propagation models describing the theoretical relationship between RSS and distance to be logarithmic [27] for distances up to 20 meters. This is a rather interesting result since the Bluetooth specification describes that Class 2 devices have an effective range of 10 meters and that useful RSSI values cannot be expected if this range is increased. However, as evidenced by the graph in Figure 1, the experiment yielded results that are in accordance with the radio propagation model. Hence, we can conclude, that our Class 2 devices do not exhibit unusual behavior at distances up to 20 meters thereby proving them amenable for comparison with Class 1 devices that theoretically, according to the Bluetooth specification, can yield useuful RSSI values up to a range of 100 meters.

3.2   RSSI Cache

Through observations [15], we determined that the implementation of RSSI value measurement on our mobile phone updates an internal RSSI value cache whenever it receives a signal in the form of a packet from one of its peers, and it is this value which is returned when enquired. If multiple enquiries are made for the RSSI value in-between updates, the same value is returned multiple times. It is subject to further research whether this problem applies to all Bluetooth hardware, or only for particular models. However, it indicates that the problem requires some attention when developing indoor position systems. A simple solution is to force the cache to be updated whenever RSSI values are enquired, by setting up a small communication which ensures that the internal RSSI cache is updated.

3.3   Measuring RSSI Values

An important characteristic of Bluetooth is its usage of frequency hopping in order to reduce the impact of interference from other wireless communication using the same frequency [26]. Specifically, Bluetooth hops every 625 microseconds between 78 frequencies with 1 MHz intervals above 2.4 GHz. This impacts the time required for discovering visible devices and connecting to devices since the frequencies must be searched to synchronise. As a consequence, a search that with high probability detects all discoverable Bluetooth devices takes up to 10 seconds [28], hence doing this will only allow position estimates to be made with this interval. However, [29] have shown that the search time can be optimised by manually decreasing the search time to 5 seconds since this does not decrease the quality of the search. That is, Bluetooth position estimates can be made with 5 seconds intervals.

In a deployment, we suggest to reduce the search time to 5 seconds, and, if supported by the hardware, make a discovery search return the RSSI values for the discovered devices. However, in our research, we want to test in a controlled environment, and therefore, we make manual connections to the access points.

4   Method

Currently, Bluetooth devices are classified into three power-classes denoting the power of the transmitted signals and thereby the effective range of these. Class 1, 2, and 3 have theoretical ranges of 100, 10, and 1 meters, respectively. Class 1 devices, given their communication range, offer the lowest installation cost of a Bluetooth-based IPS, but Class 2 devices theoretically have the potential to perform better due to RSSI values being distributed in a lower range which produces higher diversity in fingerprints on smaller distances [15]. Class 3 devices are not considered applicable for an IPS due to the short communication range and thereby high cost.

The applicability of Class 1 and 2 Bluetooth location fingerprinting IPSs has been evaluated by performing experiments in three different environments. As part of the evaluation, it has been examined how the *average method*, *four directions*, and *single direction* approaches influence the results.

4.1   Hardware

The experiments were conducted using three types of hardware: A mobile phone, access points, and Bluetooth adapters for the access points. Table 1 specifies the hardware used and the corresponding price.

Table 1. Specification and price of used hardware in the experiments

Mobile Phone	
Device	HTC Touch Diamond
Operating System	Windows Mobile 6.1
Access point	
Device	Asus WL-500gP V2
Firmware	OpenWRT Kamikaze 8.09.1
Price	93.96 Eur
Bluetooth Class 1 Adapter	
Device	Belkin Bluetooth USB
Price	10.00 Eur
Bluetooth Class 1 Adapter	
Device	Deltaco BT-108 USB
Price	14.63 Eur
Bluetooth Class 2 Adapter	
Device	Kensington BT Micro USB
Price	29.99 Eur

4.2   Test Bed and Setup

The experiments are conducted in three different environments (denoted Cluster 1 to 3) at the Department of Computer Science at Aalborg University. The floor plans of the clusters are shown in Figure 2 and 3. Note that the floor plans of Cluster 1 and 2 are equal.

Following is a description of each of the clusters:

Cluster 1. Is primarily a corridor environment containing many offices that leads to a common room. The experiments were conducted in the summer period, and this cluster remained empty from people and Wi-Fi activity was at a minimum. This environment was chosen to provide a picture of how Bluetooth applies under nearly optimal conditions.

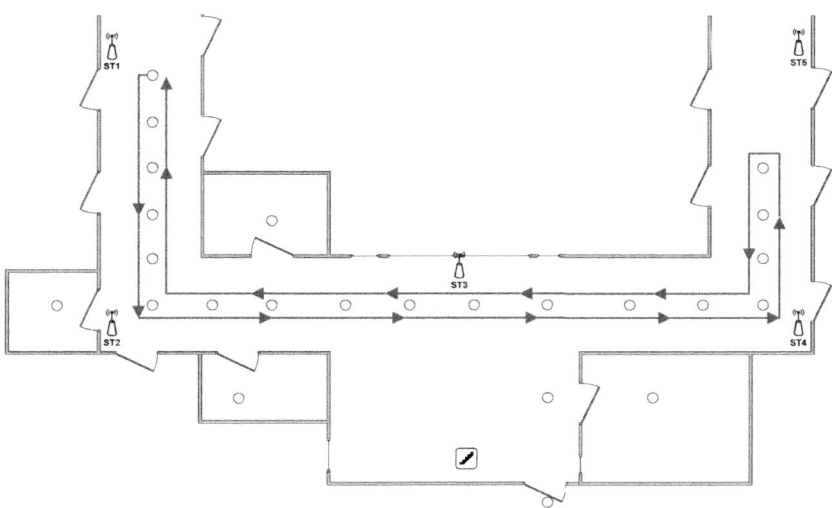

Fig. 2. Floor plan of Cluster 1 and Cluster 2. Circles represent fingerprinted locations. The marked line indicates the walked route during the online phase.

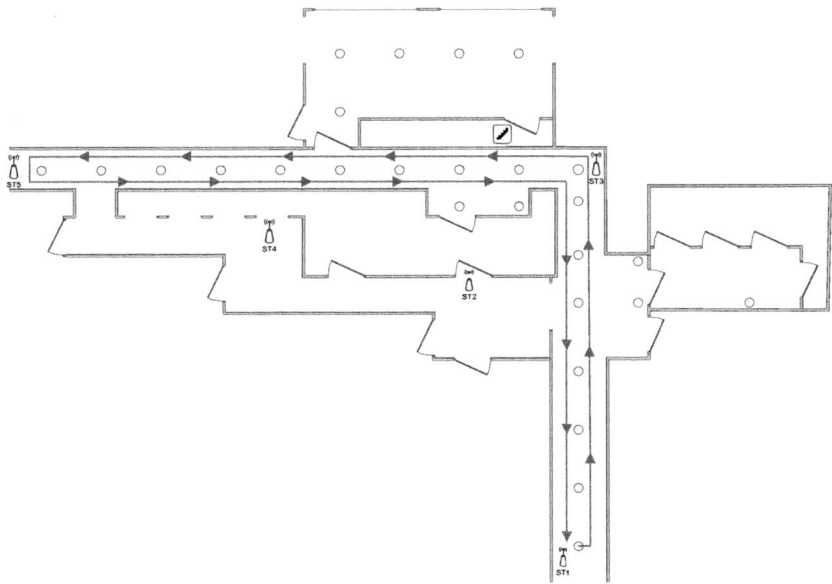

Fig. 3. Floor plan of Cluster 3

Cluster 2. The floor plan of this environment is equivalent to Cluster 1. However, the significant difference is that many negatively impacting factors were present. These include people in the majority of the offices, high Wi-Fi activity, and other temporal differences in the environment caused by the presence of people. This cluster was chosen to provide evidence for whether or not Bluetooth is sufficiently robust to sustain many temporal differences in the environment.

Cluster 3. This environment is not equivalent to the two aforementioned. It resembles medium Wi-Fi activity and people density and is generally a more open environment containing a wide range of different surface materials and obstacles. The environment was primarily chosen to determine how well Bluetooth performs in a different environment than the others.

Fingerprints in the offline phase were collected according to the *four directions method*, since data collected using this method can be used to derive radio maps for the remaining two methods. For each fingerprint, 20 RSSI values for each detectable access point were collected in a round-robin fashion 10 times. This gives a total of 200 RSSI values per access point collected over a period of time, reducing the impact of the natural fluctuation of RSSI values [15].

This approach is consistent with the fact that the values are normally distributed. We have demonstrated this distribution experimentally by sampling RSSI values at a constant distance. The distribution of these is shown in Figure 4. Hence a certain amount of values can be averaged to indicate which RSSI value is most likely to be obtained on the specific position from a specific access point.

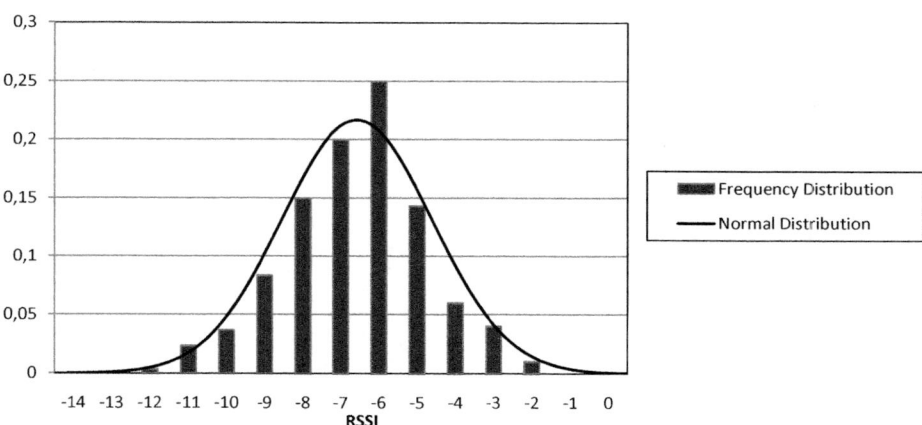

Fig. 4. Histogram showing the frequency of each measured RSSI value at a 4 meters distance

Fingerprinted locations were spaced three meters apart, and access points were placed such that three access points were detectable at each fingerprinted location.

In the online phase, a path covering the majority of the given cluster was walked in both directions. Online fingerprints were collected by measuring RSSI values when moving between two adjacent positions along the path. When the latter position was reached, the measurements were stored. During all measurements, the mobile phone was positioned in the hand of the user to resemble normal usage and the walking speed was approximately $1m/s$.

A benchmarking tool was developed which calculates the cumulated accuracy of a walked path in the online phase given a radio map and online fingerprints [15]. The tool uses K-Nearest Neighbour (KNN) and the Manhattan distance to match online fingerprints against the radio map. In our case, we have chosen $K = 3$. The results from the *average method* were obtained by comparing the online fingerprints against the averaged values in the radio map. The results from the *four directions method* were obtained by comparing online fingerprints against the radio map entries in all directions. Finally, the *single direction* results were obtained by comparing online fingerprints against the subset of fingerprints that matches the user's orientation.

5   Results

This section summarises the results of the Class 1 and 2 Bluetooth experiments in the three different environments using the three different radio map construction methods. Figures 5 and 6 show the results of the construction methods with Class 1 and 2 devices, respectively.

Figure 5 shows the results of using the three different methods together with Class 1 devices. As can be seen, Class 1 differences are negligible with different construction methods. In contrast, Class 2 devices indicate greater variations (see Figure 6). Here, the *single direction method* tends to be better up until the 6 meter accuracy mark. This behaviour is likely attributed to the fact the Class 2 devices use a lower power output which means that larger fluctuations occur if a person breaks the line of sight between a sender and receiver.

Comparing the overall performance of Class 1 and 2 devices from the graphs, it can be seen that they have a similar median accuracy (around 2 meters) with the single direction method. The performance remains similar up until the 85% precision mark (at ca. 6 meters). After that, the accuracy curve of the Class 2 devices grows more steeply, and 8 meters accuracy is achieved with 95% precision compared to ca. 11 meters for Class 1 devices. Of course, the question for a given deployment is whether the final performance boost of Class 2 devices warrants the increased hardware cost. To cover very large areas, approximately ten times more Class 2 devices are needed due to the shorter communication ranges. Motion models, such as the weighted graph model suggested in [1] may prove to even out the differences at no extra hardware cost.

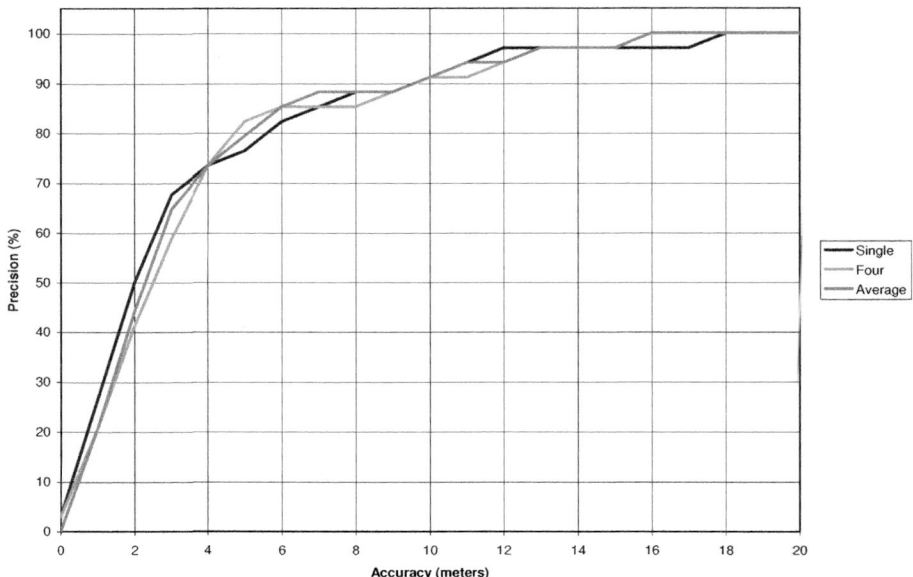

Fig. 5. Accuracy for the different map construction methods using Class 1

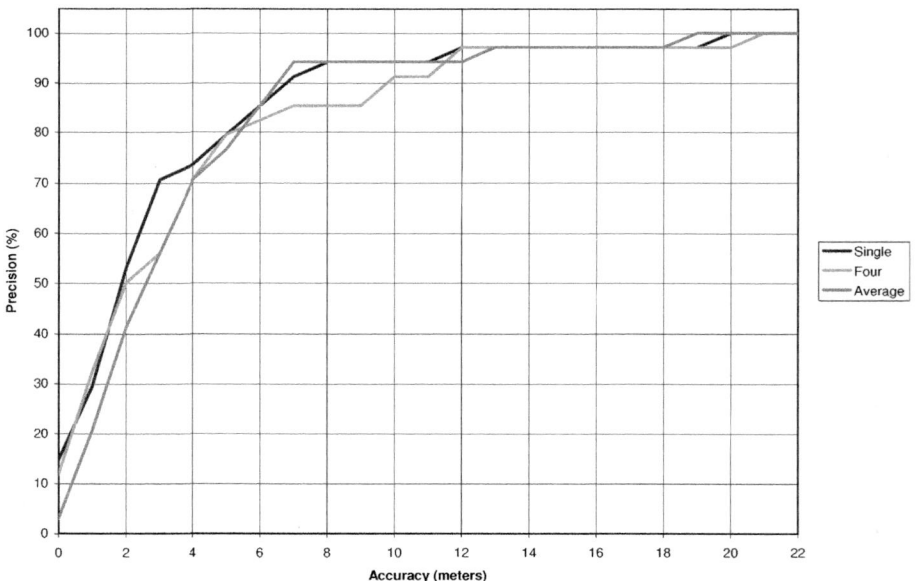

Fig. 6. Accuracy for the different map construction methods using Class 2

5.1   Class 1 and 2 Performance in Different Environments

Figure 7 illustrates the differences caused by environmental factors. The results indicate that the best accuracy is observed in Cluster 1 - the office environment where relatively few people were present at the time of the experiment. In comparison, Cluster 2 - the same office plan but with more people present - illustrates the effect on positioning accuracy in a more dynamic environment. The main difference between Cluster 2 and Cluster 3 is that Cluster 3 contains a significantly more open environment with a diverse set of obstacles such as continuous pillars in each side of the corridor; factors that are known to impact the Bluetooth signals [4]. However, the median accuracy in Cluster 3 with Class 1 devices is still below three meters, thus, demonstrating the applicability of Bluetooth location fingerprinting with low-cost equipment.

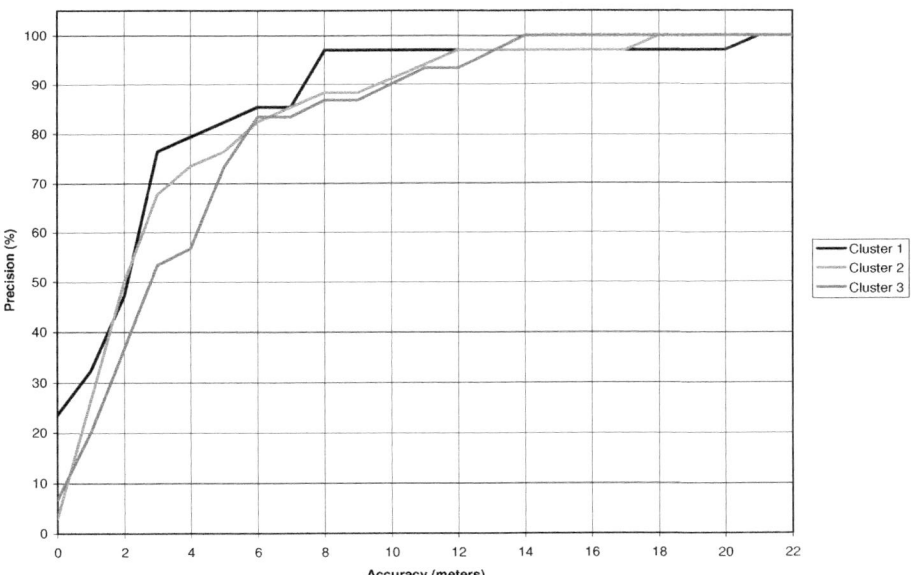

Fig. 7. Class 1 accuracy in the different environments using the *single direction method*

5.2   Summary

To better distinguish the different configurations, Table 2 summarises the accuracy for different configurations at certain percentiles interpreted from the graphs.

As shown, the accuracy is 2-3 meters at the 50th percentile for all the configurations. A similar constant accuracy is observed at the 80th percentile where the accuracy is 5-6 meters. Finally, at the 90th percentile the accuracy ranges from 7 to 10 meters. From table 2 it is clear that Class 2 devices give better accuracy, especially at the higher percentiles.

Table 2. The achievable accuracy at given percentiles using Class 1 and 2 devices. Furthermore, whether the accuracy is for the *single direction, four directions* or *average direction* construction method is denoted as Single, Four, and Average, respectively.

Percentile	Class 1			Class 2		
	Single	Four	Average	Single	Four	Average
50	2m	3m	3m	2m	2m	3m
80	6m	5m	6m	6m	6m	6m
90	10m	10m	10m	7m	10m	7m

6   Conclusion

In this paper, we have examined the positioning accuracy of Class 1 and 2 Bluetooth IPSs using the location fingerprinting technique and low-cost Bluetooth dongles that function as Bluetooth infrastructure. The impact of orientation was examined by comparing three different radio map construction methods. The *single direction method*, where a user's current orientation is used as a parameter in location determination, indicates to have a positive effect on the achievable positioning accuracy with Class 2 devices. Overall, the experiments showed that both Class 1 and 2 devices are able to achieve 2 meters median accuracy. These results, coupled with the fact that the signal strength measurement process can be conducted simultaneously for Bluetooth and Wi-Fi, means that accurate IPSs that target the maximum number of end-users can be developed easily and cheaply. Furthermore, using Bluetooth in isolation in IPSs can potentially reduce the power consumption of the mobile phones.

Using Bluetooth as foundation for an IPS, we observed some challenges which must be accounted for. Initially, one must be aware of the usage of RSSI values and how they are affected by increasing the distance from transmitter and receiver. Through an analysis, we showed that the equipment used in our experiments, distributes the RSSI values sufficiently over increasing distances, hence allowing for fingerprints with fine granularity.

Acknowledgements. This work was partly supported by the SmartCampusAAU project funded by the North Jutland Region - Vækstforum/Den Europæiske Fond for Regionaludvikling (European Regional Development Fund).

References

1. Hansen, R., Thomsen, B.: Efficient and Accurate WLAN Positioning with Weighted Graphs. Mobile Lightweight Wireless Systems, 372–386 (2009)
2. Kaemarungsi, K., Krishnamurthy, P.: Modeling of indoor positioning systems based on location fingerprinting. In: Twenty-third Annual Joint Conference of the IEEE Computer and Communications Societies, INFOCOM 2004, vol. 2, pp. 1012–1022 (2004)

3. Honkavirta, V., Perala, T., Ali-loytty, S., Piche, R.: A Comparative Survey of WLAN Location Fingerprinting Methods. In: Proceedings of the Wpnc: 2009 6th Workshop on Positioning, Navigation and Communication, pp. 243–251 (2009)
4. Bose, A., Foh, C.H.: A practical path loss model for indoor WiFi positioning enhancement. In: 2007 6th International Conference Information, Communications Signal Processing, pp. 1–5 (2007)
5. Brunato, M., Battiti, R.: Statistical learning theory for location fingerprinting in wireless LANs. Comput. Netw. ISDN Syst., 825–845 (2005)
6. Wu, C., Fu, L., Lian, F.: WLAN location determination in e-home via support vector classification. In: IEEE International Conference on Networking, Sensing and Control, pp. 1026–1031 (2004)
7. Battiti, R., Villani, A., Le Nhat, T.: Location-Aware Computing: A Neural Network Model For Determining Location In Wireless Lans. Techical report (2002)
8. Laoudias, C., Eliades, D., Kemppi, P., Panayiotou, C., Polycarpou, M.: Indoor Localization Using Neural Networks with Location Fingerprints. In: Alippi, C., Polycarpou, M., Panayiotou, C., Ellinas, G. (eds.) ICANN 2009. LNCS, vol. 5769, pp. 954–963. Springer, Heidelberg (2009)
9. Coda: Wi-Fi enabled handset penetration in the US to quadruple by 2015 (2010), http://www.codaresearch.co.uk/wificellular/pressrelease.htm
10. CTIA: CTIA Semi-Annual Wireless Industry Survey (2011), http://www.ctia.org/advocacy/research/index.cfm/AID/10316
11. Nokia: Nokia mobile phones in UK (2011), http://www.nokia.co.uk/find-products/all-phones
12. Breeze Tech: Bluetooth enabled mobile phones (2011), http://www.breeze-tech.com.au/FAQs.htm
13. BLIP: Marketing and Tracking Solutions (2010), http://www.blipsystems.com/
14. Frost, C., Jensen, C.S., Luckow, K.S.: Easy clocking - a system for automatically clocking in and out employees. Technical report, Aalborg University (2009), http://www.cs.aau.dk/~luckow/easyclocking-2009.pdf
15. Frost, C., Jensen, C.S., Luckow, K.S.: Bluecaml - bluetooth, collaborative, and maintainable location system. Technical report, Aalborg University (2010), http://www.cs.aau.dk/~luckow/bluecaml-2010.pdf
16. Pering, T., Agarwal, Y., Gupta, R., Want, R.: Coolspots: Reducing the power consumption of wireless mobile devices with multiple radio interfaces. In: Proceedings of the 4th International Conference on Mobile Systems, Applications and Services, pp. 220–232. ACM (2006)
17. Bahl, P., Padmanabhan, V.N.: RADAR: an in-building RF-based user location and tracking system. In: Nineteenth Annual Joint Conference of the IEEE Computer and Communications Societies, INFOCOM 2000, vol. 2, pp. 775–784. IEEE (2000)
18. Haeberlen, A., Flannery, E., Ladd, A.M., Algis, R., Wallach, D.S., Kavraki, L.E.: Practical robust localization over large-scale 802.11 wireless networks. In: Proceedings of the 10th Annual International Conference on Mobile Computing and Networking, pp. 70–84. ACM (2004)
19. Roos, T., Myllymki, P., Tirri, H., Misikangas, P., Sievnen, J.: A Probabilistic Approach to WLAN User Location Estimation. International Journal of Wireless Information Networks, 155–164 (2002)
20. Bahl, P., Padmanabhan, V.N., Balachandran, A.: Enhancements to the RADAR user location and tracking system. Citeseer. Microsoft Research (2000)
21. Liu, H., Darabi, H., Banerjee, P., Liu, J.: Survey of Wireless Indoor Positioning Techniques and Systems. IEEE Transactions on Systems, Man and Cybernetics, 1067–1080 (2007)

22. Kjærgaard, M.B.: A Taxonomy for Radio Location Fingerprinting. In: Hightower, J., Schiele, B., Strang, T. (eds.) LoCA 2007. LNCS, vol. 4718, pp. 139–156. Springer, Heidelberg (2007)
23. King, T., Haenselmann, T., Effelsberg, W.: Deployment, Calibration, and Measurement Factors for Position Errors in 802.11-Based Indoor Positioning Systems. In: Hightower, J., Schiele, B., Strang, T. (eds.) LoCA 2007. LNCS, vol. 4718, pp. 17–34. Springer, Heidelberg (2007)
24. Elnahrawy, E., Li, X., Martin, R.P.: The limits of localization using signal strength: a comparative study. In: IEEE Communications Society Conference on Sensor, Mesh and Ad Hoc Communications and Networks, pp. 406–414 (2004)
25. Thapa, K., Case, S.: An Indoor Positioning Service for Bluetooth Ad Hoc Networks. In: MISC 2003: The 36th Annual Midwest Instruction and Computing Symposium, Duluth, MN, USA (2003)
26. Sig, B.: Specification of the Bluetooth System Version 2.1+ EDR (2007)
27. Papamanthou, C., Preparata, F.P., Tamassia, R.: Algorithms for Location Estimation Based on RSSI Sampling. In: Fekete, S.P. (ed.) ALGOSENSORS 2008. LNCS, vol. 5389, pp. 72–86. Springer, Heidelberg (2008)
28. Woodings, R., Joos, D., Clifton, T., Knutson, C.: Rapid heterogeneous connection establishment: Accelerating Bluetooth inquiry using IrDA. In: Wireless Communications and Networking Conference, pp. 342–349 (2002)
29. Fresnedo, O., Iglesia, D.I., Escudero, C.J.: Bluetooth inquiry procedure: optimization and influence of the number of devices. In: Proceedings of the Sixth IASTED International Conference on Communication Systems and Networks, pp. 205–209. ACTA Press (2007)
30. Bolliger, P., Partridge, K., Chu, M., Langheinrich, M.: Improving Location Fingerprinting through Motion Detection and Asynchronous Interval Labeling. In: Choudhury, T., Quigley, A., Strang, T., Suginuma, K. (eds.) LoCA 2009. LNCS, vol. 5561, pp. 37–51. Springer, Heidelberg (2009)

User Movements Forecasting by Reservoir Computing Using Signal Streams Produced by Mote-Class Sensors

Claudio Gallicchio[1], Alessio Micheli[1], Paolo Barsocchi[2], and Stefano Chessa[1,2]

[1] Computer Science Department, University of Pisa,
Largo B. Pontecorvo 3, 56127 Pisa, Italy
[2] ISTI-CNR, Pisa Research Area, Via Morruzzi , 56124 Pisa, Italy

Abstract. Real-time, indoor user localization, although limited to the current user position, is of great practical importance in many Ambient Assisted Living (AAL) applications. Moreover, an accurate prediction of the user next position (even with a short advice) may open a number of new AAL applications that could timely provide the right services in the right place even before the user request them. However, the problem of forecasting the user position is complicated due to the intrinsic difficulty of localization in indoor environments, and to the fact that different paths of the user may intersect at a given point, but they may end in different places. We tackle with this problem by modeling the localization information stream obtained from a Wireless Sensor Network (WSN) using Recurrent Neural Networks implemented as efficient Echo State Networks (ESNs), within the Reservoir Computing paradigm. In particular, we have set up an experimental test-bed in which the WSN produces localization information of a user that moves along a number of different paths, and in which the ESN collects localization information to predict a future position of the user at some given mark points. Our results show that, with an appropriate configuration of the ESN, the system reaches a good accuracy of the prediction also with a small WSN, and that the accuracy scales well with the WSN size. Furthermore, the accuracy is also reasonably robust to variations in the deployment of the sensors. For these reasons our solution can be configured to meet the desired trade-off between cost and accuracy.

Keywords: Movement Forecasting, Sensor Stream Analysis, Received Signal Strength, Echo State Networks, Wireless Sensor Networks, Ambient Assisted Living.

1 Introduction

Wireless Sensor Networks (WSN) [7] are a recent development for unattended monitoring, which resulted particularly useful in many different application fields. In a typical deployment, a WSN is composed by a number of wireless sensors: small micro-systems that embed a radio transceiver and a set of transducers

J. Del Ser et al. (Eds.): MOBILIGHT 2011, LNICST 81, pp. 151–168, 2012.
© Institute for Computer Sciences, Social Informatics and Telecommunications Engineering 2012

suitable to monitor different environmental parameters. In many applications sensors are battery powered. One of the most promising application is Ambient Assisted Living (AAL) [12], an innovation funding program issued by the European Commission. AAL seeks for solutions integrating different technologies suitable for the improvement of the quality of life of elders and disabled in the environments where these people live (primarily in their houses) and work. In AAL spaces WSN play an important role as they are generally the primary source of context information about the user. For example WSN can monitor physiological parameter of the user, the environmental conditions and his/her movements and activities [34]. In most cases, raw data acquired by the WSN is given in input to software components that refine this information and that forecast the behavior or needs of the user in order to supply the user with appropriate services. In this paper we consider a scenario related to forecasting of user movements. In this scenario the user is localized in real time by a WSN (composed by low cost, low power sensors such as those of mote class [1]), and localization information is used to predict (with a short advance) whether the user will enter in a room or not, in order to timely supply the user with some services available in the room where the user is entering in. To this purpose the user wears a sensor, whose position is computed by a number of static sensors (also called *anchors*) deployed in the house. The sensor on the user and the environmental sensors exchange packets in order to compute the Received Signal Strength (RSS) for each packet, and use this information to evaluate the position of the user in real time. Although simple, this approach faces two main problems. The first is that indoor user localization is not sufficient by itself, since the current user position is not sufficient to predict the future behavior of the user. The second is that RSS measurements in indoor environments are rather noisy and this fact makes localization information imprecise. This latter problem is due both to multipath effects of indoor environments, and to the fact that the body of the user affects the radio signal propagation with irregular patterns, depending on the orientation of the user, orientation of the antenna etc. Overall, the considered scenario requires an approach which is adaptive, efficient and robust to the input noise. For this reasons, we take into consideration Machine Learning models in general, and neural networks in particular. More specifically, we exploit Recurrent Neural Networks (RNNs) [22], a class of dynamical neural network models that can work directly on the streams of RSS produced by the environmental sensors rather than on a localization information. The RNN takes into account also past measurements that reflect the history of previous movements, in order to overcome the fact that the current user position does not provide enough information. In particular, we consider the Reservoir Computing (RC) [26, 37] approach for modeling RNNs. Featured by extreme efficiency, RC models represent ideal candidates for approaching the problem in the considered scenario. Efficiency of the learning model used is in fact a critical factor, in particular in view of its deployment within the sensors themselves. In this work we present the results of a set of experiments in a real indoor environment aimed at producing a sufficiently large dataset to be used for the learning and evaluation of the RC model, and we evaluate our solution in terms of predictive classification accuracy and cost. In particular, we evaluate the cost in terms of the number of anchors that are necessary to achieve the desired accuracy and in terms of independence from the actual deployment of the anchors (that we evaluate by comparing the accuracy of the predictions depending on the position of the anchors) which has a direct impact on deployment costs. In our experiments we show that our approach provides optimal accuracy with 4 anchors, but it can already provide a good accuracy even with a single anchor. Furthermore our approach scales with the number of anchors, hence it can be easily tuned in order to attain the desired trade-off between accuracy and cost of the solution.

2   Related Work

In the past years, many developed indoor positioning systems extract the location-dependent parameters such as time of arrival, time difference of arrival and angle of arrival [35] from the received radio signal transmitted by the mobile station. Such a measurement needs to be estimated accurately and it requires line of sight (LOS) between the transmitter and the receiver. Furthermore, it requires specialized and expensive hardware integrated into the existing equipments. Due to the high implementation cost, the indoor positioning system based on the use of RSS thus gets more and more interests. Since the deployments of WLAN infrastructures are widespread and the RSS sensor function is available in every 802.11 interface, the RSS-based positioning system is obviously a more cost-effective solution.

The model-based positioning approach is one of the most widely used technology seen in the literature since it expresses the radio frequency signal attenuation using a path loss model [8,9]. From an observed RSS, these methods triangulate the person based on a distance calculation from multiple access points. However, the relationship between position and RSS relationship is highly complex due to multipath, metal reflection, and interference noise. Thus, the RSS propagation may not be adequately captured by a fixed invariant model. In contrast to model-based positioning, fingerprinting based RSS approaches are used [6,23,31,39]. Fingerprints are generated during an offline training phase, where RSS data is collected at a set of marked training locations. The most challenging aspect of the fingerprinting based method is to formulate a distance calculation that can measure similarity between the observed RSS and the known RSS fingerprints. Various Machine Learning techniques can be applied to the location estimation problem [21]. Probabilistic method [28], k-nearest-neighbor [6], neural networks [29], and Support Vector Machines [23] are exploited in popular positioning techniques based on the location fingerprinting. Euclidean distance based calculation has been used in [20] to measure the minimum distance between the observed RSS and the mean of the fingerprints collected at each training point. RADAR [6] uses a k-nearest-neighbors method in order to find the closest match between fingerprints and RSS observation. Recently, research efforts have concentrated on developing a better distance measure that can take into account the variability of the RSS training vectors. These methods estimate probability density for the training RSS and then compute likelihood/a posteriori estimates during the tracking phase using the observed RSS and the estimated densities [39]. User localization is then performed using a maximum-likelihood (ML) or maximum a posteriori (MAP) estimate of position. All these location determination methods do not solve the problem to forecast the user behaviors leveraging on empirical RSS measures. The Machine Learning approach can take advantage of training RSS data to capture characteristics of interest of their unknown underlying probability distribution. In this paper we consider the Echo State Network (ESN) [17, 18] model within the RC paradigm for modeling of RNNs. ESNs are dynamical neural networks used for sequence processing tasks. One of the main characteristics of ESNs is the efficiency of the approach. Learning is indeed restricted to a simple linear output tool, while the dynamical part of the network (the *reservoir*) is left untrained after initialization. The contractive reservoir dynamics provides a fading memory of past inputs, allowing the network to intrinsically discriminate among different input histories [17] in a suffix-based fashion [13, 14, 36], even in the absence of training. Despite the extreme efficiency of the approach, ESNs have been successfully applied to many common tasks in the area of sequence processing, often outperforming other state-of-the-art learning models for sequence domains (e.g. [17, 18]). In particular, in the last years ESN models have shown good potentialities in a range of tasks related to autonomous systems modeling. Examples of such tasks include event detection and localization in autonomous robot navigation [4, 5], multiple robot behavior modeling and switching [3, 38], robot behavior acquisition [16] and robot control [30]. However, such applications are mostly focused on modeling robot behaviors and often use artificial data obtained by simulators (e.g. [3–5, 38]). In this paper we apply the ESN approach to a real-world scenario for user indoor movements forecasting, characterized by real and noisy RSS input data, paving the way for potential applications in the field of AAL.

3   Experimental Setup

We carried out a measurement campaign on the first floor of the the ISTI institute of CNR in the Pisa Research Area, in Italy. The environment is composed of 2 rooms (namely Room 1 and Room 2), which are typical office environments with overall dimensions of approximately 12 m by 5 m divided by an hallway. The rooms contain typical office furniture: desks, chairs, cabinets, monitors that are asymmetrically arranged. This is a harsh environment for wireless communications because of multi-path reflections due to walls and interference due to electronic devices. For the experiments we used a sensor network of 5 IRIS nodes [1] (4 sensors, in the following *anchors*, and one sensor placed on the user, hereafter *mobile*), embedding a Chipcon AT86RF230 radio subsystem that implements the IEEE 802.15.4 standard. The experiments consisted in a set of measures between anchors and mobile. Figure 1 shows the anchors deployed in the environment as well as the movements of the user. The height of the anchors

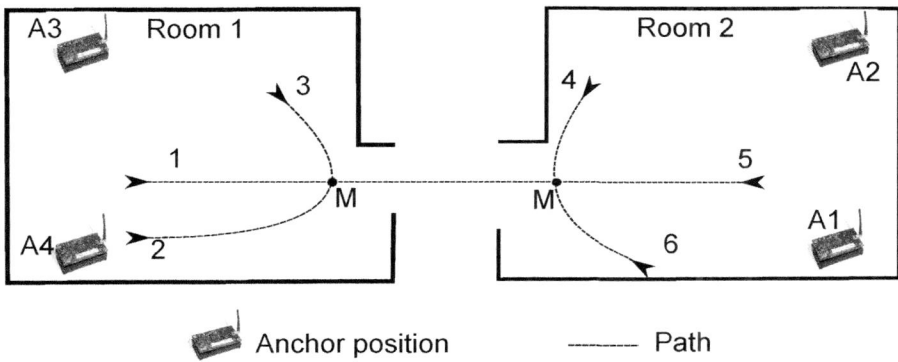

Fig. 1. Test-bed environment where the measurements have been done. The positions of the anchors, and the 6 user movements are shown.

was 1.5 m from the ground and the mobile was worn on the chest. The measurements were carried out in empty rooms to facilitate a constant speed of the user of about 1 m/s. Each measure collected about 200 RSS samples (integer values ranging from 0 to 100), where every sample was obtained by sending a beacon packet from the anchors to the mobile at regular intervals, 10 times per second, using the full transmission power of the IRIS. During the measures the user performs two types of movements: straight and curved, for a total of 6 paths (2 of which straight) that are shown in Fig. 1 with arrows numbered from 1 to 6. The straight movement runs from Room 1 to Room 2 or viceversa (paths 1 and 5 in Fig. 1) for 50 times in total. The curved movement is executed 25 times in Room 1 and 25 times in Room 2 (paths 2, 3, 4 and 6 in Fig. 1). Each path produces a trace of RSS measurements that begins from the corresponding arrow and that is marked when the user reaches a point (denoted with M in Fig. 1) located at 60 cm from the door. Overall, the experiment produced about 5000 RSS samples from each of the 4 anchors. The marker M is the same for all the movements, therefore the different path can not be only distinguished from the RSS values collected in M. The scenario and the collected RSS measures described so far can naturally lead to the definition of a binary classification task on time series for movements forecasting. The RSS samples from the four anchors are organized in 100 input sequences, corresponding to the conducted measures until the marker (M) is reached. The RSS traces can be freely downloaded in [2]. Each single trace is stored in a separate file that contains one row for each RSS measurement. Each row has 4 columns corresponding to: anchor ID, sequence number of the beacon packet, RSS value, and the boolean marker (1 if that measurement is done in point M, 0 otherwise). The resulting input sequences have length varying between 16 and 101. A target classification label is then associated to each input sequence, namely +1 for entering movements (paths 1 and 5 in Fig. 1) and −1 for non-entering ones (paths 2, 3, 4 and 6 in Fig. 1). The constructed dataset is therefore balanced, i.e. the number of sequences with positive classification is equal to the number of sequences with negative classification.

3.1   Slow Fading Analysis

The wireless channel is affected by multipath fading that causes fluctuations in the receiver signals amplitude and phase. The sum of the signals can be constructive or destructive. This phenomenon, together with the shadowing effect, may strongly limit the performance of wireless communication systems and makes the RSS values unstable. Most of the recent research works in wireless sensor networks, modeled wireless channel with Rayleigh fading channel model [11,25], which is suitable channel model for wireless communications in urban areas where dense and large buildings act as rich scatterers. In indoor environments Nakagami or Ricean fading channel model works well, because it contains both non-LOS and LOS components. But Nakagami-m distribution function, proposed by Nakagami [27], is a more versatile statistical representation that can model a variety of fading scenarios including those modeled by Rayleigh and one-sided Gaussian distributions. Furthermore, in [33] the authors demonstrated that Nakagami-m distribution is more flexible and fits more accurately with experimental data for many propagation channels than the other distributions. We observe that the received signal

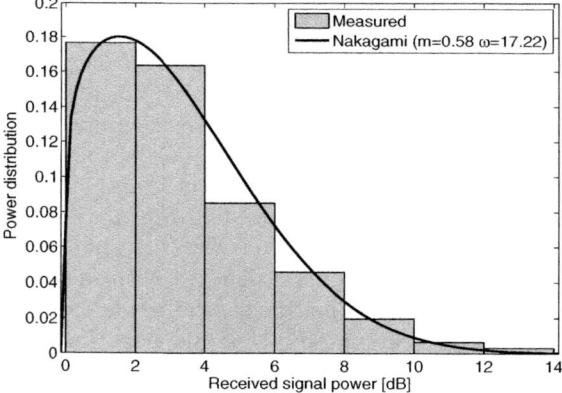

Fig. 2. Distribution of received power level from the anchor A4 and Nakagami-m distribution with $\mu = 0.58$ and $\omega = 17.22$

envelope is modulated by a slow fading process that produce an oscillation of RSS with respect its mean. This is due to multipath effects caused by scattering of the radio waves on office furniture. In order to verify this hypothesis, we look for the signature of multipath fading, by considering the distribution of the received power. The fading distribution approximates a Nakagami-m distribution around the mean received power. The Nakagami-m distribution has two parameters: a shape parameter m and a controlling spread ω. ω and m, lie in the range from 17 to 50 and from 0.5 to 0.8 for most of the measurements, respectively. The distribution observed on measured data (Fig. 2 shows an example for the same measurement

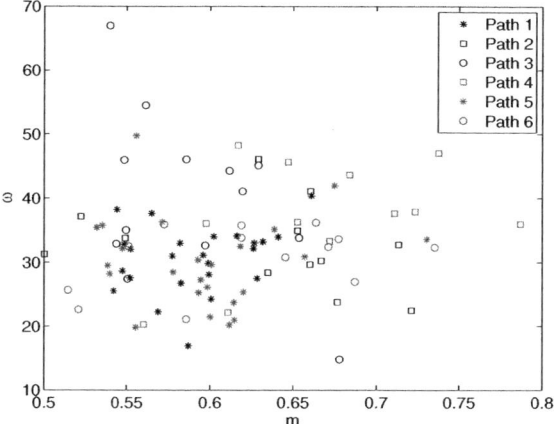

Fig. 3. Nakagami parameters of the different user paths

as above and for the anchor A4) are consistent with what is observed in [33]. As far as the dependence of the fading statistics on measurement parameters is concerned, we observe that the power spectrum is similar for all the measurements. In fact, m and ω are similar for all the measurements. For the arc movements, the Nakagami parameters are more widely spread, as shown in Fig. 3 with respect to the straight ones. As highlighted in Fig. 3 all the paths produce similar RSS traces, making it hard to forecasting the user behavior. Despite the similar RSS distributions, in Section 5 we will show that the proposed system is able to forecast the user behavior. It is also interesting to note that the traces collected in Room 1 can be modeled with m parameter values more close together with respect to the traces collected in Room 2. Consequently, we expect that the proposed system will mis-classify these paths more frequently.

4   Model

An ESN is a RNN with an input layer of N_U units, a large and sparsely connected hidden *reservoir* layer of N_R recurrent non-linear units and a *readout* layer of N_Y feed-forward linear units (see Fig. 4).

The untrained reservoir acts as a *fixed* non-linear temporal expansion function, implementing an encoding process of the input sequence into a state space where the trained linear readout is applied. More formally, given an input sequence $\mathbf{s} = [\mathbf{u}(1), \dots, \mathbf{u}(n)]$ over the input space $\mathbb{R}^{N_U}$, at each time step $t = 1, \dots, n$ the reservoir computes the following state transition function:

$$\mathbf{x}(t) = f(\mathbf{W}_{in}\mathbf{u}(t) + \hat{\mathbf{W}}\mathbf{x}(t-1)) \tag{1}$$

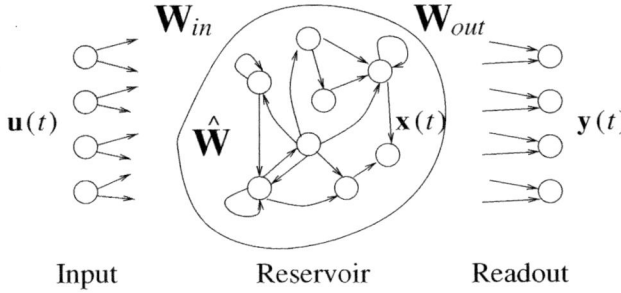

Fig. 4. The architecture of an ESN

where $\mathbf{x}(t) \in \mathbb{R}^{N_R}$ denotes the reservoir state (i.e. the output of the reservoir units) at time step t, $\mathbf{W}_{in} \in \mathbb{R}^{N_R \times N_U}$ is the input-to-reservoir weight matrix (possibly including a bias term), $\hat{\mathbf{W}} \in \mathbb{R}^{N_R \times N_R}$ is the (sparse) recurrent reservoir weight matrix and f is the component-wise applied activation function of the reservoir units (we use $f \equiv tanh$). A null initial state is used, i.e. $\mathbf{x}(0) = \mathbf{0} \in \mathbb{R}^{N_R}$. For the case of sequence binary classification tasks, which is of interest for this paper, the linear readout is applied only after the encoding process computed by the reservoir is terminated, according to the equation:

$$\mathbf{y}(\mathbf{s}) = sgn(\mathbf{W}_{out}\mathbf{x}(n)) \tag{2}$$

where sgn is a sign threshold function returning $+1$ for non-negative arguments and -1 otherwise, $\mathbf{y}(\mathbf{s}) \in \{-1, +1\}$ is the output classification computed for the input sequence $\mathbf{s}$ and $\mathbf{W}_{out} \in \mathbb{R}^{N_Y \times N_R}$ is the reservoir-to-output weight matrix (possibly including a bias term).

In this paper we also consider leaky integrator ESNs (LI-ESNs) [19], in which leaky integrator reservoir units are used. In this case, the state transition function of equation 1 is replaced by the following:

$$\mathbf{x}(t) = (1 - a)\mathbf{x}(t - 1) + af(\mathbf{W}_{in}\mathbf{u}(t) + \hat{\mathbf{W}}\mathbf{x}(t - 1)) \tag{3}$$

where $a \in [0, 1]$ is a *leaking rate* parameter, which is used to control the speed of the reservoir dynamics, with small values of a resulting in reservoirs that react slowly to the input [19, 26]. Compared to the standard ESN model, LI-ESN applies an exponential moving average to the state values produced by the reservoir units (i.e. $\mathbf{x}(t)$), resulting in a low-pass filter of the reservoir activations that allows the network to better handle input signals that change slowly with respect to the sampling frequency [5, 26]. LI-ESN state dynamics are therefore more suitable for representing the history of input signals. Note that for $a = 1$ equation 3 reduces to equation 1 and standard ESNs are obtained. In the following, we thereby use equation 3 to refer the reservoir computation for both LI-ESN and ESN (with $a = 1$).

The reservoir is initialized to satisfy the so called *Echo State Property* (ESP) [17]. The ESP simply states that the reservoir state of an ESN driven by a long input sequence does only depend on the input sequence itself. Dependencies on the initial states are progressively forgotten after an initial *transient* (the reservoir provides an echo of the input signal). A sufficient and a necessary condition for the reservoir initialization are given in [17]. Usually, only the necessary condition is used for reservoir initialization, whereas the sufficient condition is often too restrictive [10, 17]. The necessary condition for the ESP is that the system governing the reservoir dynamics of equation 3 is locally asymptotically stable around the zero state $\mathbf{0} \in \mathbb{R}^{N_R}$. Setting $\tilde{\mathbf{W}} = (1-a)\mathbf{I} + a\hat{\mathbf{W}}$, where a is the leaking rate parameter of equation 3, the necessary condition is satisfied whenever the following constraint holds:

$$\rho(\tilde{\mathbf{W}}) < 1 \tag{4}$$

where $\rho(\tilde{\mathbf{W}})$ is the *spectral radius* of $\tilde{\mathbf{W}}$. In the following, for the ease of notation, we simply use ρ to refer the spectral radius of matrix $\tilde{\mathbf{W}}$. Matrices $\mathbf{W}_{in}$ and $\hat{\mathbf{W}}$ are therefore randomly initialized from a uniform distribution, and then $\hat{\mathbf{W}}$ is scaled such that equation 4 holds. Values of ρ close to 1 are commonly used in practice, leading to reservoir dynamics close to the edge of chaos [24], often resulting in the best performance in applications (e.g. [17]).

For sequence classification tasks, each training sequence is presented to the reservoir a number of $N_{transient}$ consecutive times, to account for the initial transient. The final reservoir states corresponding to the training sequences are collected in the columns of matrix $\mathbf{X}$, while the vector $\mathbf{y}_{target}$ contains the corresponding target classifications. The linear readout is therefore trained to solve the least squares linear regression problem

$$\min \| \mathbf{W}_{out}\mathbf{X} - \mathbf{y}_{target} \|_2^2 \tag{5}$$

Usually, Moore-Penrose pseudo-inversion of matrix $\mathbf{X}$ or ridge regression are used to train the readout [26].

The most striking feature of ESNs is efficiency. Indeed, training is restricted only to a linear output part and is very efficient, whereas the dynamic part of the network is fixed and the cost of its encoding procedure scales linearly with the length of the input for both training and test. In this regard, the ESN approach compares extremely well with competitive state-of-the-art learning models for sequence domains, including RNNs (in which the dynamic recurrent part is trained, e.g. [22]), Hidden Markov Models (with the additional cost for the inference also at test time, e.g. [32]) and Kernel Methods for sequences (whose cost can scale quadratically or more with the length of the input, e.g. [15]).

5   Computational Experiments

5.1   Experimental Settings

Accordingly to the classification task defined in Section 3, the t-th element $\mathbf{u}(t)$ of an input sequence $\mathbf{s}$ consists in the t-th set of RSS samples from the different anchors considered in the corresponding measure, rescaled into the real interval $[-1, 1]$. Each input sequence was presented $N_{transient} = 3$ times to account for the initial transient. For our purposes, we considered different experimental settings in which the RSS from one or more of the four anchors is non-available. Therefore, the dimension of each $\mathbf{u}(t)$ can vary from 1 to 4 depending on the setting considered, i.e. on the number $N_{anchors}$ of anchors used.

In our experiments, we used reservoirs with $N_R = 500$ units and 10% of connectivity, spectral radius $\rho = 0.99$ and input weights scaled in the interval $[-1, 1]$. For LI-ESNs, we used the leaking rate $a = 0.1$. A number of 10 independent (random guessed) reservoirs was considered for each experiment (and the results presented are averaged over the 10 guesses). The performances of ESNs and LI-ESNs were evaluated by 5-fold cross validation, with stratification on the movement types, resulting in a test set of 20 sequences for each fold. For model selection, in each fold the training sequences were split into a training and a (33%) validation set. To train the readout, we considered both pseudo-inversion and ridge regression with regularization parameter $\lambda \in \{10^{-3}, 10^{-5}, 10^{-7}\}$. The readout regularization was chosen by model selection on the validation set.

5.2   Experimental Results

Number of Anchors. In this subsection we present the performance results obtained by ESNs and LI-ESNs corresponding to the different experimental settings considered, with a number of anchors $N_{anchors}$ varying from 1 to 4. For every value of $N_{anchors}$, the results are averaged (and standard deviations are computed) over the possible configurations of the anchors. The accuracies on the test set achieved by ESNs and LI-ESNs are graphically shown in Fig. 5. It is evident that the performances of both ESNs and LI-ESNs scale gracefully (and almost linearly) with the number of anchors used, i.e. with the cost of the WSN. The accuracy of ESNs varies from 0.53 (for $N_{anchors} = 1$) to 0.66 (for $N_{anchors} = 4$), whereas the accuracy of LI-ESNs varies from 0.81 (for $N_{anchors} = 1$) to 0.96 (for $N_{anchors} = 4$). Thus, the performance of the LI-ESN model is excellent for $N_{anchors} = 4$, scaling to acceptable values even for $N_{anchors} = 1$. In this regard it is also interesting that ESNs are consistently outperformed by LI-ESNs for every value of $N_{anchors}$. This result enlightens the better suitability of LI-ESNs for appropriately emphasizing the overall input history of the RSS signals considered with respect to the noise. The ROC plot in Fig. 6 provides a further graphical comparison of the test performances of ESNs and LI-ESNs.

Tables 1, 2, 3 and 4 detail the mean accuracy, sensitivity and specificity of ESNs and LI-ESNs, respectively, on the training and test sets, for increasing $N_{anchors}$. For both ESNs and LI-ESNs, sensitivity is slightly higher than specificity on the test set.

The nice scaling behavior of the performance with the decreasing number of anchors used, thus with the decreasing cost of the WSN, is also apparent from Tables 5 and 6, which provide the confusion matrices for ESNs and LI-ESNs, respectively, averaged over all the test set folds, with 20 sequences each (10 with positive target, 10 with negative target).

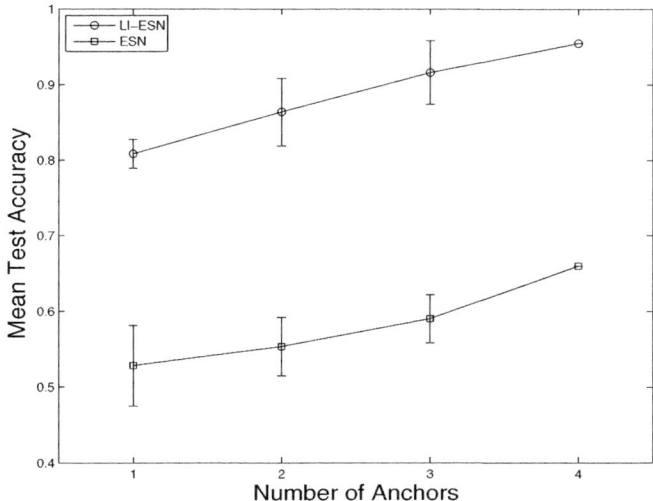

Fig. 5. Mean accuracy of ESNs and LI-ESNs on the test set, varying the number of anchors considered

Table 1. Mean training accuracy, sensitivity and specificity of ESNs, varying the number of anchors considered

$N_{anchors}$	Accuracy	Sensitivity	Specificity
1	0.96($\pm$0.01)	0.96($\pm$0.02)	0.97($\pm$0.01)
2	1.00($\pm$0.00)	1.00($\pm$0.00)	1.00($\pm$0.00)
3	1.00($\pm$0.00)	1.00($\pm$0.00)	1.00($\pm$0.00)
4	1.00($\pm$0.00)	1.00($\pm$0.00)	1.00($\pm$0.00)

Table 2. Mean test accuracy, sensitivity and specificity of ESNs, varying the number of anchors considered

$N_{anchors}$	Accuracy	Sensitivity	Specificity
1	0.53($\pm$0.05)	0.53($\pm$0.05)	0.53($\pm$0.06)
2	0.55($\pm$0.04)	0.55($\pm$0.06)	0.56($\pm$0.02)
3	0.59($\pm$0.03)	0.59($\pm$0.04)	0.58($\pm$0.02)
4	0.66($\pm$0.00)	0.69($\pm$0.00)	0.63($\pm$0.00)

The distribution of LI-ESN classification errors occurring in correspondence of each of the path types (see Fig. 1) is provided in Table 7, for the case of $N_{anchors} = 4$. Interestingly, the classification errors mainly occur for input sequences which correspond to movements in the Room 1, i.e. paths 1, 2 and 3 in Fig. 1. This actually confirms the coherence of the LI-ESN model with respect

Table 3. Mean training accuracy, sensitivity and specificity of LI-ESNs, varying the number of anchors considered

$N_{anchors}$	Accuracy	Sensitivity	Specificity
1	0.92($\pm$0.02)	0.97($\pm$0.01)	0.87($\pm$0.04)
2	0.99($\pm$0.01)	1.00($\pm$0.00)	0.98($\pm$0.02)
3	1.00($\pm$0.00)	1.00($\pm$0.00)	1.00($\pm$0.00)
4	1.00($\pm$0.00)	1.00($\pm$0.00)	1.00($\pm$0.00)

Table 4. Mean test accuracy, sensitivity and specificity of LI-ESNs, varying the number of anchors considered

$N_{anchors}$	Accuracy	Sensitivity	Specificity
1	0.81($\pm$0.02)	0.86($\pm$0.04)	0.76($\pm$0.02)
2	0.86($\pm$0.04)	0.88($\pm$0.05)	0.85($\pm$0.04)
3	0.92($\pm$0.04)	0.93($\pm$0.04)	0.90($\pm$0.04)
4	0.96($\pm$0.00)	0.98($\pm$0.00)	0.93($\pm$0.00)

Table 5. Averaged confusion matrix on the test set (with 10 positive samples and 10 negative samples for each fold) for ESNs, varying the number of anchors considered

$N_{anchors}$	True Positives	True Negatives	False Positives	False Negatives
1	5.29($\pm$0.52)	5.28($\pm$0.57)	4.73($\pm$0.57)	4.71($\pm$0.52)
2	5.46($\pm$0.59)	5.60($\pm$0.24)	4.40($\pm$0.24)	4.54($\pm$0.59)
3	5.90($\pm$0.42)	5.84($\pm$0.22)	4.17($\pm$0.22)	4.10($\pm$0.42)
4	6.90($\pm$0.00)	6.30($\pm$0.00)	3.70($\pm$0.00)	3.10($\pm$0.00)

Table 6. Averaged confusion matrix on the test set (with 10 positive samples and 10 negative samples for each fold) for LI-ESNs, varying the number of anchors considered

$N_{anchors}$	True Positives	True Negatives	False Positives	False Negatives
1	8.57($\pm$0.40)	7.60($\pm$0.15)	2.40($\pm$0.15)	1.43($\pm$0.40)
2	8.80($\pm$0.52)	8.47($\pm$0.43)	1.53($\pm$0.43)	1.20($\pm$0.52)
3	9.31($\pm$0.45)	9.01($\pm$0.40)	0.99($\pm$0.40)	0.69($\pm$0.45)
4	9.84($\pm$0.00)	9.26($\pm$0.00)	0.74($\pm$0.00)	0.16($\pm$0.00)

to the RSS input signals. Indeed (see Section 3), the movement paths in Room 1 are very similar and more hardly distinguishable among each other (in particular paths 1 and 2, see Fig. 1) than the path types in Room 2.

Actual Deployment of the Anchors. In this sub-section, we detail the performance results of ESNs and LI-ESNs for each possible configuration of the set of anchors used, with $N_{anchors}$ varying from 1 to 4. For each configuration

Table 7. Distribution of test errors for LI-ESNs in the case $N_{anchors} = 4$ (with a total test error of 4%) occurring for each of the path types in Fig. 1

	Test Error (%)				
Path 1	Path 2	Path 3	Path 4	Path 5	Path 6
10.76%	51.86%	32.52%	1.43%	3.43%	0%

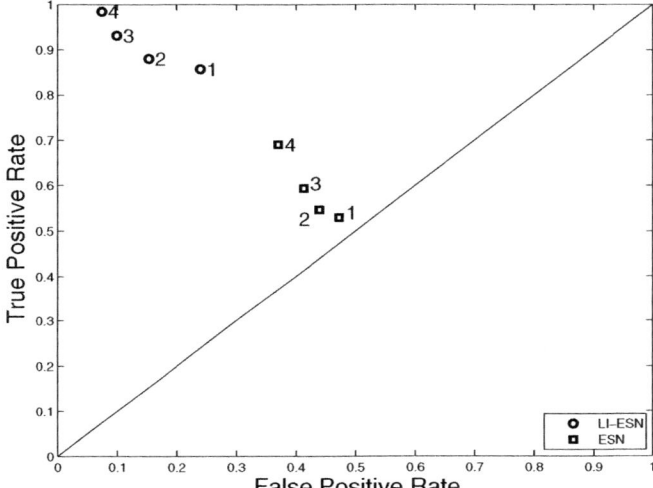

Fig. 6. ROC plot of ESNs and LI-ESNs on the test set, varying the number of anchors considered (indicated beside each point in the graph)

considered, the results are averaged (and the standard deviations are computed) over the 10 reservoir guesses. Tables 8 and 9 show the mean test accuracy, sensitivity and specificity for ESNs and LI-ESNs, respectively, in correspondence of every configuration of the anchors. Although the performances achieved in correspondence of the different choices for the same value of $N_{anchors}$ are quite similar, Tables 8 and 9 indicate that specific configurations can result in better performances. Despite the fact that different combination of anchors could give different performance was expected (since the disturbance and quality of signal is clearly affected by the position of the anchors in the environment), we observe from the tables that the accuracy of the prediction is reasonably robust to the position of the available anchors, which means that the deployment of the anchors does not need to be extremely accurate (thus reducing deployment costs).

On the other hand, the results show clearly that it is better to distribute the anchors as much as possible, e.g. in Table 9 the worse results are obtained when the available anchors are in the same room, while with two anchors displaced in different rooms the system already achieves accuracy in the range 87% - 93%.

Table 8. Mean test accuracy, sensitivity and specificity of ESNs for the possible configurations of the considered anchors

Anchors	Accuracy	Sensitivity	Specificity
A1	0.49($\pm$0.06)	0.51($\pm$0.10)	0.47($\pm$0.09)
A2	0.47($\pm$0.05)	0.45($\pm$0.08)	0.48($\pm$0.08)
A3	0.59($\pm$0.06)	0.58($\pm$0.10)	0.61($\pm$0.07)
A4	0.57($\pm$0.06)	0.57($\pm$0.08)	0.56($\pm$0.10)
A1, A2	0.52($\pm$0.06)	0.47($\pm$0.10)	0.56($\pm$0.08)
A1, A3	0.51($\pm$0.07)	0.50($\pm$0.11)	0.52($\pm$0.08)
A1, A4	0.58($\pm$0.07)	0.60($\pm$0.08)	0.57($\pm$0.09)
A2, A3	0.61($\pm$0.07)	0.62($\pm$0.09)	0.60($\pm$0.08)
A2, A4	0.53($\pm$0.06)	0.50($\pm$0.11)	0.55($\pm$0.08)
A3, A4	0.58($\pm$0.06)	0.60($\pm$0.08)	0.56($\pm$0.10)
A1, A2, A3	0.57($\pm$0.07)	0.57($\pm$0.10)	0.56($\pm$0.10)
A1, A2, A4	0.58($\pm$0.06)	0.57($\pm$0.08)	0.59($\pm$0.08)
A1, A3, A4	0.57($\pm$0.06)	0.56($\pm$0.09)	0.57($\pm$0.08)
A2, A3, A4	0.64($\pm$0.06)	0.66($\pm$0.08)	0.62($\pm$0.08)
A1, A2, A3, A4	0.66($\pm$0.07)	0.69($\pm$0.10)	0.63($\pm$0.09)

Table 9. Mean test accuracy, sensitivity and specificity of LI-ESNs for the possible configurations of the considered anchors

Anchors	Accuracy	Sensitivity	Specificity
A1	0.84($\pm$0.01)	0.90($\pm$0.01)	0.77($\pm$0.01)
A2	0.80($\pm$0.03)	0.85($\pm$0.06)	0.75($\pm$0.03)
A3	0.81($\pm$0.03)	0.88($\pm$0.03)	0.74($\pm$0.02)
A4	0.78($\pm$0.03)	0.79($\pm$0.06)	0.78($\pm$0.03)
A1, A2	0.83($\pm$0.03)	0.87($\pm$0.06)	0.78($\pm$0.02)
A1, A3	0.87($\pm$0.03)	0.89($\pm$0.04)	0.85($\pm$0.05)
A1, A4	0.88($\pm$0.02)	0.91($\pm$0.02)	0.85($\pm$0.02)
A2, A3	0.89($\pm$0.02)	0.89($\pm$0.03)	0.88($\pm$0.03)
A2, A4	0.93($\pm$0.01)	0.95($\pm$0.02)	0.91($\pm$0.02)
A3, A4	0.79($\pm$0.03)	0.78($\pm$0.04)	0.80($\pm$0.05)
A1, A2, A3	0.87($\pm$0.03)	0.89($\pm$0.04)	0.86($\pm$0.04)
A1, A2, A4	0.94($\pm$0.03)	0.96($\pm$0.04)	0.92($\pm$0.03)
A1, A3, A4	0.88($\pm$0.03)	0.89($\pm$0.04)	0.87($\pm$0.03)
A2, A3, A4	0.98($\pm$0.02)	0.99($\pm$0.01)	0.96($\pm$0.03)
A1, A2, A3, A4	0.96($\pm$0.03)	0.98($\pm$0.03)	0.93($\pm$0.03)

6   Conclusions

We have discussed the problem of forecasting the user movements in indoor environments. This problem cannot be tackled with by using mere user localization, since the information about the current position of the user does not necessary provide an indication about his future position. In our approach we have combined localization information obtained by a wireless sensor network of MicaZ sensors with a RNN (implemented as an ESN) that takes in input a stream of RSS data produced by the sensors and signals when the user is about to enter in/exit from a given room. The problem is also made complex due to the intrinsic difficulty of localization in indoor environments, since presence of walls and objects disturb the radio propagation and makes RSS data imprecise. We have considered a scenario in which a user enters in and exits from two rooms according to different paths, which intersect in a marker point at the time in which the ESN is requested to make the prediction. We also have considered different numbers and combinations of anchors in order to investigate the trade-off between the cost of the WSN, the cost of deployment (that is dependent on the sensibility of the solution to the position of the anchors) and accuracy of prediction. The results confirmed the potentiality of our appproach. In particular we have observed that our solution obtains good precisions also with a single anchor, and it scales gracefully with the number of anchors (with 4 anchors it reaches a test accuracy of 96%). Furthermore it is reasonably robust to the position of the anchors, although the experiments gave a clear indication that the anchors should be placed as distant from each other as possible, in order to guarantee a better coverage. Concerning the ESN models considered, results in Section 5.2 have shown that LI-ESNs consistently lead to better performances than standard ESNs for every experimental setting (i.e. for varying the number and the deployment of the used anchors). The bias of in the LI-ESN model, acting as a low-pass filter of the reservoir states, has therefore revealed to be suitable for approaching the characteristics of the problem considered. LI-ESNs have indeed shown a good ability to appropriately represent the history of the noisy RSS input signals used in experiments. Standard ESNs, on the other hand, would need a larger dataset and less noise in the input signals in order to achieve better generalization performances. Moreover, the experimental results have enlightened the coherence of the learning models used with respect to the known difficulties of the problem. In fact, the great part of the classification errors of LI-ESNs (for the 4 anchors setting) has occurred in correspondence of movements in Room 1, where the possible path types are much more similar among each other than the corresponding paths in Room 2. Finally, we observe that the ESN-based solution is potentially suitable for its embedding in wireless sensors of the mote class, due to its efficiency and low requirements of memory and processing. This embedding is matter of ongoing work. Future work also include the investigation of trade-off between accuracy and energy spent by the sensors to produce localization information, and the extension of our approach to environments of different nature (e.g. public buildings, building with different composition of walls, room size etc.). As final remark, we stress that the dataset used in our experiments is openly available for download in our website [2].

Acknowledgments. This work is partially supported by the EU FP7 RUBI-CON project (contract n. 269914). The authors also wish to thank their colleague Alberto Gotta for his patience and help during the measurements.

References

1. Crossbow technology inc., http://www.xbow.com
2. Experimental dataset (February 2011),
   http://wnlab.isti.cnr.it/paolo/index.php/dataset/forecasting
3. Antonelo, E.A., Schrauwen, B., Stroobandt, D.: Modeling multiple autonomous robot behaviors and behavior switching with a single reservoir computing network. In: Proceedings of the IEEE International Conference on Systems, Man and Cybernetics, pp. 1843–1848 (October 2008)
4. Antonelo, E.A., Schrauwen, B., Campenhout, J.M.V.: Generative modeling of autonomous robots and their environments using reservoir computing. Neural Processing Letters 26(3), 233–249 (2007)
5. Antonelo, E.A., Schrauwen, B., Stroobandt, D.: Event detection and localization for small mobile robots using reservoir computing. Neural Networks 21(6), 862–871 (2008)
6. Bahl, P., Padmanabhan, V.: Radar: an in-building rf-based user location and tracking system. In: Proceedings of the Nineteenth Annual Joint Conference of the IEEE Computer and Communications Societies, INFOCOM 2000, vol. 2, pp. 775–784. IEEE (2000)
7. Baronti, P., Pillai, P., Chook, V.W.C., Chessa, S., Gotta, A., Hu, Y.F.: Wireless sensor networks: A survey on the state of the art and the 802.15.4 and zigbee standards. Comput. Commun. 30(7), 1655–1695 (2007)
8. Barsocchi, P., Lenzi, S., Chessa, S., Giunta, G.: A novel approach to indoor rssi localization by automatic calibration of the wireless propagation model. In: IEEE 69th Vehicular Technology Conference, VTC Spring 2009, pp. 1–5 (April 2009)
9. Barsocchi, P., Lenzi, S., Chessa, S., Giunta, G.: Virtual calibration for rssi-based indoor localization with ieee 802.15.4. In: IEEE International Conference on Communications, ICC 2009, pp. 1–5 (June 2009)
10. Buehner, M., Young, P.: A tighter bound for the echo state property. IEEE Transactions on Neural Networks 17(3), 820–824 (2006)
11. Cui, S., Goldsmith, A., Bahai, A.: Energy-efficiency of mimo and cooperative mimo techniques in sensor networks. IEEE Journal on Selected Areas in Communications 22(6), 1089–1098 (2004)
12. Ducatel, K., Bogdanowicz, M., Scapolo, F., Leijten, J., Burgelman, J.C.: Scenarios for Ambient Intelligence in 2010. Tech. rep., IST Advisory Group (February 2001)
13. Gallicchio, C., Micheli, A.: A markovian characterization of redundancy in echo state networks by PCA. In: Proceedings of the ESANN 2010, pp. 321–326. d-side (2010)
14. Gallicchio, C., Micheli, A.: Architectural and markovian factors of echo state networks. Neural Networks 24(5), 440–456 (2011)
15. Gärtner, T.: A survey of kernels for structured data. SIGKDD Explorations Newsletter 5, 49–58 (2003)
16. Hartland, C., Bredeche, N.: Using echo state networks for robot navigation behavior acquisition. In: IEEE International Conference on Robotics and Biomimetics (ROBIO 2007), pp. 201–206. IEEE Computer Society Press (2007)

17. Jaeger, H.: The "echo state" approach to analysing and training recurrent neural networks. Tech. rep., GMD - German National Research Institute for Computer Science (2001)
18. Jaeger, H., Haas, H.: Harnessing nonlinearity: Predicting chaotic systems and saving energy in wireless communication. Science 304(5667), 78–80 (2004)
19. Jaeger, H., Lukosevicius, M., Popovici, D., Siewert, U.: Optimization and applications of echo state networks with leaky- integrator neurons. Neural Networks 20(3), 335–352 (2007)
20. Kaemarungsi, K., Krishnamurthy, P.: Modeling of indoor positioning systems based on location fingerprinting. In: Twenty-third Annual Joint Conference of the IEEE Computer and Communications Societies, INFOCOM 2004, vol. 2, pp. 1012–1022 (March 2004)
21. Kjærgaard, M.B.: A Taxonomy for Radio Location Fingerprinting. In: Hightower, J., Schiele, B., Strang, T. (eds.) LoCA 2007. LNCS, vol. 4718, pp. 139–156. Springer, Heidelberg (2007)
22. Kolen, J., Kremer, S. (eds.): A Field Guide to Dynamical Recurrent Networks. IEEE Press (2001)
23. Kushki, A., Plataniotis, K.N., Venetsanopoulos, A.N.: Kernel-based positioning in wireless local area networks. IEEE Transactions on Mobile Computing 6(6), 689–705 (2007)
24. Legenstein, R.A., Maass, W.: Edge of chaos and prediction of computational performance for neural circuit models. Neural Networks 20(3), 323–334 (2007)
25. Liu, W., Li, X., Chen, M.: Energy efficiency of mimo transmissions in wireless sensor networks with diversity and multiplexing gains. In: Proceedings of the IEEE International Conference on Acoustics, Speech, and Signal Processing (ICASSP 2005), vol. 4, pp. iv/897–iv/900 (March 2005)
26. Lukosevicius, M., Jaeger, H.: Reservoir computing approaches to recurrent neural network training. Computer Science Review 3(3), 127–149 (2009)
27. Nakagmi, M.: The m-distribution. a general formula of intensity distribution of rapid fading. Statistical methods in radio wave propagation. Pergamon, Oxford (1960)
28. Madigan, D., Einahrawy, E., Martin, R., Ju, W.H., Krishnan, P., Krishnakumar, A.: Bayesian indoor positioning systems. In: INFOCOM 2005, vol. 2, pp. 1217–1227 (March 2005)
29. Martínez, E.A., Cruz, R., Favela, J.: Estimating User Location in a WLAN Using Backpropagation Neural Networks. In: Lemaître, C., Reyes, C.A., González, J.A. (eds.) IBERAMIA 2004. LNCS (LNAI), vol. 3315, pp. 737–746. Springer, Heidelberg (2004)
30. Oubbati, M., Kord, B., Palm, G.: Learning Robot-Environment Interaction Using Echo State Networks. In: Doncieux, S., Girard, B., Guillot, A., Hallam, J., Meyer, J.-A., Mouret, J.-B. (eds.) SAB 2010. LNCS, vol. 6226, pp. 501–510. Springer, Heidelberg (2010)
31. Pan, J.J., Kwok, J., Yang, Q., Chen, Y.: Multidimensional vector regression for accurate and low-cost location estimation in pervasive computing. IEEE Transactions on Knowledge and Data Engineering 18(9), 1181–1193 (2006)
32. Rabiner, L.R.: A tutorial on hidden markov models ans selected applications in speech recognition. Proceedings of the IEEE 77(2) (1989)
33. Rubio, L., Reig, J., Cardona, N.: Evaluation of nakagami fading behaviour based on measurements in urban scenarios. AEU - International Journal of Electronics and Communications 61(2), 135–138 (2007)

34. Rui, C., Yi-bin, H., Zhang-qin, H., Jian, H.: Modeling the ambient intelligence application system: Concept, software, data, and network. IEEE Transactions on Systems, Man, and Cybernetics, Part C: Applications and Reviews 39(3), 299–314 (2009)
35. Sun, G., Chen, J., Guo, W., Liu, K.: Signal processing techniques in network-aided positioning: a survey of state-of-the-art positioning designs. IEEE Signal Processing Magazine 22(4), 12–23 (2005)
36. Tiño, P., Hammer, B., Bodén, M.: Markovian Bias of Neural-based Architectures with Feedback Connections. In: Hammer, B., Hitzler, P. (eds.) Perspectives of Neural-Symbolic Integration. SCI, vol. 77, pp. 95–133. Springer, Heidelberg (2007)
37. Verstraeten, D., Schrauwen, B., D'Haene, M., Stroobandt, D.: An experimental unification of reservoir computing methods. Neural Networks 20(3), 391–403 (2007)
38. Waegeman, T., Antonelo, E.A., Wyffels, F., Schrauwen, B.: Modular reservoir computing networks for imitation learning of multiple robot behaviors. In: Proceedings of the 8th IEEE International Symposium on Computational Intelligence in Robotics and Automation, pp. 27–32. IEEE (2009)
39. Youssef, M., Agrawala, A.: The horus wlan location determination system. In: MobiSys 2005: Proceedings of the 3rd International Conference on Mobile Systems, Applications, and Services, pp. 205–218. ACM, New York (2005)

Bayesian Filtering Methods for Target Tracking in Mixed Indoor/Outdoor Environments

Katrin Achutegui[1], Javier Rodas[2], Carlos J. Escudero[2], and Joaquín Míguez[1]

[1] Department of Signal Theory and Communications,
Universidad Carlos III de Madrid, Spain
{kachutegui,jmiguez}@tsc.uc3m.es
[2] Department of Electronics and Systems, Universidade da Coruña, Spain
{jrodas,escudero}@udc.es

Abstract. We propose a stochastic filtering algorithm capable of integrating radio signal strength (RSS) data coming from a wireless sensor network (WSN) and location data coming from the global positioning system (GPS) in order to provide seamless tracking of a target that moves over mixed indoor and outdoor scenarios. We adopt the sequential Monte Carlo (SMC) methodology (also known as particle filtering) as a general framework, but also exploit the conventional Kalman filter in order to reduce the variance of the Monte Carlo estimates and to design an efficient importance sampling scheme when GPS data are available. The superior performance of the proposed technique, when compared to outdoor GPS-only trackers, is demonstrated using experimental data. Synthetic observations are also generated in order to study, by way of simulations, the performance in mixed indoor/outdoor environments.

Keywords: Bayesian filtering, indoor/outdoor tracking, Kalman filter, particle filter, switching models.

1 Introduction

The existing outdoor and indoor systems for target positioning and/or tracking have evolved in rather different ways. The global positioning system (GPS) is the most common technology in outdoor scenarios. It provides broad coverage, essentially ubiquitous except for a few "tough" environments, such as urban canyons [11], yet it has a poor accuracy, in the order of 10 meters [8,16]. Positioning based on cellular networks yields a similar precision and the coverage, even if not global [15], can include urban areas where GPS fails. Combinations of both technologies [15] are attractive but do not resolve the accuracy problem. During recent years, localization systems based on wireless sensor networks (WSNs) have gained momentum, specially for indoor applications [15,12]. In outdoors environments, WSNs providing radio signal strength (RSS), time of arrival (ToA) or angle of arrival (AoA) data can potentially beat GPS and cellular networks in terms of accuracy, but they are *ad hoc* systems to be deployed only in small areas [15].

J. Del Ser et al. (Eds.): MOBILIGHT 2011, LNICST 81, pp. 169–185, 2012.
© Institute for Computer Sciences, Social Informatics and Telecommunications Engineering 2012

Another major difference between positioning in outdoor and indoor environments is the need to model very different kind of signals. Let us focus hereafter in systems that use RSS observations, although similar arguments could be put forward for ToA and AoA measurements. In an "open" outdoor area, with no obstacles, it is relatively easy to extract range information (i.e., estimate the distance between the transmitter and the receiver of the signal) which can then be used for positioning. Unfortunately, such information is much harder to extract in indoor environments, due to the multipath propagation of the radio signals [15]. As a consequence the kind of models that are needed outdoors, with a direct line of sight (LOS) between the transmitter and receiver, and indoors, with strong multipath and non LOS transmission, can be very different.

In order to deal with the nonlinearities inherent to RSS (but also AoA and ToA) observations, sequential Monte Carlo (SMC) methods, also known as particle filters (PFs) [9,6,5], have been proposed as tools for positioning and tracking, both outdoors and (specially) indoors [10]. These methods rely on the simulation of candidate positions and tracks for the target of interest, which are later weighted and combined using a statistical procedure, and differ substantially from the (much simpler) Kalman filtering methods that are often used with GPS data [14].

In this paper, we tackle the design of a tracking algorithm that can work both indoors and outdoors, using GPS and/or RSS data collected from a WSN. The basic methodology that we adopt is particle filtering, which enables us to deal with variety of different models (representing various indoor and outdoor scenarios) both for the observations and for the target motion, possibly switching among them using the general scheme of [1]. However, we also exploit the availability of GPS data and the ability to process them using Kalman filtering in order to (a) simplify the complexity of the tracker (when GPS alone is available) and (b) design efficient particle filtering algorithms for the online fusion of GPS and RSS observations. The superior performance of the resulting methods, when compared to outdoor GPS-only trackers, is demonstrated using experimental data. Synthetic observations are also generated in order to study, by way of simulations, the performance in mixed indoor/outdoor environments.

The rest of the paper is organized as follows. In Section 2 we describe three environment-specific tracking models. The proposed algorithms are introduced in Section 3. In Section 4 we describe the experimental setup used to collect GPS and RSS data in an outdoor environment and we illustrate the performance of the proposed tracker, both with synthetic and experimental data.

2   System Model

2.1   Outdoor Linear Model

The dynamics of the target can be described using a linear and Gaussian state-space system [10]. Let $\mathbf{x}_{4,t} = [\mathbf{r}_t^\top \mathbf{v}_t^\top]^\top \in \mathbb{R}^4$ be the state of the system. The state vector contains the position, $\mathbf{r}_t \in \mathbb{R}^2$ and the velocity, $\mathbf{v}_t \in \mathbb{R}^2$, of the object to be tracked in a 2-dimensional plane and the subscript $t \in \mathbb{N}$ denotes discrete time. The state of the dynamic system evolves according to the stochastic model

$$
\underbrace{\begin{bmatrix} r_{1,t} \\ r_{2,t} \\ v_{1,t} \\ v_{2,t} \end{bmatrix}}_{\mathbf{x}_{4,t}} = \underbrace{\begin{bmatrix} 1 & 0 & T & 0 \\ 0 & 1 & 0 & T \\ 0 & 0 & 1 & 0 \\ 0 & 0 & 0 & 1 \end{bmatrix}}_{\mathbf{A}} \underbrace{\begin{bmatrix} r_{1,t-1} \\ r_{2,t-1} \\ v_{1,t-1} \\ v_{2,t-1} \end{bmatrix}}_{\mathbf{x}_{4,t-1}} + \underbrace{\begin{bmatrix} \frac{1}{2}T^2 & 0 \\ 0 & \frac{1}{2}T^2 \\ T & 0 \\ 0 & T \end{bmatrix}}_{\mathbf{Q}} \underbrace{\begin{bmatrix} u_{1,t} \\ u_{2,t} \end{bmatrix}}_{\mathbf{u}_t}, \tag{1}
$$

where $\mathbf{A}$ is a transition matrix that depends on the period T, $\mathbf{x}_{4,t-1}$ is the state vector of the previous time instant and $\mathbf{u}_t$ is a 2×1 real Gaussian vector of zero mean and diagonal covariance matrix, $\sigma_u^2 \mathbf{I}_2$. As a result, the process noise is a 4×1 real Gaussian vector, $\mathbf{Q}\mathbf{u}_t$, with zero mean and covariance matrix $\sigma_u^2 \mathbf{Q}\mathbf{Q}^\top$, which represents the effect of unknown accelerations. This model is often termed constant velocity model [10].

More often, tracking in a purely outdoor scenario can be carried out using GPS data. The GPS observations give the position of the object in geodetic coordinates (latitude, longitude and altitude) and are easily converted to carte-sian local coordinates. Thus, mathematically, we model the GPS observations as a linear function of the state, $\mathbf{x}_{4,t} \in \mathbb{R}^4$, with an added noise term ε_t, which accounts for errors in the measurements

$$
\underbrace{\begin{bmatrix} y_{1,t} \\ y_{2,t} \end{bmatrix}}_{\mathbf{y}_t} = \underbrace{\begin{bmatrix} 1 & 0 & 0 & 0 \\ 0 & 1 & 0 & 0 \end{bmatrix}}_{\mathbf{B}} \underbrace{\begin{bmatrix} r_{1,t} \\ r_{2,t} \\ v_{1,t} \\ v_{2,t} \end{bmatrix}}_{\mathbf{x}_{4,t}} + \varepsilon_t. \tag{2}
$$

Note that ε_t is modelled as Gaussian noise vector of zero mean and known covariance matrix $\varepsilon_t \sim N(\varepsilon_t; 0, \Sigma_\varepsilon)$. When experimental data are available, the covariance parameter, Σ_ε, can be adjusted from the sample (empirical) variance.

2.2   Indoor Non-linear System Model

For an indoor environment we seek a more flexible system model capable of capturing rapidly changing movements and capable of modeling measurements with a high variance. On one hand, the movements which we are going to track are closer to maneuvers than to a linear motion. On the other hand, we assume that the observations available for tracking the target are RSS measurements collected by a sensor network. Unfortunately, the severe multipath propagation effects in indoor scenarios make the modeling of RSS data a challenging task.

Let $\omega_t \in \mathbb{R}$ be the change, in radians, of the angle of the velocity at time $t+1$, and redefine the state vector as $\mathbf{x}_{5,t} = [\omega_t, \mathbf{r}_t, \mathbf{v}_t]^\top$, that evolves according to

$$\omega_t \quad \sim \quad p(\omega_t | \omega_{t-1})$$

$$\underbrace{\begin{bmatrix} r_{1,t} \\ r_{2,t} \\ v_{1,t} \\ v_{2,t} \end{bmatrix}}_{\mathbf{x}_{4,t}} = \underbrace{\begin{bmatrix} 1 & 0 & \frac{\sin(\omega_{t-1}T)}{\omega_{t-1}} & -\frac{\cos(\omega_{t-1}T)-1}{\omega_{t-1}} \\ 0 & 1 & \frac{1-\cos(\omega_{t-1}T)}{\omega_{t-1}} & \frac{\sin(\omega_{t-1}T)}{\omega_{t-1}} \\ 0 & 0 & \cos(\omega_{t-1}T) & -\sin(\omega_{t-1}T) \\ 0 & 0 & \sin(\omega_{t-1}T) & \cos(\omega_{t-1}T) \end{bmatrix}}_{\mathbf{A}(\omega_{t-1})} \underbrace{\begin{bmatrix} r_{1,t-1} \\ r_{2,t-1} \\ v_{1,t-1} \\ v_{2,t-1} \end{bmatrix}}_{\mathbf{x}_{4,t-1}} + \mathbf{Q}\mathbf{u}_t, \tag{3}$$

where the transition matrix $\mathbf{A}(\omega_{t-1})$ is now a function of the angle ω_{t-1} as well as the period T and the conditional probability density function (pdf) $p(\omega_t | \omega_{t-1})$ is known. If we select different distributions for ω_t, we can create different motion models. Note also that in the extreme case where $\omega_t = 0$ for all t, (3) becomes the constant velocity model (1).

Let us assume that, the objective can move according to one out of L models of motion, identified by the indices $\{1, 2, \ldots, L\}$. Each one of the motion models corresponds to a different transition pdf for the Markovian process $\{\omega_t\}_{t\in\mathbb{N}}$. Thus, to identify the different densities, we introduce a new state variable, denoted a_t. This is a discrete random indicator, $a_t \in \{1, \ldots, L\}$, so that $a_{t-1} = l$ implies that ω_t is generated according to the l-th model. Therefore we need to write $\omega_t \sim p(\omega_t | \omega_{t-1}, a_{t-1})$ to make the dependence explicit. The probability mass function (pmf) $p(a_t | a_{t-1})$ is part of the model and therefore is assumed known. This type of dynamic model description, where the are various sub-models to describe different type of motion, is often denoted interacting multiple models (IMM) [10]. Incorporating the indicator a_t to the state, we obtain a 6×1 vector, $\mathbf{x}_{6,t} = [a_t, \omega_t, \mathbf{r}_t, \mathbf{v}_t]^\top$ which evolves in time according to the equations

$$a_t \sim p(a_t | a_{t-1}), \quad \omega_t \sim p(\omega_t | \omega_{t-1}, a_{t-1}), \quad \mathbf{x}_{4,t} = \mathbf{A}(\omega_{t-1})\mathbf{x}_{4,t-1} + \mathbf{Q}\mathbf{u}_t. \tag{4}$$

For the observation model, we assume that at time t we obtain J RSS measurements. The datum obtained from sensor j at time t is denoted as $y_{j,t}$. The relationship between the received observation, $y_{j,t}$, and the position of the target, $\mathbf{r}_t$, depends on the environment in which the measurement is taken and can change with time [13]. In order to model this uncertainty in the observations we are going to use an interacting multiple model (IMM) approach the same as for the dynamic model. Specifically, we represent the observation $y_{j,t}$ using one out of K different models that describe the different environments. Finally, we model the observations as

$$y_{j,t} = f_{m_{j,t}}(\mathbf{r}_t) + \varepsilon_{m_{j,t}}, \tag{5}$$

where $m_{j,t} \in \{1, \ldots, K\}$ is a random index with a known probability mass function (pmf), $p(m_{j,t})$, which identifies the observation model at time t for each sensor j, $f_{m_{j,t}}$ is the function that describes the propagation conditions in model $m_{j,t}$, and $\varepsilon_{m_{j,t}} \sim N(\varepsilon_{m_j}; 0, \sigma^2_{m_{j,t}})$ is Gaussian noise with zero mean and a known variance $\sigma^2_{m_{j,t}}$, which is also associated with the model $m_{j,t}$. The form of the functions $\{f_1, f_2, \ldots, f_K\}$ and variances $\{\sigma^2_1, \sigma^2_2, \ldots, \sigma^2_K\}$ should be determined from a bank of empirical observations collected in the scenarios in which the tracking will be performed. In [1] we give full details of the functions and variances obtained when the RSS is measured using a network of ZigBee models. We assume the same indoor environment and hardware setup in the present paper.

We write the measurement-model indicators together in a $J \times 1$ vector $\mathbf{m}_t = [m_{1,t}, \ldots, m_{J,t}]^\top$, hence the full target state has $J + 6$ components, $\mathbf{x}_{J+6,t} = [\mathbf{m}_t, a_t, \omega_t, \mathbf{r}_t, \mathbf{v}_t]^\top$. The observations are put together in a $J \times 1$ vector $\mathbf{y}_t = [y_{1,t}, \ldots, y_{J,t}]^\top$. The indices in $\mathbf{m}_t$ are assumed independent, but not necessarily identically distributed. The probability mass functions $p(m_{j,t}), j = 1, \ldots, J$ are assumed known and independent of time.

2.3   Outdoor Nonlinear Model

In this section we describe an outdoor system model that includes ZigBee observations as well as GPS observations. The dynamic model is the same as described in Section 2.1.

The observation vector is composed of J measurements received from ZigBee sensors and S GPS measurements, that is

$$\mathbf{y}_t = [y_{1,t}, \ldots, y_{J,t}, y_{J+1,t}, \ldots, y_{S+J,t}]. \tag{6}$$

The GPS observation model is the same as described in Section 2.1.

The outdoor ZigBee observation models we propose are based on the log distance path-loss model described in [13]

$$y_{j,t} = f_j(\mathbf{r}_t) + \varepsilon_j = L_{j,0} + \gamma_j 10 \log_{10}\left(\frac{d_{j,0}}{d_{j,t}}\right) + \varepsilon_j, \tag{7}$$

where $d_{j,t}$ is the distance between sensor j and the target at time instant t, $d_{j,0}$ is a reference distance for sensor j, $L_{j,0}$ is the path loss corresponding to the reference distance, γ_j is the path loss exponent and $\varepsilon_j \sim N(\varepsilon_j; 0, \sigma_{\varepsilon_j}^2)$ has a normal distribution with zero mean and variance $\sigma_{\varepsilon_j}^2$. The parameters $L_{j,0}$, γ_j, $d_{j,0}$ and $\sigma_{\varepsilon_j}^2$ should be adjusted using experimental data.

In particular, assume that we collect k RSS measurements for a sensor-to-target distance d_i and this is repeated for l different distances, $d_1, \ldots, d_l$. Then, the parameters L_0 and γ are selected as the solution to the optimization problem

$$(L_{0,j}, \gamma_j) = \arg\min_{L_0, \gamma} = \left\{ \sum_{i=1}^{l} \sum_{n=1}^{k} \left(y_{n,i} - L_0 - \gamma 10 \log_{10}\left(\frac{d_0}{d_i}\right) \right)^2 \right\},$$

where $\mathbf{y}_{n,i}$ is the n-th observation measured at distance d_i in the experiments, k is the number of measurements we have at distance d_i and l is the number of distances at which we have measurements.

The variance parameter is fitted as

$$\hat{\sigma}_{\varepsilon,j}^2 = \frac{1}{l}\frac{1}{k} \sum_{i=1}^{l} \sum_{n=1}^{k} \left(y_{n,i} - L_{0,j} - \gamma_j 10 \log_{10}\left(\frac{d_0}{d_i}\right) \right)^2.$$

Note that $L_{0,j}$, γ_j and $\sigma_{\varepsilon,j}^2$ are adjusted for each sensor separately.

3   Tracking Algorithms

3.1   Kalman Filter

When both the dynamic and observation models are linear and Gaussian, the density $p(\mathbf{x}_t|\mathbf{y}_{1:t})$, $t = 1, 2, \ldots$ is also Gaussian and can be exactly computed using the Kalman filter [14]. Specifically if we assume
Specifically, $p(\mathbf{x}_t|\mathbf{y}_{1:t})$ is Gaussian if we assume that

- $\mathbf{u}_t$ y $\boldsymbol{\varepsilon}_t$ are independent and have known Gaussian distributions,
- the dynamic model is a linear function of $\mathbf{x}_{t-1}$ and $\mathbf{u}_t$, and
- the observation model is a linear function of $\mathbf{x}_t$ and $\boldsymbol{\varepsilon}_t$.

Obviously these requirements are satisfied by the outdoor linear model of Section 2.1.

Assume that at time $t = 0$ the prior distribution of $\mathbf{x}_t$ is Gaussian with mean $\hat{\mathbf{x}}_{0|0}$ and covariance matrix $\mathbf{P}_{0|0}$ denoted $\mathbf{x}_0 \sim N(\mathbf{x}_0; \hat{\mathbf{x}}_{0|0}, \mathbf{P}_{0|0})$. The Kalman filter consists on the set of recursive equations [14]

$$\left.\begin{aligned}
\hat{\mathbf{x}}_{t|t-1} &= \mathbf{A}\hat{\mathbf{x}}_{t-1|t-1} \\
\mathbf{P}_{t|t-1} &= \mathbf{Q}_{u,t-1} + \mathbf{A}\mathbf{P}_{t-1|t-1}\mathbf{A}^\top \\
\hat{\mathbf{x}}_{t|t} &= \hat{\mathbf{x}}_{t|t-1} + \mathbf{K}_t(\mathbf{y}_t - \mathbf{B}\hat{\mathbf{x}}_{t|t-1}) \\
\mathbf{P}_{t|t} &= \mathbf{P}_{t|t-1} - \mathbf{K}_t\mathbf{S}_t\mathbf{K}_t^\top
\end{aligned}\right\} \tag{8}$$

where $\mathbf{Q}_{u,t} = \sigma_u^2 \mathbf{Q}\mathbf{Q}^\top$ is the covariance matrix of the process noise,

$$\mathbf{S}_t = \mathbf{B}_{t-1}\mathbf{P}_{t|t-1}\mathbf{B}_{t-1}^\top + \mathbf{Q}_{v,t} \tag{9}$$

is the covariance matrix of the innovation $\boldsymbol{\nu}_t = \mathbf{y}_t - \mathbf{B}_t\hat{\mathbf{x}}_{t|t-1}$, $\mathbf{Q}_{v,t} = \boldsymbol{\Sigma}_\varepsilon$ is the covariance matrix of the observation noise and

$$\mathbf{K}_t = \mathbf{P}_{t|t-1}\mathbf{B}_t^\top \mathbf{S}_t^{-1} \tag{10}$$

is the Kalman gain. The recursive application of these equations gives us the mean and covariance matrix of the posterior probability distribution function $p(\mathbf{x}_t|\mathbf{y}_{1:t})$, namely $p(\mathbf{x}_t|\mathbf{y}_{1:t}) = N(\mathbf{x}_t; \hat{\mathbf{x}}_{t|t}, \mathbf{P}_{t|t})$.

Table 1 summarizes the Kalman filter algorithm for outdoor tracking using GPS observation data.

3.2   Particle Filter for the Indoor Nonlinear Model

Sequential Monte Carlo approximation. From a Bayesian point of view, the smoothing pdf

$$p(\mathbf{r}_{0:t}, \omega_{0:t}, a_{0:t}|\mathbf{y}_{1:t}) = \sum_{\mathbf{m}_{0:t}} \int_{\mathbf{v}_{0:t}} p(\mathbf{x}_{J+4,0:t}|\mathbf{y}_{1:t})d\mathbf{v}_{0:t} \tag{11}$$

Table 1. Kalman filter for online tracking in outdoor environments using GPS data

---

1. Initialization, at $t = 0$:
   - Assign a mean vector $\hat{\mathbf{x}}_0$ and a covariance matrix $\mathbf{P}_0$ to the first time instant $t = 0$, all taken from the prior distribution $p(\mathbf{x}_0; \hat{\mathbf{x}}_0, \mathbf{P}_0)$.
2. Recursive step, for $t > 0$:
   - Obtain a new GPS observation $\mathbf{y}_t$.
   - Apply the recursive formulae (8), (9) and (10) to obtain $p(\mathbf{x}_t|\mathbf{y}_{1:t}) = N(\mathbf{x}_t; \hat{\mathbf{x}}_{t|t}, \hat{\mathbf{P}}_{t|t})$.

---

contains all relevant statistical information for the estimation of $\mathbf{r}_{0:t}$. The elimination of $\mathbf{v}_{0:t}$ and $\mathbf{m}_{0:t}$ by marginalization is often termed Rao-Blackwellization and reduces the estimation variance [7,3]. Unfortunately, the density of (11) cannot be obtained analytically and we have to resort to numerical approximation techniques. Our approach, is to build a point-mass approximation of the distribution with density $p(\mathbf{r}_{0:t}, \omega_{0:t}, a_{0:t}|\mathbf{y}_{1:t})$, consisting of M random samples in the space of $\{\mathbf{r}_{0:t}, \omega_{0:t}, a_{0:t}\}$, denoted $\{\mathbf{r}_{0:t}^{(i)}, \omega_{0:t}^{(i)}, a_{0:t}^{(i)}\}_{i=1}^{M}$, and associated importance weights, $\{w_t^{(i)}\}_{i=1}^{M}$. Each pair $\left\{\left(\mathbf{r}_{0:t}^{(i)}, \omega_{0:t}^{(i)}, a_{0:t}^{(i)}\right), w_t^{(i)}\right\}$ is called a *particle* and we can use them to build the random measure

$$p_M(\mathbf{r}_{0:t}, \omega_{0:t}, a_{0:t}|\mathbf{y}_{1:t}) = \sum_{i=1}^{M} \delta_i(\mathbf{r}_{0:t}, \omega_{0:t}, a_{0:t})w_t^{(i)}, \qquad (12)$$

where δ_i is a unit delta measure located at $\left(\mathbf{r}_{0:t}^{(i)}, \omega_{0:t}^{(i)}, a_{0:t}^{(i)}\right)$ and the weights are assumed normalized, i.e., $\sum_{i=1}^{M} w_t^{(i)} = 1$. If the approximation is properly constructed, meaning that the moments of $p_M(\mathbf{r}_{0:t}, \omega_{0:t}, a_{0:t}|\mathbf{y}_{1:t})$ converge to those of $p(\mathbf{r}_{0:t}, \omega_{0:t}, a_{0:t}|\mathbf{y}_{1:t})$ in some adequate sense [4], then it is straightforward to use (12) in order to approximate any estimators of $\mathbf{r}_{0:t}$ or $\mathbf{r}_t$. In particular, since

$$p_M(\mathbf{r}_t|\mathbf{y}_{1:t}) = \sum_{a_{0:t}} \int_{\omega_{0:t}} \int_{\mathbf{r}_{0:t-1}} p_M(\mathbf{r}_{0:t}, \omega_{0:t}, a_{0:t}|\mathbf{y}_{1:t})d\mathbf{r}_{0:t-1}d\omega_{0:t} = \sum_{i=1}^{M} \delta_i(\mathbf{r}_t)w_t^{(i)},$$

$$(13)$$

where δ_i is the delta unit measure located at $\mathbf{r}_t^{(i)}$, we readily calculate the (approximate) minimum mean square error (MMSE) estimate of $\mathbf{r}_t$ as

$$\hat{\mathbf{r}}_t^{mmse} = \int \mathbf{r}_t p_M(\mathbf{r}_t|\mathbf{y}_{1:t})d\mathbf{r}_t = \sum_{i=1}^{M} \mathbf{r}_t^{(i)}w_t^{(i)}. \qquad (14)$$

The generation of samples and the computation of weights is carried out by means of the sequential importance sampling (SIS) principle [7]. Specifically, we can decompose teh smoothing pdf using a Bayess theorem

$$p(\mathbf{r}_{0:t}, \omega_{0:t}, a_{0:t} | \mathbf{y}_{1:t}) \propto p(\mathbf{y}_t | \mathbf{r}_t) p(\mathbf{r}_t | \mathbf{r}_{0:t-1}, \omega_{0:t-1}) p(a_t | a_{t-1})$$
$$\times\ p(\omega_t | \omega_{t-1}, a_{t-1}) p(\mathbf{r}_{0:t-1}, \omega_{0:t-1}, a_{0:t-1} | \mathbf{y}_{1:t-1}), \quad (15)$$

and, if we draw the particles using the transition pdf's

$$a_t^{(i)} \sim p(a_t | a_{t-1}^{(i)})$$
$$\omega_t^{(i)} \sim p(\omega_t | \omega_{t-1}^{(i)}, a_{t-1}^{(i)})$$
$$\mathbf{r}_t^{(i)} \sim p(\mathbf{r}_t | \mathbf{r}_{0:t-1}^{(i)}, \omega_{0:t-1}^{(i)}, \mathbf{y}_{1:t-1}) \quad (16)$$

then the importance weight becomes

$$w_t^{(i)} \propto w_{t-1}^{(i)} p(\mathbf{y}_t | \mathbf{r}_t^{(i)}). \quad (17)$$

Eqs. (16) and (17) together yield a sequential IS (SIS) type of algorithm for the construction of $p_M(\mathbf{r}_{0:t}, \omega_{0:t}, a_{0:t} | \mathbf{y}_{1:t})$ [7].

It is well known, however, that the sequential application of (16) and (17) with a finite number of samples, $M < \infty$, quickly leads to a degenerate set of particles [7]. Indeed, the variance of the weights increases stochastically with time and, after a few time steps, one single particle tends to accumulate all the weight and the approximation $p_M(\mathbf{r}_{0:t}, \omega_{0:t}, a_{0:t} | \mathbf{y}_{1:t})$ becomes useless. This difficulty is commonly overcome by adding a resampling step [7,2] which, intuitively, consists in stochastically discarding the particles with low weights while the particles with higher weights are replicated. Although several resampling schemes exist (and all of them can be plugged into the tracking algorithm without any added difficulty), in this paper we adopt the conceptually simple multinomial resampling method [7,4]. A resampling step can be taken every time the approximate effective sample size [7] $\hat{M}_{eff} = \frac{1}{\sum_{i=1}^{M} w_t^{(i)2}}$ falls below a user-defined threshold. Since $\hat{M}_{eff} \leq M$, typical threshold values are λM for some $0 < \lambda < 1$.

Evaluation of the weights. In order to ensure that the weights of (17) can be computed, we must be able to draw from $p(\mathbf{r}_t | \mathbf{r}_{0:t-1}, \omega_{0:t-1})$ and to evaluate the factors $p(a_t | a_{t-1})$, $p(\omega_t | \omega_{t-1}, a_{t-1})$ and $p(\mathbf{y}_t | \mathbf{r}_t)$. The transition densities $p(a_t | a_{t-1})$ and $p(\omega_t | \omega_{t-1}, a_{t-1})$ are part of the model, hence known by assumption. The prior density of the position at time t, $p(\mathbf{r}_t | \mathbf{r}_{0:t-1}, \omega_{0:t-1})$, is Gaussian and can be obtained in closed form for each particle. Indeed, given $\mathbf{r}_{0:t-1}^{(i)}$ and $\omega_{0:t-1}^{(i)}$, the system

$$\begin{bmatrix} v_{1,t} \\ v_{2,t} \\ r_{1,t}^{(i)} \\ r_{2,t}^{(i)} \end{bmatrix} = \begin{bmatrix} \cos(\omega_{t-1}^{(i)} T) & -\sin(\omega_{t-1}^{(i)} T) & 0 & 0 \\ \sin(\omega_{t-1}^{(i)} T) & \cos(\omega_{t-1}^{(i)} T) & 0 & 0 \\ \frac{\sin(\omega_{t-1}^{(i)} T)}{\omega_{t-1}^{(i)}} & -\frac{\cos(\omega_{t-1}^{(i)} T)-1}{\omega_{t-1}^{(i)}} & 1 & 0 \\ \frac{1-\cos(\omega_{t-1}^{(i)} T)}{\omega_{t-1}^{(i)}} & \frac{\sin(\omega_{t-1}^{(i)} T)}{\omega_{t-1}^{(i)}} & 0 & 1 \end{bmatrix} \begin{bmatrix} v_{1,t-1} \\ v_{2,t-1} \\ r_{1,t-1}^{(i)} \\ r_{2,t-1}^{(i)} \end{bmatrix} + \begin{bmatrix} T & 0 & 0 & 0 \\ 0 & T & 0 & 0 \\ 0 & 0 & \frac{1}{2}T^2 & 0 \\ 0 & 0 & 0 & \frac{1}{2}T^2 \end{bmatrix} \begin{bmatrix} u_{3,t} \\ u_{4,t} \\ u_{1,t} \\ u_{2,t} \end{bmatrix}$$
$$(18)$$

is linear and Gaussian, with known parameters, and all posterior pdf's, including $p(\mathbf{r}_t | \mathbf{r}_{0:t-1}^{(i)}, \omega_{0:t-1}^{(i)})$, are Gaussian and can be computed exactly using a Kalman filter [3,10]. In the sequel, we denote

$$p(\mathbf{r}_t | \mathbf{r}_{0:t-1}^{(i)}, \omega_{0:t-1}^{(i)}) = N(\mathbf{r}_t; \overline{\mathbf{r}}_{t|t-1}^{(i)}, \overline{\mathbf{\Sigma}}_{t|t-1}^{(i)}). \tag{19}$$

The pdf $p(\mathbf{y}_t | \mathbf{r}_t)$ is usually referred to as the likelihood of $\mathbf{r}_t$. If we write $p(y_{j,t} | \mathbf{r}_t)$ as a marginal of the joint density $p(y_{j,t}, m_{j,t} | \mathbf{r}_t)$, then it is straightforward to obtain the expression

$$p(\mathbf{y}_t | \mathbf{r}_t) = \prod_{j=1}^{J} p(y_{j,t} | \mathbf{r}_t) = \prod_{j=1}^{J} \sum_{m_{j,t}} p(y_{j,t} | \mathbf{r}_t, m_{j,t}) p(m_{j,t}), \tag{20}$$

(note that the observations are conditionally independent given the position $\mathbf{r}_t$) where both

$$p(y_{j,t} | \mathbf{r}_t, m_{j,t}) = N(y_{j,t}; f_{m_{j,t}}(\mathbf{r}_t), \sigma_{m_{j,t}}^2) \tag{21}$$

and $p(m_{j,t})$ are known from the model, for all $j = 1, ..., J$.

Table 2 summarizes the proposed SIS algorithm for the system model described in Section 2.2.

Table 2. SIS for indoor tracking with RSS data

---

1. Initialization, at $t = 0$:
   - For $i = 1, \ldots, M$, sample $\mathbf{r}_0$, $\mathbf{v}_0$, ω_0 and a_0 from the priors $p(\mathbf{r}_0)$, $p(\mathbf{v}_0)$, $p(\omega_0)$ and $p(a_0)$. Initialize weights to $w_0^{(i)} = \frac{1}{M}$.
2. Recursive step, for $t > 0$:
   - For $i = 1, \ldots, M$, sample $a_t^{(i)} \sim p(a_t | a_{t-1}^{(i)})$, $\omega_t^{(i)} \sim p(\omega_t | \omega_{t-1}^{(i)}, a_{t-1}^{(i)})$ and obtain $\mathbf{r}_t^{(i)}$ from the distribution $p(\mathbf{r}_t | \mathbf{r}_{0:t-1}^{(i)}, a_{0:t-1}^{(i)}, \omega_{0:t-1}^{(i)}) = N(\mathbf{r}_t; \mathbf{r}_{t|t-1}^{(i)}, \mathbf{\Sigma}_{t|t-1}^{(i)})$ obtained with the Kalman filter.
   - For $i = 1, \ldots, M$, update weights, $w_t^{(i)} \propto w_{t-1}^{(i)} p(\mathbf{y}_t | \mathbf{r}_t^{(i)})$
   - Compute the particle effective size $\hat{M}_{eff} = 1 / \sum_{k=1}^{M} w_t^{(k)^2}$. If $\hat{M}_{eff} < \lambda M$ perform resampling and set $w_t^{(i)} = \frac{1}{M} \forall i$.

3.3 Particle Filter for the Outdoor Nonlinear Model

The algorithm that we are going to use for outdoor tracking with GPS and RSS data is a particle filter similar to the one used for indoor tracking. The main difference is the use of a different proposal pdf for the generation of particles and the subsequent change in the computation of the weights. Recall that the state vector for our nonlinear outdoor model is defined as $\mathbf{x}_{4,t} = [\mathbf{r}_t^\top \mathbf{v}_t^\top]^\top$ and, as a consequence, the probability density function of interest is

$$p(\mathbf{r}_{0:t} | \mathbf{y}_{1:t}) = \int_{\mathbf{v}_{0:t}} p(\mathbf{x}_{0:t} | \mathbf{y}_{1:t}) d\mathbf{v}_{0:t},$$

where the velocity is integrated to reduce the variance in the estimation.

In order to obtain an efficient proposal pdf, we note that even though the complete vector of observations is not linear, the part of the vector that corresponds to the GPS observations is linear, therefore if we take only this part of the vector we can construct a posterior Gaussian pdf of $\mathbf{r}_t$ given partial data. To be specific, if we define the vector of GPS observations as

$$\mathbf{y}_{S,t} = \mathbf{r}_t + \varepsilon_{S,t},$$

then we can derive an analytic expression for the posterior density $p(\mathbf{r}_t|\mathbf{r}_{0:t-1}, \mathbf{y}_{S,t})$, namely

$$p(\mathbf{r}_t|\mathbf{r}_{0:t-1}, \mathbf{y}_{S,t}) \propto p(\mathbf{y}_{S,t}|\mathbf{r}_t)p(\mathbf{r}_t|\mathbf{r}_{0:t-1}) =$$
$$\frac{1}{2\pi|\boldsymbol{\Sigma}_{t|t-1}|^{\frac{1}{2}}|\boldsymbol{\Sigma}_\varepsilon|^{\frac{1}{2}}} \exp\{-\frac{1}{2}[(\mathbf{y}_{S,t} - \mathbf{r}_t)^\top \boldsymbol{\Sigma}_\varepsilon^{-1}(\mathbf{y}_{S,t} - \mathbf{r}_t) +$$
$$(\mathbf{r}_t - \mathbf{r}_{t|t-1})^\top \boldsymbol{\Sigma}_{t|t-1}^{-1}(\mathbf{r}_t - \mathbf{r}_{t|t-1})]\}, \tag{22}$$

where $\boldsymbol{\Sigma}_\varepsilon$ is the covariance matrix of $\mathbf{y}_{S,t}$ and $p(\mathbf{r}_t|\mathbf{r}_{0:t-1}) = N(\mathbf{r}_t|\mathbf{r}_{t|t-1}, \boldsymbol{\Sigma}_{t|t-1})$.

Equations 22 can be written in a more compact form if we use the equality

$$(\mathbf{y} - \mathbf{Br})^\top \mathbf{V}(\mathbf{y} - \mathbf{Br}) + (\mathbf{r} - \mathbf{r}_a)^\top \mathbf{U}(\mathbf{r} - \mathbf{r}_a) = (\mathbf{r} - \mathbf{r}_p)^\top \mathbf{C}(\mathbf{r} - \mathbf{r}_p) +$$
$$+ (\mathbf{y}^\top \mathbf{Vy} + \mathbf{r}_a^\top \mathbf{Ur}_a) - (\mathbf{B}^\top \mathbf{Vy} + \mathbf{Ur}_a)^\top \mathbf{C}^{-1}(\mathbf{B}^\top \mathbf{Vy} + \mathbf{Ur}_a) \tag{23}$$

where all vectors have compatible dimensions, $\mathbf{C} = \mathbf{B}^\top \mathbf{VB} + \mathbf{U}$ and $\mathbf{r}_p = \mathbf{C}^{-1}(\mathbf{B}^\top \mathbf{Vy} + \mathbf{Ur}_a)$. if we identify $\mathbf{y} = \mathbf{y}_{S,t}$, $\mathbf{V} = \boldsymbol{\Sigma}_\varepsilon^{-1}$, $\mathbf{B} = \mathbf{I}$, $\mathbf{r} = \mathbf{r}_t$, $\mathbf{U} = \boldsymbol{\Sigma}_{t|t-1}^{-1}$, $\mathbf{r}_a = \mathbf{r}_{t|t-1}$, $\tilde{\boldsymbol{\Sigma}}_t = \mathbf{C}^{-1}$ and $\tilde{\mathbf{r}}_t = \mathbf{r}_p$, then we can see that substitution of (23) into (22) gives a Gaussian density

$$p(\mathbf{r}_t|\mathbf{r}_{0:t-1}, \mathbf{y}_t) \propto \frac{1}{2\pi|\tilde{\boldsymbol{\Sigma}}^{-1}|^{1/2}} \exp\{-\frac{1}{2}\left[(\mathbf{r}_t - \tilde{\mathbf{r}}_t)^\top \tilde{\boldsymbol{\Sigma}}_t^{-1}(\mathbf{r}_t - \tilde{\mathbf{r}}_t)\right]\}.$$

Therefore, the proposed pdf that incorporates GPS observations can be characterized as a Gaussian density, $N(\mathbf{r}_t; \tilde{\mathbf{r}}_t, \tilde{\boldsymbol{\Sigma}}_t)$, where the vector of means and the covariance matrix are computed as

$$\tilde{\boldsymbol{\Sigma}}_t^{-1} = \boldsymbol{\Sigma}_\varepsilon^{-1} + \boldsymbol{\Sigma}_{t|t-1}^{-1} \quad \text{and} \quad \tilde{\mathbf{r}}_t = \tilde{\boldsymbol{\Sigma}}_t(\boldsymbol{\Sigma}_\varepsilon^{-1}\mathbf{y}_{S,t} + \boldsymbol{\Sigma}_{t|t-1}^{-1}\mathbf{r}_{t|t-1}), \tag{24}$$

respectively. The weight update equation becomes

$$w_t^{(i)} \propto w_{t-1}^{(i)} \frac{p(\mathbf{y}_t|\mathbf{r}_t^{(i)})p(\mathbf{r}_t^{(i)}|\mathbf{r}_{0:t-1}^{(i)})}{N(\mathbf{r}_t; \tilde{\mathbf{r}}_t^{(i)}, \tilde{\boldsymbol{\Sigma}}_t^{(i)})}. \tag{25}$$

Note that the pdf $p(\mathbf{y}_t|\mathbf{r}_t)$ incorporates both GPS and RSS data as

$$p(\mathbf{y}_t|\mathbf{r}_t) = \prod_{n=1}^{J+S} p(\mathbf{y}_{n,t}|\mathbf{r}_t) = \prod_{s=1}^{S} p(\mathbf{y}_{s,t}|\mathbf{r}_t) \prod_{j=1}^{J} p(y_{j,t}|\mathbf{r}_t), \tag{26}$$

where both $p(\mathbf{y}_{s,t}|\mathbf{r}_t) = N(\mathbf{y}_{s,t}; \mathbf{r}_t, \sigma_v^2 \mathbf{I}_2)$ and $p(y_{j,t}|\mathbf{r}_t) = N(y_{j,t}; f_j(\mathbf{r}_t), \sigma_{\varepsilon_j}^2)$ are known, for $j = 1, \ldots, J$ y $s = 1, \ldots, S$.

Table 3 shows a summary of the SIS algorithm for the nonlinear outdoor model.

Table 3. SIS algorithm with a more efficient proposal function for tracking in an outdoor environment with GPS and RSS data

---

1. Initialization, at $t = 0$:
   - For $i = 1, \ldots, M$, sample $\mathbf{r}_0$ and $\mathbf{v}_0$ from the prior functions $p(\mathbf{r}_0)$ and $p(\mathbf{v}_0)$. Initialize weights $w_0^{(i)} = \frac{1}{M}$.
2. Recursive step, for $t > 0$:
   - For $i = 1, \ldots, M$, compute the vector of means $\mathbf{r}_{t|t-1}^{(i)}$ and the covariance matrices $\boldsymbol{\Sigma}_{t|t-1}^{(i)}$ with the Kalman filter.
   - With the current GPS observation, $\mathbf{y}_{S,t}$, compute the vector of means $\tilde{\mathbf{r}}_t^{(i)}$ and the covariance matrices $\tilde{\boldsymbol{\Sigma}}_t^{(i)}$ of the proposal pdf as defined in (24).
   - For $i = 1, \ldots, M$, draw $\mathbf{r}_t^{(i)}$ from the distribution $p(\mathbf{r}_t|\mathbf{r}_{0:t-1}^{(i)}, \mathbf{y}_{S,t}) = N(\mathbf{r}_t; \tilde{\mathbf{r}}_t^{(i)}, \boldsymbol{\Sigma}_t^{(i)})$.
   - For $i = 1, \ldots, M$, update the weights, $w_t^{(i)} \propto w_{t-1}^{(i)} \frac{p(\mathbf{y}_t|\mathbf{r}_t^{(i)})p(\mathbf{r}_t^{(i)}|\mathbf{r}_{0:t-1}^{(i)})}{N(\mathbf{r}_t; \tilde{\mathbf{r}}_t^{(i)}, \tilde{\boldsymbol{\Sigma}}_t)}$ with the complete vector of observations of the current time instant $\mathbf{y}_t$.
   - Compute the effective sample size $\hat{M}_{eff} = 1/\sum_{k=1}^{M} w_t^{(k)^2}$. If $\hat{M}_{eff} < \lambda M$ resample and set $w_t^{(i)} = \frac{1}{M} \; \forall i$.

3.4 Switching between Algorithms

As the targets move from an indoor environment to an outdoor environment, or vice versa, we have to switch between different tracking algorithms. There are essentially three cases. If the target moves form indoors to outdoors, or outdoors to indoors, but RSS data are available in *both* environments, then the tracking algorithms are particle filters and it is straightforward to go from one to another.

In order to switch from an outdoor environment with GPS data alone to an indoor environment with RSS data, we have to generate a collection of particles from the Gaussian distribution computed by the Kalman filter immediately before the transition, which plays the role of a prior pdf for the particle filter. Specifically if $p(\mathbf{x}_{t-1}|\mathbf{y}_{1:t-1}) = N(\mathbf{x}_{t-1}; \hat{\mathbf{x}}_{t-1|t-1}, \mathbf{P}_{t-1|t-1})$ then we can draw M samples with equal weights,

$$\mathbf{x}_{t-1}^{(i)} \sim N(\mathbf{x}_t; \mathbf{x}_{t-1|t-1}, \mathbf{P}_{t-1|t-1}), \quad w_{t-1}^{(i)} = \frac{1}{M} \quad i = 1, \ldots, M, \qquad (27)$$

from which the positions of the particles, $\mathbf{r}_{t-1}^{(i)}, i = 1, \ldots, M$, are extracted.

The prior $p(\mathbf{v}_{t-1}|\mathbf{y}_{1:t-1})$ is a Gaussian marginal of $N(\mathbf{x}_{t-1}; \hat{\mathbf{x}}_{t-1|t-1}, \mathbf{P}_{t-1|t-1})$ that is straightforward to compute.

$$\hat{\mathbf{x}}_{t-1|t-1} = \sum_{i=1}^{M} w_{t-1}^{(i)} \mathbf{x}_{t-1}^{(i)}$$

$$\mathbf{P}_{t-1|t-1} = \sum_{i=1}^{M} w_{t-1}^{(i)} (\mathbf{x}_{t-1}^{(i)} - \hat{\mathbf{x}}_{t-1|t-1})(\mathbf{x}_{t-1}^{(i)} - \hat{\mathbf{x}}_{t-1|t-1})^{\top}. \tag{28}$$

where $\mathbf{x}_t^{(i)} = [\mathbf{r}_t^{(i)\top} \mathbf{v}_t^{(i)\top}]^{\top}$ and $\mathbf{v}_t^{(i)}$ is the mean of $p(\mathbf{v}_t|\mathbf{r}_{0:t}^{(i)})$ which is obtained from a Kalman filter for the i-th particle (this is the *same* Kalman filter that is used to compute $p(\mathbf{r}_t|\mathbf{r}_{0:t-1}^{(i)}, \omega_{0:t-1}^{(i)})$.

The algorithms require of a variable that will indicate them the available technology at each instant t. To do so, we introduce a new variable $K_t \in \{0, 1, 2\}$ that can take 3 values: $K_t = 0$ indicates that we only have ZIgbee observations and that we must use particle filters for indoor tracking, $K_t = 1$ indicates us that we only have GPS observations available and that we must use the Kalman filter, and lastly $K_t = 2$ indicates that we have GPS and ZigBee observations and we therefore use the data fusion particle filter for outdoor.

4 Experimental Setup and Results

4.1 Experimental Setup

We have developed two kind of hardware nodes using standard parts. For the collection of RSS data, we have set up J= 8 anchors, acting as ZigBee transmitters, built using Arduino boards and XBee (series 1) modules (IEEE 802.15.4 compliant). The mobile device, that acts as a target, uses an Arduino Mega board, a XBee (series 1) module, and adds an small OEM GPS receiver. The antennas for 2.4 GHz band were all equal, 2 dBi monopole omnidirectional antennas. For the GPS receiver we used a RHCP external antenna. Each node was mounted inside a plastic enclosure and attached to a 2 m long non-metal pole. This was done to minimize the interferences caused by the person who carries the device. Figure 1 shows the setup of the mobile device. The anchors look really similar, with a smaller Arduino inside and without the GPS receiver (and its antenna).

The ZigBee and GPS measurements were taken in an outdoor scenario, in the middle of a clear area (without walls or trees) of about 600 m^2. We deployed an 8 ZigBee node network (anchors) covering a 6×10 m area and, on the other hand, another mobile node, with ZigBee and GPS receivers. This mobile node acted as a gateway, receiving both the ZigBee packets (from the 8 anchors) and the GPS signal every 1 s. The measurement were tagged by this node and sent to a small laptop using the USB port, where they were stored for future off line process.

In order to test the performance of the proposed algorithms and to be able to build realistic observation models we collected a large number of RSS and GPS observations in the mentioned area. The GPS observations are modelled as the position of the target with an added Gaussian noise. Therefore once known the

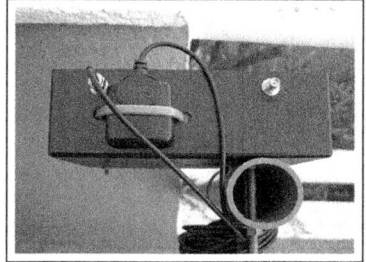

Fig. 1. Mobile device setup, with ZigBee and GPS receivers

true position where the GPS data are being taken, $\mathbf{x}_{n,real}$, we may translate the GPS global coordinates onto ENU coordinates (east north up) [14] and we may compute the covariance matrix applying a maximum likelihood criteria

$$\boldsymbol{\Sigma}_\varepsilon = \frac{1}{N} \sum_{n=1}^{N} (\mathbf{x}_n - \mathbf{x}_{n,real})(\mathbf{x}_n - \mathbf{x}_{n,real})^\top, \qquad (29)$$

where N is the total number of GPS observations taken in the experiment, n is the index of a unique observation and $\mathbf{x}_{n,real}$ is the true position where the n-th observation has been taken. The resulting covariance matrix is

$$\boldsymbol{\Sigma}_\varepsilon = \begin{bmatrix} 7.5 & -0.58 \\ -0.58 & 11.3 \end{bmatrix}. \qquad (30)$$

For the modeling of the ZigBee observation functions we have followed the criteria described in 2.2 to fit the real data taken in the experiments. Figure 2 shows experimental data taken from sensors 1 and 8 and the log-distance path loss models adjusted to them. Note we have fitted one model for each sensor.

The details of the hardware and experiments performed for the indoor model construction can be found in [1].

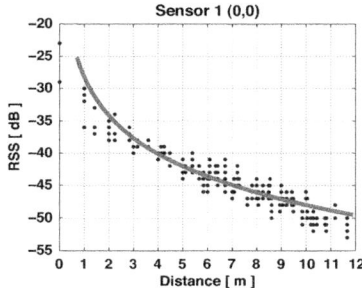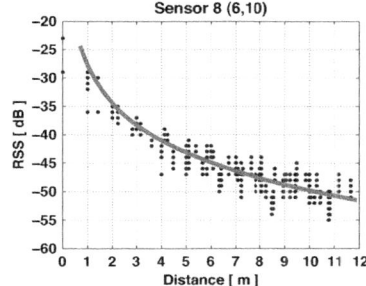

Fig. 2. Real ZigBee outdoor measurements taken in an outdoor environment for Sensors 1 and 8 and the observation functions adjusted to them

4.2 Results

In order to illustrate the performance of the outdoor algorithms we have taken experimental data of 8 ZigBee sensors and experimental GPS data from a moving target following three specific trajectories (see Figures 3, 4 and 5).

Figures 3, 4 and 5 show a tracking example of the two outdoor algorithms: the Kalman filter with GPS data alone and SIS algorithm. The figures in the left show the estimation of the target trajectory with the Kalman filter and the GPS observations and the figures in the right show the estimated trajectories with the SIS filter that uses GPS and RSS data. The true trajectory is drawn with a dark colored line, the estimated trajectory is drawn in a light colored line and the ZigBee nodes are depicted with squares.

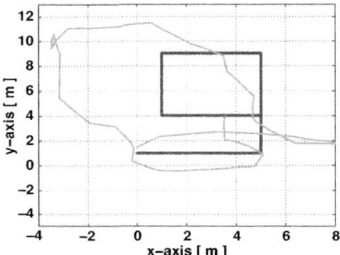 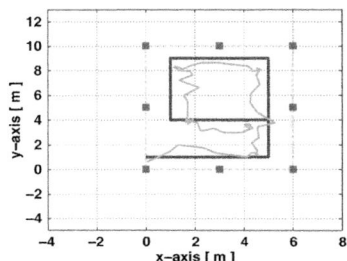

Fig. 3. Outdoor tracking example using ZigBee and GPS real data. The figure in the left uses GPS only for the estimation (via de Kalman filter) and the figure in the right uses both GPS and ZigBee data (vis de SIS algorithm).

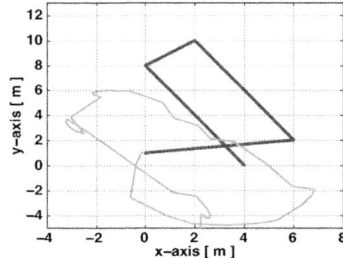 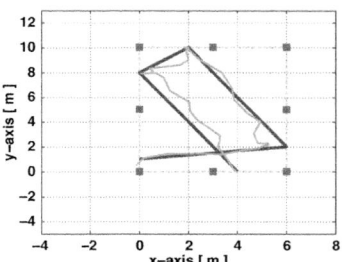

Fig. 4. Outdoor tracking example using ZigBee and GPS real data. The figure in the left uses GPS only for the estimation (via de Kalman filter) and the figure in the right uses both GPS and ZigBee data (vis de SIS algorithm).

As we can observe the trajectories with GPS data have a greater error and the improvement we obtain by incorporating RSS data is high. This is as expected, because the precision that the civil GPS system can obtain is from 5 to 10 meters and depends on open sky conditions [16,8]. The area in which we perform the

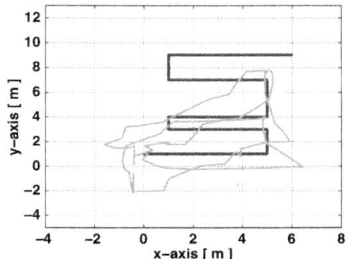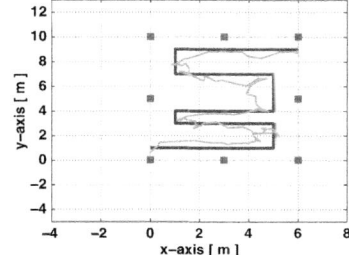

Fig. 5. Outdoor tracking example using ZigBee and GPS real data. The figure in the left uses GPS only for the estimation and the figure in the right uses both GPS and ZigBee real data.

tracking is, therefore, too small for the precision that the technology provides. The ZigBee technology, on the other hand, has a much lower range but achieves a greater precision.

In order to illustrate the performance of the tracking scheme we have generated synthetic trajectories that switch between environments (from indoor to outdoor) and we have synthetically generated observations that switch between technologies (from indoor ZigBee to outdoor ZigBee with GPS, to outdoor with GPS only).

We have simulated a scenario consisting of an indoor room of dimension 6×10 meters, linked to a 2 meter long and 2 meters wide corridor that leads on to an outdoor environment, which is also an area of 6×10 meters. This is shown in Figure 6. In the 6×10 meters area on the right and in the corridor we assume we are in an indoor environment. In the 6×10 meter area of the left we assume we are in an outdoor environment.

A random trajectory that switches between environments, generated according to the corresponding dynamic models, has been randomly generated for ex-emplification. At each time instant t, we check the position of the target in the previous time instant, $\mathbf{r}_{t-1}$, and the new position, $\mathbf{r}_t$, is generated according to the environment-specific dynamic model. In order to generate the observations that switch between technologies we also check where the target is at every time instant, and we generate the observations according to the environment-specific model. On the other hand, as we have two observation models for the outdoor environment (GPS only and GPS and ZigBee) we have introduced a new random variable which in outdoor environments chooses the available technology. When $K_t = 1$ we assume we only have GPS technology available and when $K_t = 2$ we assume we have the two technologies and we, therefore, simulate both types of observations. We have also associated a transition probability for the technology selection variable so that we do not change technologies too fast, that is, $p(K_t = i | K_{t-1} = i) = 0.95$, $i = 1, 2$.

Figure 6 shows a tracking example in this mixed scenario. The dark line is the true trajectory, the light line is the estimated trajectory and the squares are the ZigBee nodes. The beginning of the simulation is situated in the right *indoor* room, and the simulation lasts 300 seconds. The observation period is $T = 0.4$ seconds. In the first 200 seconds the target moves in an indoor environment and we have used the indoor specific algorithm. From the 200-th second, the target has moved on to the corridor and then on to the outdoor environment. In the corridor we assume we have both GPS and outdoor ZigBee measurements available. Then, in the left outdoor environment we have allowed variable K_t to choose which technology is available in each time instant.

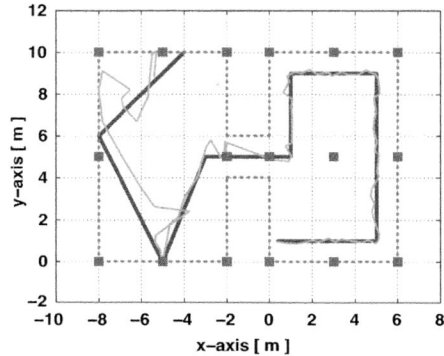

Fig. 6. Example of tracking a simulated trajectory using different technologies in different time instants and commuting between environment and technology specific algorithms. The beginning of the trajectory is situated in the right *indoor* room and the trajectory end in an outdoor environment (in the left area). For the indoor environment we have used ZigBee measurements to perform the tracking and for the outdoor environment we have used for some instant GPS measurements only and others GPS and ZigBee outdoor measurements.

4.3   Conclusions

We have introduced a target tracking algorithm for mixed indoor and outdoor scenarios based on the particle filtering methodology and a class of generalized interacting multiple-model state-space systems. The resulting method enables the fusion of GPS and RSS data in such a way that a target moving between indoor and outdoor environments, and therefore switching between different available data (GPS alone, RSS alone or both), can be seamlessly tracked. Within the particle filter, we use Kalman filtering techniques to reduce the variance of Monte Carlo estimates and to design efficient importance sampling schemes when GPS data are available. The performance improvement, compared to outdoor trackers that use only GPS data, is demonstrated using experimental data. Computer simulation results are presented to illustrate the application of the method in mixed indoor/outdoor scenarios.

Acknowledgments. The authors acknowledge the support of *Centro para el De-sarrollo Tecnológico e Industrial* (CDTI) through project CEN-2002-2007 TIMI. K. A. and J. M. acknowledge the support of *Ministerio de Ciencia e Innovación* (program CONSOLIDER-INGENIO 2010 CSD2008-00010 COMONSENS, and project TEC2009-14504-C02-01 DEIPRO) and *Ministerio de Industria, Turismo y Comercio* (project TSI-020400-2009-103 uService). J. R. and C. J. E. acknowledge the support of *Ministerio de Ciencia e Innovación* (project IPT-020000-2010-35).

References

1. Achutegui, K., Martino, L., Rodas, J., Escudero, C.J., Míguez, J.: A Multi-Model Particle Filtering algorithm for indoor tracking of mobile terminals using RSS data. In: 18th IEEE International Conference on Control Applications, pp. 1702–1707 (July 2009)
2. Carpenter, J., Clifford, P., Fearnhead, P.: Improved particle filter for nonlinear problems. IEE Proceedings - Radar, Sonar and Navigation 146(1), 2–7 (1999)
3. Chen, R., Liu, J.S.: Mixture Kalman filters. Journal of the Royal Statistics Society B 62, 493–508 (2000)
4. Crisan, D.: Particle filters - a theoretical perspective. In: Doucet, A., de Freitas, N., Gordon, N. (eds.) Sequential Monte Carlo Methods in Practice, ch. 2, pp. 17–42. Springer, Heidelberg (2001)
5. Djurić, P.M., Kotecha, J.H., Zhang, J., Huang, Y., Ghirmai, T., Bugallo, M.F., Míguez, J.: Particle filtering. IEEE Signal Processing Magazine 20(5), 19–38 (2003)
6. Doucet, A., de Freitas, N., Gordon, N. (eds.): Sequential Monte Carlo Methods in Practice. Springer, New York (2001)
7. Doucet, A., Godsill, S., Andrieu, C.: On sequential Monte Carlo Sampling methods for Bayesian filtering. Statistics and Computing 10(3), 197–208 (2000)
8. Reza Ehsani, M., Sullivan, M.D., Zimmerman, T.L., Stombaugh, T.: Evaluating the dynamic accuracy of low-cost gps receivers. American Society of Agricultural and Biological Engineers, ASAE Meeting Paper No. 031014. St. Joseph, Mich.: ASAE (2003)
9. Gordon, N., Salmond, D., Smith, A.F.M.: Novel approach to nonlinear and non-Gaussian Bayesian state estimation. IEE Proceedings-F 140, 107–113 (1993)
10. Gustafsson, F., Gunnarsson, F., Bergman, N., Forssell, U., Jansson, J., Karlsson, R., Nordlund, P.-J.: Particle filters for positioning, navigation and tracking. IEEE Transactions Signal Processing 50(2), 425–437 (2002)
11. Kaplan, E.D.: Understanding GPS: Principles and Applications, 2nd edn. Artech House Publishers (2005)
12. Patwari, N., Ash, J.N., Kyperountas, S., Hero III, A.O., Moses, R.L., Correal, N.S.: Locating the nodes. IEEE Signal Processing Magazine 22(4), 54–69 (2005)
13. Rappaport, T.S.: Wireless Communications: Principles and Practice, 2nd edn. Prentice-Hall, Upper Saddle River (2001)
14. Ristic, B., Arulampalam, S., Gordon, N.: Beyond the Kalman Filter. Artech House, Boston (2004)
15. Sun, G., Chen, J., Guo, W., Liu, K.J.R.: Signal processing techniques in network-aided positioning. IEEE Signal Processing Magazine 22(4), 12–23 (2005)
16. Wing, M.G., Eklund, A., Kellogg, L.D.: Consumer-Grade Global Positioning System (GPS) Accuracy and Reliability. Society of American Foresters (2005)

Social Acceptance and Usage Experiences from a Mobile Location-Aware Service Environment

Bernhard Klein[1], Jorge Perez[2], Christian Guggenmos[1], Olli Pihlajamaa[3], Immo Heino[3], and Javier Del Ser[2]

[1] DeustoTech - Deusto Institute of Technology, 48007 Bilbao, Spain
{bernhard.klein,christian.guggenmos}@deusto.es
[2] TECNALIA Research and Innovation, 48170 Bilbao, Spain
{jorge.perez,javier.delser}@tecnalia.com
[3] VTT Technical Research Centre of Finland, Espoo, Finland
{olli.pihlajamaa,immo.heino}@vtt.fi

Abstract. MUGGES is a European research project with the goal of evaluating peer-to-peer service concepts based on GNSS systems for mobile phones. MUGGES provides an infrastructure to create, publish, provide and consume mobile micro-services directly from mobile devices. As part of the project four application prototypes have been developed, which allow for the description and sharing of places and routes between users. This paper reports about a user trial conducted in real environments with early adopters. The goal has been to identify benefits and best practices for the type of applications envisioned in the MUGGES project. The obtained results indicate that users like to share information about preferred places or routes, and see it as a complementary application to already existing applications such as Facebook or Twitter. These types of applications are often used in time-killing situations, e.g. at the bus stop. Their value lies on the highlighting of important places or non-everyday events, and on making daily coordinations simpler. People feel that the application is less intrusive, since no information such as the whereabouts of persons are shared anyhow.

Keywords: Prosumer concept, peer-to-peer, semantic location model, micro-services, user trial.

1 Introduction

Social services such as Facebook, YouTube or Twitter have empowered users by making them not only consumers of information but also producers. This trend is now moving to mobile social services, which offer a more natural way of social interaction – anywhere at any time from the user's mobile device. Interestingly, most current mobile devices incorporate Global Navigation Satellite Systems (GNSS) technology such as GPS, thus making location-based mobile social services a very clear current trend of business innovation. This is due to the fact that location is considered a key context attribute for more optimal service filtering and recommendation in mobile domains [1,2].

J. Del Ser et al. (Eds.): MOBILIGHT 2011, LNICST 81, pp. 186–197, 2012.
© Institute for Computer Sciences, Social Informatics and Telecommunications Engineering 2012

The MUGGES project (Mobile User Generated Geo Services, [3]), funded by the European Commission's 7th Framework Programme [4], goes one step beyond current mobile location-based services (LBS) by providing such location-aware services and their contents directly from the user's mobile device, i.e. the mobile device evolves to be a server. Thus, mobile users turn into location-aware service super-prosumers, i.e. producers, providers, and consumers of services and associated contents from their mobile devices. Besides, this project also focuses on the development and real-world testing of a new and sophisticated heterogeneous location model which combines GNSS-based positioning and user-provided social positioning.

This paper deals with on the analysis of the information gathered from the testing of the MUGGES concept through a trial. The main objective of the user trials is to analyze how the prosumer concept is accepted, the feasibility of mobile peer-to-peer approach and the benefits for mobile location-based micro-services. The structure of the paper is divided into two main parts. The first part (Section 2) defines the MUGGES system by providing an explanation of the key topics behind this project: the super-prosumer role of users, the MUGGES location-aware services, the peer-to-peer architecture and the location management. The second part of the document (Section 3) describes the trial held in Finland, tackling the planning and the first experiences collected from the assessment of the information gathered therein.

2   The MUGGES System

The rationale behind the MUGGES project is to demonstrate that there is significant market potential for location-based services in Europe, and that the technology associated to this kind of services is currently mature enough to deliver real benefits to users. The main challenge of MUGGES is to find a new perspective to introduce massive mobile applications focused on mobile users' instant needs and interests, i.e. user- generated services, where the location is the enabling feature.

So far location has been exploited in the mobile environment mainly as a key service enabler for personal usage in client-server schemes, e.g. navigation, guidance and tourism. The underlying idea beneath MUGGES is to provide users of mobile devices the right tools that allow them to create focused, knowledge-based, mobile location-based micro-services *on the go*, converting the mobile device not only into a "player", but also into a "server".

In this context mobile users are not just seen as mere passive consumers of contents, but they are actively involved in the process of producing, creating and distributing location-based services and contents from their own mobile devices. Therefore, users in MUGGES are considered *super-prosumers*. By super-prosumer, we understand that users act as providers, producers and consumers of micro-services by using only the mobile device. In other words, prosumers provide contents directly from their mobile devices (or other user-owned electronic devices), where these contents are stored, and thus transforming the mobile device into a server. Within MUGGES these micro-services are called *Mugglets*;

they are small and independent location-based social services hosted in the mobile terminal and provided from the mobile terminal to a distant mobile device.

2.1   Location-Aware Micro-services in MUGGES

Before delving into the roots of the MUGGES system, let us first elaborate on the application of the MUGGES project to a short real-world example by commenting on one specific Mugglet, the "MUGGES Note":

Bob is visiting Lisbon's most traditional nightspot, Barrio Alto. After wandering around the streets for some time, he decides he may need some advices on the best spots. He takes his mobile phone and searches for MUGGES Note on the MUGGES Mugglet Search Engine. Shortly thereafter, he is looking at a list of notes posted around his current location and decides to check out "Maria Caxuxa" on the "Rua da Barroca", where people seem to be having a lot of fun! When he reaches the bar, he posts a photo of himself showing off how much fun he is having. This photo is now accessible by all the MUGGES Note users walking by and maybe attracts them to join Bob at the bar.

This example illustrates the benefits of attaching a location attribute to notes, achieving an improved integration between the physical and digital world. In order to demonstrate that 1) there is significant market potential for location-based services in Europe, and 2) that the LBS technology is mature enough to deliver real benefits to the users, we have implemented more complex micro-services which represent groupings of MUGGES notes. Specifically, the following four Mugglets have been designed and experimented by real users:

- "MUGGES Note": This Mugglet allows publishing short messages with a photo and attaching it to a specific location. Other users can then retrieve these messages at this location (see Figures 1.a and 1.b).
- "MUGGES Journal": The main goal of this Mugglet is to maintain a journal bound to the current position of the user. This Mugglet represents a set of semantically-related MUGGES notes maintained by a single author and ordered by date. Each note has its own location (See Figure 1.c).
- "MUGGES Trail": This Mugglet is an application which allows users to define routes with information about the places on them by adding a sequence of MUGGES notes (i.e. a starting point, intermediate points and a goal). This kind of Mugglet permits users to see the directions from their current location to the next point on the route, with the aim of guiding them to the end of the route without any trouble (See Figure 1.d).
- "MUGGES Race": This Mugglet is made for runners allowing them to follow a predefined route and compete with others in an asynchronic manner. Users must reach specific checkpoints to get to the goal and complete the race. The total time is measured and recorded by the service (See Figure 1.e). In the case of this Mugglet, each point of the race is considered and implemented as a MUGGES note.

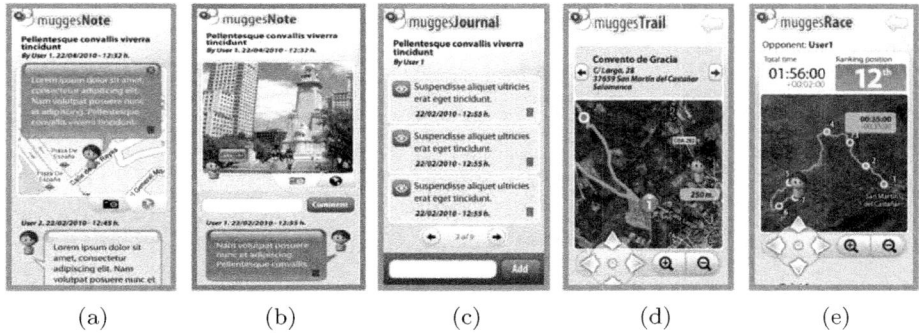

| (a) | (b) | (c) | (d) | (e) |

Fig. 1. MUGGES application interfaces: a) MUGGES Note. b) MUGGES Note photo view. c) MUGGES Journal. d) MUGGES Trail. e) MUGGES Race.

2.2 MUGGES Peer-to-Peer Architecture

As depicted in Figure 2, the MUGGES platform is modeled as a triangle where there is no central service provider, only the service infrastructure that facilitates connections between producers and consumers. The process is as follows: the service provider creates the service and publishes it through the service infrastructure. A Mugglet published through the service infrastructure can be searched and found by other Mugglet users. To consume a Mugglet, users have to query the server infrastructure. After downloading and installing the Mugglet on their device, users can execute the Mugglet. Communication is then handled directly between the Mugglet provider and consumer. Such peer-to-peer architectures have several advantages for the user:

– Service provision can be done instantly, without need for cumbersome uploads to an intermediate internet platform.
– Users maintain always full control over their services and can withdraw them at anytime.

MUGGES provides two mobile applications to manage Mugglets on the mobile phone: 1) the "creation kit", which is used to guide the user through the creation process and make Mugglet creation as convenient as possible for the user; and 2) the "execution platform", which manages Mugglet installations and provides access to mobile phone capabilities such as the embedded GPS or camera systems. On the server side, different software components ensure the correct functionality of the MUGGES system: on one hand, the service infrastructure hosts a storage system for the Mugglets and related templates. It allows users to search them based on keywords, template category and location. Since mobile devices may use different positioning technologies, a location manager part of the service infrastructure supports the translation between different location concepts. On the other hand, the user management and accounting component of the server infrastructure provides user and community profile management functions, and also records information required for billing purposes.

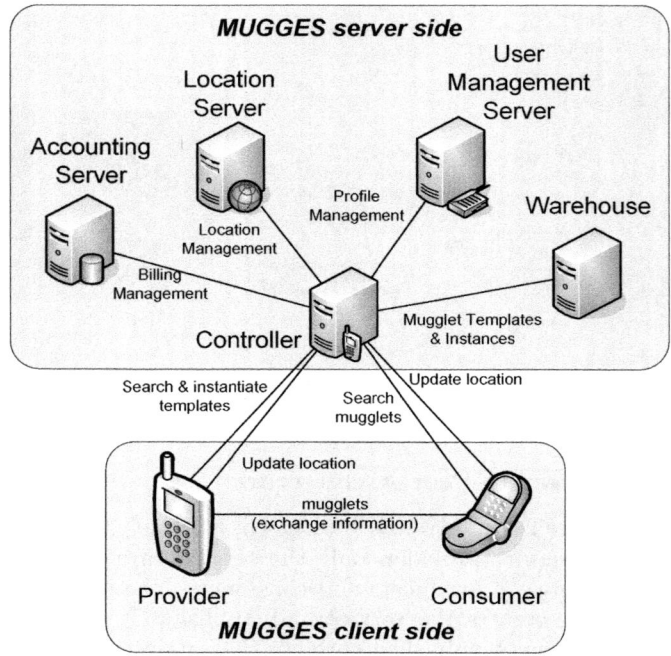

Fig. 2. The prosumer triangle

2.3   Location Management in MUGGES

Mugglets are bound to the physical world through location descriptions. Consequently, location descriptions become a crucial filter to search and access Mugglets. Managing locations can be complicated due to heterogeneous location technologies and different interpretations by humans. Different systems exist for indoor and outdoor positioning. Well-known indoor position systems are visual markers, Wi-Fi or Bluetooth based systems. Visual marker technologies translate symbols in locations, whereas Wi-Fi and Bluetooth positioning are based on triangulation, a methodology which calculates the position from different signal strengths issued by senders placed in the environment. For outdoor positioning the most commonly used system is GPS. GPS expresses locations through coordinates in latitude, longitude and altitude, e.g. in the World Geodetic System [5]. For instance, the city of Bilbao is located at latitude 4315'25", longitude -255'24" and altitude 19 m.

The Location Server (LS) component of the service infrastructure hides the underlying technology used to determine the user's location. This may be anything from a GNSS receiver's coordinates, to the phone's Cell-ID, the surrounding Wi-Fi base stations, a RFID reader, 2-dimensional barcodes or even plain text. For the verification of this location management concept we implemented GPS and Cell-ID for outdoor positioning and visual markers as indoor positioning system.

It is the LSs responsibility to identify the correct position based on one or a combination of these pieces of location information. The significant advantage of this approach is a stronger independence of heterogeneous location technologies, and makes the MUGGES system well suited for a wider range of usage scenarios.

As opposed to machines, people use conceptual language to represent location data. In other words, while computers use numbers, people use concepts (e.g. "near the station", "at the museum", "in the market area"). Hence, in MUGGES location-based services are accommodated to the way people need to utilize them, and correctly interpret expressions at a semantic level, just as humans do. Location information is represented and used in a humanized way. Therefore, the Location Server distinguishes between physical, symbolic and semantic locations. In MUGGES users collaboratively create, maintain and constantly update a coherent location model based on their collective contributions. Besides the location server seeks the help from external location mapping or modeling services such as GeoNames [6] or Google Maps [7].

3   MUGGES Trial and Evaluation

3.1   MUGGES Finnish Trial Setup

A functional and technical evaluation of the MUGGES system was carried out through user trials hosted in Espoo, Finland. This trial was conducted based on the Living Lab concept requiring and seeking high involvement of the users in real usage setups. MUGGES services were provided from a server installed and operated in Spain. The trial itself was held in one single phase of two weeks, one indoor and outdoor trial aiming to test heterogeneous location mechanisms for LBS applications. The base scenario consisted of:

- 8 users equipped with advanced mobile phones powered by MUGGES components and a GNSS receiver.
- 10 hot areas for interaction limited by geographic coordinates.
- 4 deployed social LBS applications that promote location-based interactions among different user types.

The trial was held close to VTT headquarters, mainly in a roughly 1 km^2 area within the Otaniemi technology campus. VTT's technical campus was furnished with symbolic locations with the help of the MUGGES location model. Trial users, however, were not restricted to experiment with the application in the main trial area, but encouraged to test the application elsewhere as well, in order to gain as much information as possible regarding the technical and operational evaluation of the system.

Regarding the recruited group, 8 IT professionals – aging from 26 to 53 years and working at VTT – were recruited. This sample group represented a group of critical, tech-savvy early adopters (enthusiastic to try new technologies). For that reason, the results of the trial cannot be identified with a layman's view on MUGGES experience. Since the trial has been held in the heart of winter, weather conditions in Finland turned out to be very demanding for testing any mobile application. The temperature was continuously below 0°C, leading to frequent heavy snowfalls (see Figure 3). Harsh conditions highlighted not only the requirements for usability, but also to fit to the context [8].

Fig. 3. Weather conditions rose exceptional challenges for mobile service use

For the trial, users were given Nokia 5800 XpressMusic touch screen smart phones with pre-installed and preconfigured MUGGES software. All the utilized phones have a small touch screen, embedded GPS and a prepaid 3G/3.5G data connectivity that allowed 0.4-6 Mbit/s downlink data transfer. Tests before the trial revealed several shortcomings of the phones and MUGGES software: the application was not easy to use due to touch screen problems, the upload bandwidth represented a bottleneck for the peer-to-peer messaging and the runtime memory was too small and slowed down the application over time or even crashed occasionally. The trial was started with a 2-hour kick-off introduction to MUGGES concepts and the functionality of the application. This intensive training session was used besides explaining the MUGGES user interface to teach trial participants to deal with the technical problems described above.

3.2 Assessment of Trial Results

Based on obtained data during the trial we analyzed MUGGES usage and its workflow. People used Mugglets over a period of two weeks. Mugglet usage was the highest in the initial days of the trial and dropped slightly in the remaining time. During weekends MUGGES was usually not used, since technical support was not available. The drop of the usage rate during the week can be explained by the bad weather conditions in Finland (heavy snowfall). Besides the weather, trial users mentioned distractions through the environment and the unstable implementation as the biggest and major reasons to dismiss the usage of MUGGES. Such reasons are graphically shown in Figure 4.b. Based on an interview performed at the end of the trial, it is inferred that most of the trial users would use Mugglets once a day (see Figure 4.a). This can be deemed as a value lying in the range of similar third-party applications.

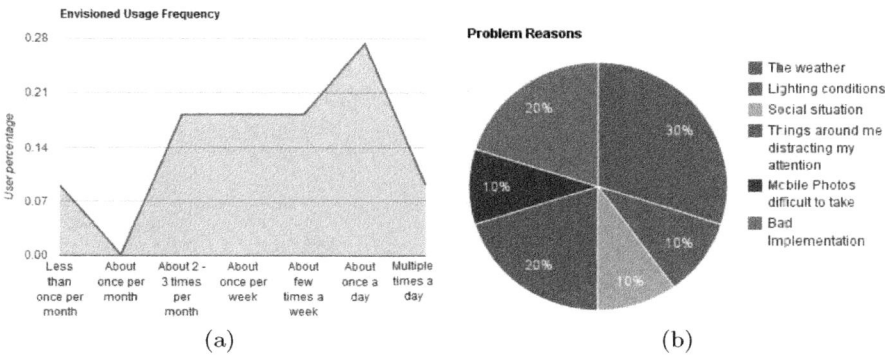

(a)   (b)

Fig. 4. a) Mugglet usage prediction; b) envisioned distractions for not using MUGGES

In the following we take a closer look on the Finnish trial results as examples of spatial distribution of the occurrences of MUGGES activities, as well as we discuss the findings and related experiments from trial conducted in Finland. In the spatial examination, we concentrate on the following three basic MUGGES tasks: creation of Mugglets, provision of existing Mugglet and consumption of Mugglets, i.e. downloads of Mugglets from another user. Execution of each task by the trial user created an event with the location, which enabled us to visualize the MUGGES activities of trial users on maps. In Figure 5 occurrences of each task type are visualized on a pair of maps. The first map in the pair renders a view on the area of interest in larger geographical scale, while the second one zooms into the Otaniemi campus, which was the main trial area. Task occurrences are represented as circles of different yet proportional sizes in the maps, in the sense that the more events bound to the same location there are, the bigger the circle representing the number of events in that location is.

• **Creation of Mugglets:** During the Finnish trial a total of 149 Mugglets were created (published) and most of them (84 %) were basic MUGGES Note Mugglets. The proportions of mash-up Mugglets from all created Mugglets were: MUGGES Journal 7%, MUGGES Trail 6%, and MUGGES Race 3%. The small proportion of mash-up Mugglets is understandable since they acted as special-purpose containers of already existing Note Mugglets. Even though the sample of trial users and number of created Mugglets were small, clear hints of participation inequality (the 90-9-1 rule for lurkers and contributors [9]) can be observed in the Finnish trial where one of the trial users created 60% of all Mugglets. In summary, This rule means that "In most online communities, 90% of users are lurkers who never contribute, 9% of users contribute a little, and 1% of users account for almost all the action".

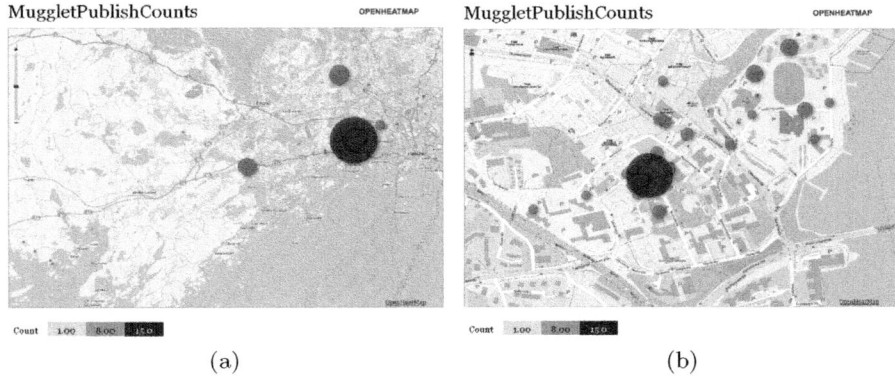

(a)   (b)

Fig. 5. Spatial distribution of Mugglet creation locations

As the left-hand side map in Figure 5 shows, Mugglets were mainly created in the Otaniemi campus area according to the initial trial assignment and due to the fact that trial participants were working in that area. Other publishing places were related mostly to users' home locations and some outdoor locations related to the leisure time activities of trial users (the trial coordinators encouraged users to test Mugglets at their leisure time as well). The right-hand side map in Figure 5 elucidates that the main publishing activity was centralized in the VTT Digitalo premises and its surroundings, including the student society building and its popular student restaurant (the Otahalli sports center) where trial users had their activities. In VTT Digitalo, QRCodes located conveniently near the coffee rooms were often used to bind the Mugglets to the location. As the users used often two-phase workflow for creating and finalizing Mugglets, there was a clear need for providing private workspace for Mugglet creation, where users are able to keep their Mugglets as long as they are finalized and ready for publishing for others.

• **Providing Mugglets:** Regarding the provision of Mugglets, it was detected and reported that mobile devices became slower over time due to the instability of the MUGGES system. By thoroughly analyzing this phenomenon we discovered a strong correlation of the provided Mugglets and the execution speed.

In a range up to 30 Mugglets per phone the execution speed was so slow that some people started to uninstall their Mugglets again. We also discovered that no more than 5 MUGGES consumers could access the same Mugglet at the same time. The reason may lie on the asymmetric upload and download bandwidth of the installed connection, which is counterproductive for the peer-to-peer concept of MUGGES. Regarding the provider-consumer ratio per Mugglet, it was seen that most Mugglets were intended for smaller groups, but some were targeted for larger audiences (i.e. up to 8 people). Since the size of the test group was 8 people, we assume that they tried to address the general public.

The peer-to-peer approach in MUGGES architecture gave rise to several technical challenges. When the user decided to either close the MUGGES application or to switch his/her phone off, the provided content disappeared from the other terminals, which was considered as a major drawback of the proposed solution. For the same reason, the searches made by the trial user often led to problems when providers closed MUGGES applications and old search results referring to the disappeared Mugglets still remained within the search results (if user did not made a re-search). Fortunately, this drawback can be solved by forcing a timed refresh of the search engine upon disconnection of a certain MUGGES user.

• **Discovery and consumption of Mugglets:** In total over 300 consumption events (i.e. the download of a given Mugglet from a peer) were logged in total during the Finnish trial. According to the compiled data, the MUGGES system was mainly utilized daily during the working hours (from 8 a.m. to 5 p.m.). The main activity gravitated on the Otaniemi campus area, but some users also examined Mugglets in their home locations outside Otaniemi. In the latter case, consumption took place also during weekends, as well as at times beyond working hours. Most of the consumption in the Otaniemi trial area was accumulated inside the VTT Digitalo premises (indoors), where users experienced with the MUGGES application during the coffee and other breaks along their work schedules. Other popular locations for consumption included the surroundings of the student society house, library and bus stops. It is remarkable that consumption of Mugglets was accumulated in indoor locations where working with mobile device was the most convenient choice. In addition, and as expected, also "deadtime" moments offered opportunities for Mugglet consumption, e.g. during the waiting time at the bus stop, or while waiting on something or someone at a public place.

Mugglets were consumed often in locations that were different from the Mugglet location. This result states that current MUGGES applications were not strongly location-aware, but just location-based applications. This result is further explained by the fact that Mugglets were searched primarily through other than location-based means (keyword and category searches). Furthermore, proactive features (e.g. those that notify users when some interesting contents are nearby the user's location) were missing from the current MUGGES implementation.

3.3   Frequently Requested Features

After the field trials, users were asked about the features they missed most during the trial. The following aspects were mentioned repeatedly:

– In order to increase the transparency of the MUGGES community, users requested the enhancement of MUGGES with social features such as Mugglet recommendations and ratings. In addition, they suggested incorporating, to the Mugglet creation procedure, the option to address specific groups of friends for their Mugglets, e.g. by extracting the contact list from the Facebook account of the Mugglet creator. Not all of the created Mugglets were intended for the public usage.
– Users also claimed that the process of creating a Mugglet is a strong effort, and consequently they are willing to get feedback from the consumers of their created Mugglets. From the consumer point of view, recommendations allow for easily distinguishing popular and important Mugglets from others, a feature which in the long run increases the value of the offered Mugglets.
– Another frequently requested property was the real-time administration of Mugglets, e.g. by notifying MUGGES users about new comments or notifications of the Mugglet itself. Some users also wanted to interact with their Mugglet provider through direct chat tools.

4   Concluding Remarks

Mugges opens new opportunities for people to create mobile micro services and share them with others. The trial conducted in Finland revealed interesting results: first, trial users found it appealing to furnish real-world locations with their own digital content. Trial users perceived the concept of mash-up Mugglets as very powerful and suggested to extend the concept by allowing the re-usage of notes from others. The creation process for Mugglets with wizards was seen as ineffective. Note Mugglets were almost always containing a photo, which emphasizes the importance of multimedia as mobile content.

Users also raised issues that were related to the discovery and search procedures of Mugglets. Especially when the number of Mugglets increases, users felt that simple Mugglet retrieval lists were not useful and better ways of grouping, sorting and restricting of search results were needed. In comparison to existing commercial products, MUGGES can be seen as a complementary offer. Facebook, Twitter and Foursquare are more concerned about social networks, and less about the physical environment. It is important to note that users are worried about their privacy, and would hence prefer to share information about their favorite places and routes instead of publishing their current position. Based on this observation, the possibility to control the content was seen as very positive. This is buttressed by the fact that people envision to use it quite frequently – at least once per day – preferable in typical time-killing situations such as bus stops. However, one user remarked that there is a potential threat if the MUGGES application is hacked and outsiders may get access to personal information on the phone. Further research will be conducted to circumvent this issue.

This paper summarizes the results obtained from the Finnish trial. Taking into account these results and the comments made by the trial users, new functionalities and improvements in the MUGGES system will be included and further tested by users in the future through new trials.

Acknowledgments. This project has been supported by the project grant no. 228297 (MUGGES [3]), funded by the European Commission's 7th Framework Programme, as well as by the Spanish Ministry of Science and Innovation through the Torres-Quevedo programme (no. PTQ-09-01-00740 and PTQ-09-01-00745).

References

1. Kaasinen, E.: User Needs for Location-Aware Mobile Services. Personal Ubiquitous Computing 7, 70–79 (2003)
2. Fisher, M.: European Mobile Mapping Trends, Directions Magazine (2008), http://www.directionsmag.com/printer.php?article_id=2756 (retrieved on March 7, 2011)
3. MUGGES (Mobile User Generated Geo Services) project, official website: http://www.mugges-fp7.org (retrieved on March 7, 2011)
4. CORDIS: FP7, European Commission, http://cordis.europa.eu/fp7/home_en.html (retrieved on March 7, 2011)
5. Department of Defense, World Geodetic System 1984, Its Definition and Relationships With Local Geodetic Systems, NIMA Technical Report TR8350.2, 3rd edn. (1997)
6. GeoNames Ontology - Geo Semantic Web, http://www.geonames.org/ontology/ (retrieved on March 7, 2011)
7. The Google Geocoding Web Service - Google Maps API Web Services, http://code.google.com/apis/maps/documentation/geocoding/index.html (retrieved on March 7, 2011)
8. Cockton, G.: From Quality In Use to Value in the World. In: Extended Abstracts on Human Factors in Computing Systems (CHI), pp. 1287–1290. ACM, New York (2004)
9. Participation Inequality: Encouraging More Users to Contribute. Jakob Nielsen's Alertbox (2006), http://www.useit.com/alertbox/participation_inequality.html (retrieved on March 7, 2011)

Mobile-Human Interaction Monitoring System

Ivan Pretel and Ana Belen Lago

DeustoTech - Deusto Institute of Technology, 48007 Bilbao, Spain
{ivan.pretel,anabelen.lago}@deusto.es

Abstract. Information and Knowledge Society is involved into a more challenging phenomenon than ever. The ability of mobile devices to access information and services from anywhere and anytime is the main reason that empowers the massive usage of this kind of technology. Software quality has to be improved by developing mobile device interaction models according to the user necessities. These necessities have more kind of users than ever. It means the quality in use improvement is vital to the interaction. By this work we aim to give a more explicit point of view of the problems that appear during mobile application quality testing. In order to do so, we have studied a context model for mobile interaction design and the existing ways to capture, analyze and evaluate the user interaction. Finally we present one software solution consisted by a tiny mobile application and a desktop application. The exposed system can capture all necessary information to calculate quality in use metrics defined within ISO/IEC 9126 standard. The contribution revealed is a new approach to quality testing methodology focused on mobile applications where it is possible to improve the reliability of its results by paying special attention to minimize the influence of external elements used to monitor the interaction.

Keywords: Quality in use, mobile services, context-awareness.

1 Introduction

The Information and Knowledge Society has evolved into a more challenging phenomenon than ever. Information and communications technologies have been introduced into all fields of human activity linked by the key technology: full connectivity mobile devices. By this key, everybody can be aware of their economics and entertainment among others.

The ability to access information and services from anywhere is the main reason that empowers the massive usage of this kind of technology, which not only focuses on the individual but also on business and social groups. These social tendencies create the need for better awareness and readiness to face these demands quickly. The software quality is becoming increasingly important due to the exigencies of the new and aggressive mobile software market. In addition, the software design and development is progressively more focused on the user. Due to this tendency, a great amount of resources are invested in the long-term ambition of finding and developing mobile device interaction models according to

J. Del Ser et al. (Eds.): MOBILIGHT 2011, LNICST 81, pp. 198–205, 2012.

the user necessities. In order to achieve this aim, knowing how the users feel using the product and what kind of problems could have is vital. Inside the quality topic, ISO 9126 [1] standard describes the quality in use. This kind of quality measures how a product can satisfy the needs of the specified user to achieve specific goals in a particular context with effectiveness, productivity, safety and satisfaction. Unfortunately, a good software quality evaluation takes more time than the companies can invert.

The contribution revealed is a new approach to quality testing methodology focused on mobile applications where it is possible to improve the reliability of its results by paying special attention to minimize the influence of external elements used to monitor the interaction. Firstly, the quality focused on mobile devices interaction is explained in Chapter 2. Secondly, the context in use is studied and the context focused on mobile interactions is defined in Chapter 3. Capture methods and existing monitoring systems are studied in Chapter 4. The mobile interaction monitoring system is presented in Chapter 5. Finally, the research is concluded and further work discussed in Chapter 6.

2   Quality Focused on the Interaction

ISO 9126 defines a quality framework by three aspects: Intern Quality, Extern Quality and Quality in Use. Internal Quality is the totality of characteristics of the software product from an internal view (i.e. cyclomatic complexity, code maintainability). This kind of quality can be improved during code implementation, reviewing and testing. External Quality is the quality when software is executed, which is measured and evaluated focusing on the software application behavior (i.e. number of wrong expected reactions of software). Finally, Quality in Use is defined within ISO/IEC 9126-4. It is the quality of the software system that the user can perceive when it is used in an explicit context of use. It measures the extent to which users can complete their tasks in a particular environment. It is measured by four main capabilities of the software product in a specified context of use:

- "Effectiveness": The capability to enable users to achieve specified goals with accuracy and completeness.
- "Productivity": The capability to enable users to expend appropriate amounts of resources in relation to the effectiveness achieved.
- "Safety": The capability to achieve acceptable levels of risk of harm to people, business, software, property or the surrounding environment.
- "Satisfaction": The capability to satisfy users.

These capabilities have to be measured in order to calculate how the quality in use of evaluated software is. Focusing on mobile devices, every software capability has to be measured per task and also per user, who is surrounded by the context in which actions are needed to be tracked. Owing to the wide range of contexts, an explicit context in use definition focused on mobile interactions has to be defined.

3   Context in Use

According to ISO 9241-11[2] standard, context in use is defined as every user, task, equipment and also physical and social environment that is affected by the interaction. In 2007, the NIST [3] institute published a new document adding every stakeholder to the context in use defined in the first standard. Other context in use definition is specified by Kankainen [4], he defines context in use as the environment that involves the user and his community. Nadav Savio and Jared Braiterman [5] explain the context by enumerating the following layers: culture, environment, activity, goals, attention, tasks, interface, device, connection and carrier. For mobile interaction, context is everything.

According to the exposed definitions, the different mobile context components are user, mobile device and environment.

The user has to be described by four main groups of attributes: personal, knowledge, skills and attitudes. Personal attributes are name, age and sex. The attributes related to knowledge are those attributes that can affect language, systems, products, work area, experience and eases with the tasks defined, culture, education level and experience using similar products. Physical abilities, mental abilities, disabilities and qualifications form the skills group. The attitudes group is formed by motivations, previous experiences and expectations.

The environment is also formed by groups of attributes: physical, ambient, technical and sociocultural groups. Inside physical group are attributes that describe the tangible environment (e.g. work area dimensions). The aim of ambient group is to keep attributes that can describe meteorological conditions, such as humidity, temperature or sound level. The sociocultural attributes group defines the cultural and social agents that can determine the user experience (e.g. cultural habits, religion). Technical group define every characteristic used during the tests excluding the mobile device, for example, connectivity attributes, hardware and software characteristics and so on.

If the work is focused on mobile environments studies, it appears the first complication. The main problem the quality in use shows is it is highly context-dependent. It is widely acknowledged that mobile environments are continuously changing. Therefore, context in use focused on mobile-human interaction (Figure 1) is formed by one mobile device, its owner, and also every environment that appears during the tasks execution.

4   Interaction Data Compilation

In order to define the best mobile-human interaction capturing we have studied the existing capture methods and the different advantages and disadvantages of the monitoring systems.

4.1   Capture Methods Classification

Firstly, we have studied the existing methods used to capture user interaction to conclude the best monitoring way. Different kinds of classification are found:

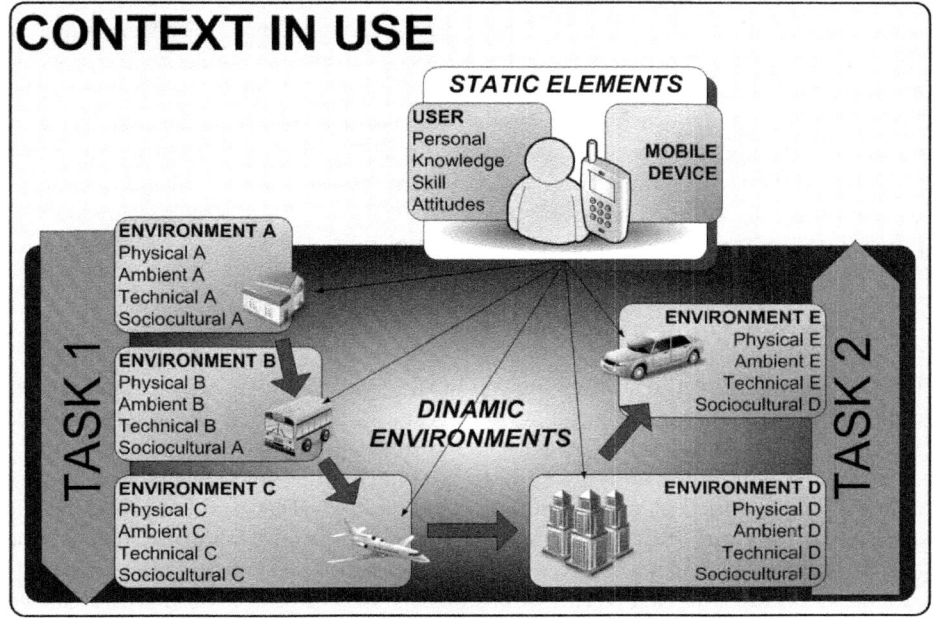

Fig. 1. Context in use definition

focused on environment, on automation level and also on the participation of users.

Focused on the environment which in the capture is developed, we can differentiate between real environments and laboratory environments. Testing executed in laboratory is easy because all influencing factors can be controlled and data can be recorded with several cameras and capturing tools. However, the context, which is the most influential factor, is not considered or it can hardly be simulated. In contrast to the laboratory methods, testing in real environment means that all data can be captured within real context influence.

Other kind of classification is by the automation level. They can be differentiated between manual and automatic methods. The manual method does not require any additional software or hardware and they are very flexible methods (it can measure specific and general parameters). The main problem this kind of method has is that it is extremely subjective. Because of that those captures could easily go wrong. However, automatic methods can capture highly objective information they are very quick.

The last method grouping is by the participation of users. Experiments done by real and direct interactions made by users (direct users and also stakeholders) can provide real data and can discover new problems. Contrary to the experiments with real users, experiments done by experts can only detect known problems and they also have more subjectivity than the first group.

After studying those methods, focusing on the objectivity and reliability of the captured information, the best method to capture user interaction is in real environments with real users in order to retrieve data as objective as possible and also the automatic methods to save time and add more objectivity to the capture. To sum up, we have explained the context in use is formed by user, mobile device and environment. Additionally, the best way to capture the most objective data is by monitoring the real user interaction in real environments by automatic systems. The next step is to study the best way to capture the information without influencing the context in use.

4.2   Existing Monitoring Systems

During the design of the system showed in this work, various interaction capture tools were studied: Morae [6], The Observer [7] and AQUA [8] among others. We have found a wide range of ways to capture the interaction.

The most common way is by camera installation in the device or added on a helmet. These added elements influence the context (the ergonomics of mobile device and the comfort of the user). When giving a sample, if a user whose phone has external capturing accessories (i.e. added camera), he will feel uncomfortable and he will change his behaviour. Consequently, this interaction will be corrupted and it will show worse quality results than without camera.

Interaction capturing by tests is a good method because they can be done after or before the interaction without influencing the context although they can add subjectivity depending on the design of the test questions.

Other way is by human observers. Although this method is manual, it has to be mentioned because is a good sample to understand the problem could lightly appear in camera installation and testing methods. The problem is that user can have a tendency to show expected (but not real) results due to the being observed and being evaluated feelings (i.e. if the user detects cameras, he will feel observed and evaluated). The most objective way to capture the interaction is by logs. The element that is altered is the mobile device because the logging software can reduce its performance.

In summary, if we capture data focusing on mobile context by the mobile device we can provide deeper and objective information without changing drastically the interaction. Therefore, the main goal to the capturer designed is to capture interaction data by registering information only using the mobile device.

5   Mobile Interaction Monitoring System

The exposed monitoring system is made up of a tiny mobile application that is able to capture the interaction and a desktop application that is able to simulate the interaction captured by the mobile application. It can capture data by saving screenshots and the user actions. Every key pressed is logged within its timestamp and its corresponding screenshot.

5.1 Methodology

In order to register the interaction, we have defined one methodology that is divided in five main steps.

- Firstly, the devices have to be configured. The configuration consists in defining tasks and contexts which in the users have to do the experiment. These tasks and contexts are defined by one xml file. This file is stored in phones to lent and ridden by the mobile application.
- Secondly, users have to borrow the configured phones and do the specified tasks. The task information is showed by the graphical user interface of the capturing system. When user chooses one task to do, he notifies to the application he is going to start. After ending the task the user notifies to the application the task is ended. The system stops capturing.
- When the user ends every task, he has to go the lent device back. The administrator has to dump the interaction data from the device to the system.
- After dumping information the system administrator has to introduce missing data (i.e. interaction errors, search times...) simulating the interaction by processing the recorded information.
- Finally, all necessary information to analyze the quality in use is stored. Therefore, the desktop application can generate graphs and reports.

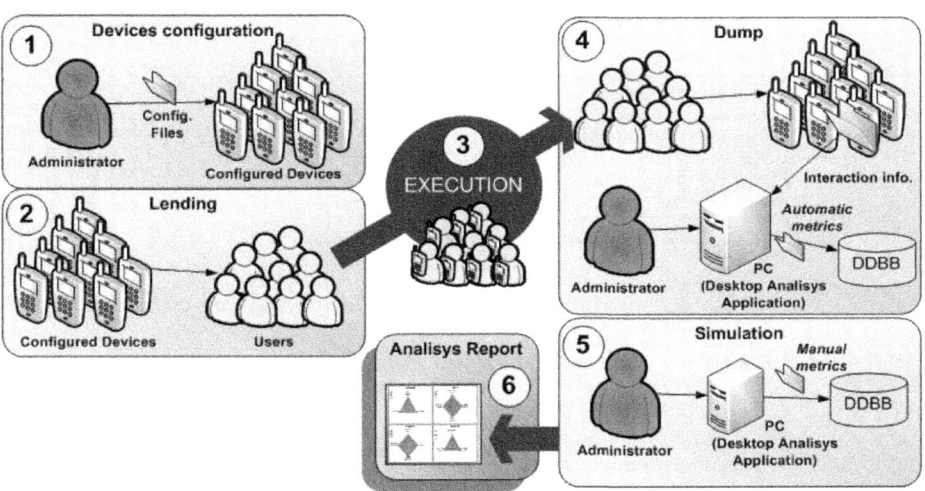

Fig. 2. Methodology

5.2   Architecture

The system is formed by two modules (see Figure 3): mobile application and desktop application. The first module is developed to be executed in mobile devices. The files which contain the configuration, context and task information are ridden by the controller. It is developed in Java and its functionality is to interact with the user by its Graphical User Interface (GUI) and to send TCP commands (Start, stop, resume and pause) to the interface interaction capturer. The controller has been developed in J2ME language. Java Virtual Machine for J2ME has screen and key access limitation. If the application has not got the focus, it cannot access to the screen and the keys of the mobile device. This limitation was solved by developing interface interaction capturer (module that can capture data interaction) in PyS60 (Python on Symbian Series 60). This capturer can save information generated by the interaction in three types of files: image files (PNG format), test answers files (XML format) and log file (text format). Due to this module, the only element affected by the interaction monitoring system is mobile phone because its performance is reduced.

The second module is the desktop application. When users finish the experiment, desktop application takes this information and normalizes every data. This data is used to calculate metrics of characteristics that define quality in use. It stores this data in the database of the system (MySQL database). Another functionality this application has is simulation. It can simulate the interaction. By this way, we can capture the data that we cannot capture automatically. After seeing simulations and introducing incomplete data manually, it can calculate every metric and show analysis reports.

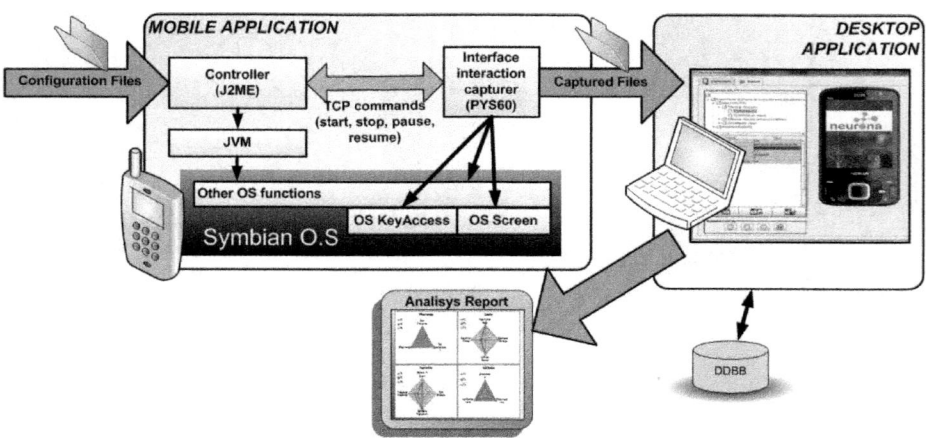

Fig. 3. Architecture

6   Conclusions and Future Work

The ability of mobile devices to access information and services from anywhere is the main reason that empowers the massive usage of this kind of technology. Owing to this tendency, software quality has to be improved by developing mobile device interaction models according to the user necessities. These necessities that have to be satisfied by the new applications have more kinds of users. It means the quality in use is getting more important than ever. The exposed system can capture all necessary information to calculate quality in use metrics defined within ISO/IEC 9126 standard. By this work we aim to give a more explicit point of view of the problems that appear during mobile application quality testing. In order to do so, we have studied a context model for mobile interaction design and ways to capture the user interaction.

To sum up, this work reveals whether the quality in use testing is focused on mobile context, we have to take care choosing the interaction monitoring methodology because the context can be easily influenced. The solution presented by this work shows limitations and problems that have to be solved in future works. The first problem is we have to trust the user is doing the tasks in the specified context. Another problem is the device influence caused by the exposed capturing tool. These problems can be solved by automatic context detection and by optimizing the mobile capturing tool. This work concludes in order to design successful mobile interactions; we must be aware and understand the context in which they take place.

References

1. ISO/IEC 9126:2001. Information Technology - Software Product Evaluation – Quality Characteristics and Guidelines for their use (2001)
2. ISO 9241-11:1998(E). Ergonomic requirements for office work with visual display terminals (VDTs) Part 11: Guidance on usability (1998)
3. NISTIR7432. Information Access Division – Information Technology Laboratory, Common Industry Specification for Usability –Requirements (2007)
4. Kankainen, A.: Thinking model and tools for understanding user experience related for information appliance product concepts. Dissertation of Anu Kankainen (2002)
5. Savio, N., Braiterman, J.: Design Sketch: The Context of Mobile Interaction, http://www.giantant.com/publications/mobile_context_model.pdf (retrieved on March 20, 2011)
6. Morae: usability testing and market research software, http://www.techsmith.com/morae.asp (retrieved on March 20, 2011)
7. The Observer XT, http://www.noldus.com/human-behavior-research/products/the-observer-xt (retrieved on March 20, 2011)
8. Covella, G.J., Olsina, L.A.: Assessing quality in use in a consistent way. In: Proceedings ICWE 2006, pp. 1–8. ACM, New York (2006)

Design and Implementation of a Simulation Environment for the Evaluation of Authentication Protocols in IEEE 802.11s Networks

Ikbel Daly, Faouzi Zarai, and Lotfi Kamoun

LETI Laboratory, University of Sfax
Sfax, Tunisia
{ikbel.daly,faouzi.zarai,lotfi.kamoun}@isecs.rnu.tn

Abstract. Mesh technology presents a great deviation in the field of wireless networks thanks to its assets brought in the level of mobility, the quality of services and distribution of radios resources. Due to its importance, a working group was formed in IEEE organization in order to work out a standard for the Mesh networks under the reference IEEE 802.11s. Consequently, a whole of tasks remain in the course of research such as security. That allows the appearance of several solutions and protocols suggested by the researchers' community in the literature. However, the existing simulators do not contain yet complete modules which allow studying this network. To remedy these problems, we thought of developing an environment of simulation for the evaluation of the authentication protocols for the IEEE 802.11s networks. This tool allows the easy development and the integration of new modules and thereafter its performances evaluation by providing appropriate measurements.

Keywords: IEEE 802.11s, Mesh network, simulator, authentication protocols, handoff.

1   Introduction

After the success of the Wireless Local Area Networks (WLANs), the world of wireless saw the birth of new technologies such as the Ad hoc networks and the Wireless Mesh Networks (WMNs) which propose a facility and a flexibility of deployment.

The Mesh network also called multi-hops network constitutes a flexible architecture ensuring the effective circulation of data between the wireless equipments by elimination of the problem of corrupted ways. This characteristic is consolidated by the automatic installation and configuration as well as the co-existence with existing networks. This technology is all the more reliable since it is dense and is by definition multi-services voice, data and video.

Moreover, WMN is characterized by its capacity to extend its coverage area. Consequently, its architecture is able to change dynamically to allow the fluent mobility of its users. This freedom during clients' movement imposes several challenges that we quote mainly the problem of security. Indeed, we must ensure the access only to authorized users by avoiding all risks of attack or intrusion. This

J. Del Ser et al. (Eds.): MOBILIGHT 2011, LNICST 81, pp. 206–218, 2012.

problem becomes increasingly serious and vulnerable with the opening of network medium towards the outside and in the case of routers movement.

A simulator network presents a fast, economic and effective solution for the networks tests. Indeed, the installation of a real network introduces a heavy operation from sides of material, time and cost. Then, it is essential to have a solution which makes it possible to solve these problems by guaranteeing the application of the various mechanisms and protocols which are necessary for the engineering of such a network. This solution reveals by the use of a network simulator which allows moreover the study of the behavior and the capacities of complex networks. Besides, it provides the possibility of developing new protocols and furthermore testing its performances and effectiveness.

In the field of networking, there are several types of network simulator. NS2 is one of the most known simulators in the researchers' community [1]. Indeed, it is open-source, fast and contains a big diversity of models. On the one hand, the complexity of the language programming used with NS2 causes a great difficulty for the development of new protocols. On the other hand, NS2 is characterized by the management of wireless networks as well as new technologies such as the Ad-hoc network and the study of some aspects in particular the mobility management, the access security and the control of congestion, etc. These aspects are treated by other simulator; we quote as an example Glomosim, Qualnet and Opnet.

Mesh technology states a subject of topicality since its standardization is under development. It is not adjusted yet in the majority of existing network simulators. In contrast, a novice version of this type of network was implemented in the new simulator NS3 [2]. This version just provides the backbone which is the first level of the WMN and is made up of the Mesh nodes bound by the routing protocol HWMP [3]. This implementation adopts the proposal IEEE 802.11s draft standard version 3.0.

Because of our need to test the performances of the new suggested solutions and protocols in the field of Mesh networks, the idea to develop an adequate simulator to this type of network is triggered. The new simulator is modular; it integrates several modules which allow the management, the control and the administration of network. We quote as an example the following modules:

- Channel Modeling
- Area modeling
- Management of the arrivals
- Deployment of the radios resources
- Radios resources management (CAC, Scheduling...)
- Neighbors discovery
- Mobility management (handoff)
- Security management
- Management of the observations.

The remainder of this paper is organized as follows. In Section 2, we give an overview about the Mesh technology. In Section 3, we explain through a class diagram the design our environment of simulation. Then, the implementation phase is illustrated in Section 4. And an example of manipulation of this new tool is described in section 5. Finally, we conclude the paper in Section 6.

2   Overview of WMN

The revolution of wireless, started by the success of the standard IEEE 802.11, pushed the community of researchers in the design, the analysis, the development, and the deployment of new wireless solutions. In particular, the Wireless Mesh Networks (WMN), have captured the interest of university research and industry, because of their capacity to meet at the same time the requirements of the suppliers of wireless access to Internet and the users.

Mesh topology is a network topology qualifying the networks (wired or not) in which all the hosts are connected gradually without central hierarchy, thus forming a structure in form of net. That makes it possible to avoid having critical points, which in the case of breakdown, cut the connection of part of the network. Indeed, if a host is out of service, its neighbors will pass by another path. The implementation of such a topology is called network Mesh [4].

Thus, Mesh network consists of nodes which communicate directly with their neighbors by removing the interconnected network between the access points. Moreover, this type of network forms an emergent class of wireless networks, which is able to be dynamically organized and configured. It takes the principle of a wireless network that is based on the multi-hop transmission. A typical example of a Mesh is presented in Figure 1. Mesh architecture of two levels concentrates the routing on the wireless part (the first level - backbone), made up of Wireless Mesh Routers (WMRs) and of the access points which offer a connectivity to mobile stations (the second level) [5].

Considering the importance of this type of networks and the various advantages brought by this technology, a working group was created within the IEEE (Institute of Electrical and Electronics Engineers) with an aim of arranging a new standard under the reference 802.11s [6], [7].

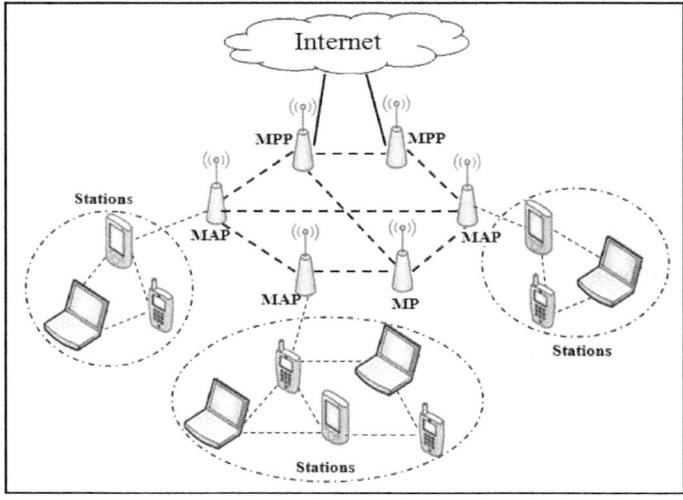

Fig. 1. Overview of wireless Mesh network

This emergent standard defines a whole of terminologies schematized in Figure 1. First of all, any node that supports Mesh services such as control, management, and the operation of the Mesh is a Mesh Point (MP). If the node supports in more the access to stations (STAs) or with the nodes which do not have the Mesh services, it is called a Mesh Access Point (MAP). Moreover, a Mesh Portal Point, noted MPP, is a Mesh point which has a non-802.11 connection with Internet and the external networks.

Although there are the significant advantages of the successful deployment of the wireless Mesh networks in the whole world, some technical limitations and problems will remain to be solved and will probably require more advanced research for the deployment of such a network [8] Moreover, the upcoming standardization of Mesh network will clarify some concepts and will put an end to some problems. We quote mainly the quality of service, the security, the mobility and the interference problem.

While these networks continue to develop, several efforts and works are provided to be ensured of the smooth running of the services offered by Mesh network. Indeed, we cannot be certain of granting the access to network only for the authorized users or the continuity of connectivity following the mobility of the various network equipments or the conservation of an acceptable quality of services even with real time applications only if we propose a set of protocols that manage all these aspects and others in order to provide a reliable Mesh network.

With the aim of solving these problems, the community of researchers suggests several solutions and protocols. And concerning the phase of validation, they were obliged to adapt their proposals to an ad hoc environment due to the absence of a Mesh network implementation in the majority of network simulators. Following this need, the existence of a simulator adapted to this type of network becomes a need which makes it possible to test the suggested protocols and to evaluate its performances.

3   Simulation Design

In this section, we start on the design of an environment of simulation for the evaluation of authentication protocols for the IEEE 802.11s networks. First of all, this developed tool is supposed to create the wireless Mesh network architecture with its various levels and nodes. Then, it handles and manages the Mesh services and nodes by the management of traffic, the management of the mobility and the mechanism of neighbors discovery, etc. Then, the simulator allows the addition of a set of control mechanism through the implementation of new protocols and in particular those of security and re-authentication.

In order to carry out all these modules in such a simulator, we start with the construction of a conceptual model for this tool in the form of a class diagram. In fact, this schema is a net of classes and associations which models the structure of an object, its role within the system in addition to its relations with the other objects.

Figure 2 presents the various modules which will be implemented in our simulator. The principal class of this diagram is the model of the WMN creation. It includes all the essential methods for the generation of such a type of network as well as the other additional methods for services control and the administration of the various network equipments.

First, we have to prepare our environment of study which is composed of the set of nodes mentioned in Figure 1. This task appears in the diagram by the reservation of classes for the creation of the different network components; the stations, the nodes MAPs and MPPs. Indeed, the MP class is formed by the whole of specific parameters to MAPs in Mesh network. Moreover, we find multiple methods to operate the various services of an access point as an intermediate point with the external networks (MPPs).

For the second network level, a station class was developed. It comprises the required information to identify a WMN user as well as the various actions which are likely to be carrying out by a simple station such as the emission and the reception of data.

After the formation of the Mesh, we have recourse to regulate the several functionalities ensured by any wireless network and those specific to IEEE 802.11s networks. Indeed, we also require specifying the type of studied architecture and the

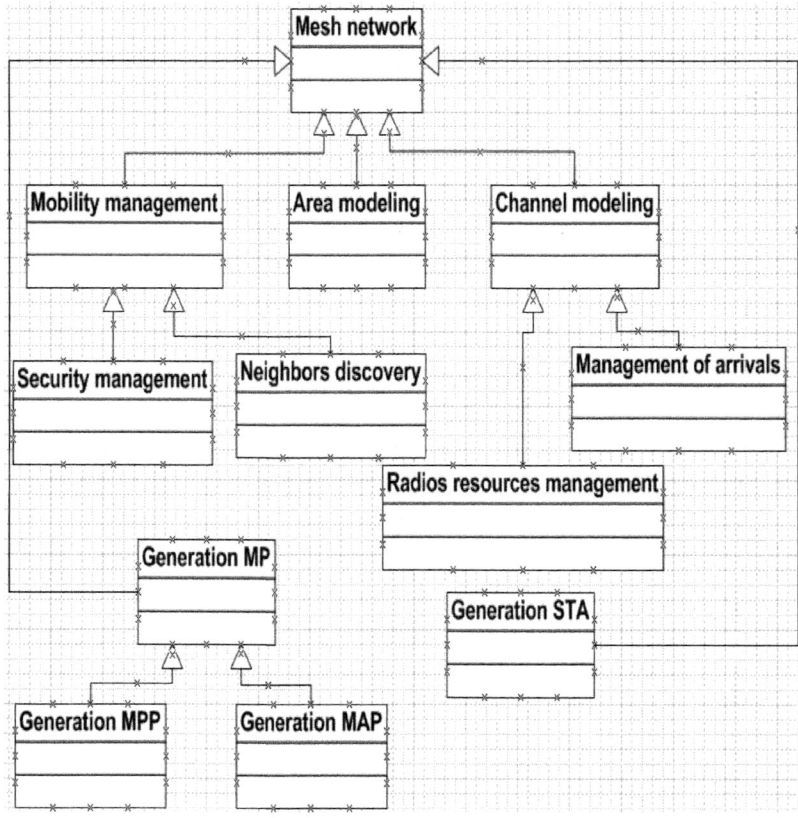

Fig. 2. Simulator class diagram

distribution of the different components (such as central, distributed or hierarchical network architecture) create by using the module "Area modeling". With an aim of ensuring and managing the communication between MPs and their users, we have recourse to a whole of models which are used to adjust the transmission of packets between the various Mesh network entities.

Indeed, to ensure the connectivity and the establishment of communications, we must manage the different channels (Channel modeling) by reserving the required radios resources (Deployment of the radios resources) and organizing the exchanged flows and packets (Management of the arrivals). Consequently, the characteristics of the services and the applications managed in a network from traffic point of view make it possible to know and control the load of a specified network. Moreover, the study of behavior and the analysis of performances of such a network require the knowledge of the characteristics and statistical measurements for the various types of applications which are active during the transmission.

Among the assets brought by an IEEE 802.11s network, we quote the effective mobility management with an aim of providing a freedom in moving for the customers who seek to communicate during their movements without any constraint of connectivity, and where the change point of attachment is completely transparent. For these reasons, the study of mobility notion is raised by the module "Mobility management". In order to establish the scan phase, the network nodes carry out the procedure of the "neighbors discovery" to browse the vicinity of every MP and STA to facilitate the connectivity between these entities.

The reliability, the adaptability and the scalability are the most significant attributes of this type of network. However, security in a Mesh network is not yet well defined which requires the adoption of powerful mechanisms to be protected from several attacks. This problem of the insecurity becomes increasingly vulnerable and critical especially during handoff phase (i.e. following the change of channel during communication). That requires the application of an effective and well defined policy of security.

Moreover, securing the transmitted information is a task whose complexity gradually grows, parallel to the increase in the number of applications and with the degree of medium opening towards the outside. Consequently, the authentication of users becomes necessary. It is significant as well on the level of ensuring security as of the facility and the safety of employment for the user. Indeed, the authentication process is one of significant measurements to confront against attacks in Mesh networks, by allowing only authorized users to obtain connections and preventing the intruders from being integrated in the network and disturbing its operation.

In our simulator, we have integrated the security aspect of the dynamic stations during channel changes, in order to evaluate the performances of such an implemented authentication protocol.

4   Simulation Implementation

A flexible and easy handling remains always among the most decisive criteria for the success of such a development project. This is why the choice of a programming language must be well studied.

The software of development on the subject of simulation is diversified and can be presented in the form of a structured programming language like the language C++ and object oriented programming language for example Java. The choice of a specific language depends on the category of the tasks which will be carried out, the requirements to satisfy and its performance compared to the other languages.

In our case, it was necessary to choose Java, an object oriented language which proves to be most adequate thanks to its rich and fertile libraries, the portability of its programs, its flexibility, its facility of use and its robustness [9].

Through the development of this environment of simulation we endeavor to:

- set up a wireless Mesh network architecture with the required entities and components,
- implement the mobility aspect in order to study the station handoff,
- implement the security aspect of the dynamic stations during their changes of channel,
- evaluate the impact of these concepts on the behavior of network.

Consequently, this tool makes it possible to specify the various parameters of this type of network and to simulate its behavior with an aim to study the performances and the effect of implemented aspects such as the protocols of mobility and security. To guarantee the correct functioning of our network generated by the simulator as well as the studied models. A model is an essential element to understand the behavior of such a phenomenon or such an object. In our case, we implemented some models which are mainly:

- Traffic model: it can be mathematical or analytical. The model of network traffic is represented by the statistical models, namely Poisson model , Exponential model or Geometrical model,
- Mobility model: it define the manner of nodes movement in the coverage area of the network,
- Propagation model: it contains the implementation of the wireless propagation environment, namely the attenuation model.

4.1   Models

4.1.1   Traffic Model

In our study, we will consider two traffic models; voice and Internet flow (Web).

• Voice Model

The voice transmission is a real time service. The constraint of time constitutes a priority that must be respected for the generation of this type of application [10].

For this type of application, we have adopted the traditional model. In this case, the arrival of the phone calls follows the "Poisson process", which is characterized by the value λ as an average rate of call and the duration of a call follows an "Exponential process" with μ average.

- **Web Model**

This model is characterized by a whole of the parameters based on the consultation of HTML pages. In Internet flow, the access to the Web sites can be split in several sessions. Each session is defined by specific characteristics such as periods of datagrams loading, periods of datagrams inter-arrived and the periods of reading (periods of silence).

The statistical characteristics of this model are [10]:

- ➢ The appearance of the sessions follows a "Poisson process",
- ➢ On the session level:
- the number of HTML pages call follows a Géométrical distribution of typical average $\mu = 5$ calls/session,
- the reading time of HTML pages follows an Exponential distribution of average μlec, with $1 / \mu lec$ between 4 and 12 s,
- the number of datagrams per call follows a Géométrical distribution of average $\mu dgm = 10$ dgm/call,
- the duration of datagrams inter-arrival is an Exponential distribution depending on the network rate of transmission.

4.1.2 Mobility Model

The mobility model is based on a non uniform and random distribution of stations movements in the coverage area of the network. The movement of a node follows a Random Waypoint model, which is expressed by [11]:

- ➢ Mean speed of the movement noted v such as $v \in [0, 20 \, m/s]$,
- ➢ Movement direction represented by the angle noted θ, such as $\theta \in [0, 360°]$.

The movement of nodes is generated in a periodic way and with a random way during simulation.

4.1.3 Propagation Model

The simulated propagation model supposes that the signals transmitted between the various network components undergo attenuation on a small scale. Indeed, if a node i sends a message with a power of emission Pe the station j receives it with a power of reception Pr (eq1) such as:

$$P_r^j = P_e^i - 10 \log\left(\frac{d}{T_r}\right)^4 \tag{1}$$

With:
  P_e^i: Power of emission of the signal of i expressed in dB,
  P_r^j: Received power of the signal of J expressed in dB,
  d : Distance between the nodes expressed in meters,
  T_r: Zone of vicinity expressed in meters.

4.2   Parameters and Results of Simulation

After having implemented the various models of a wireless Mesh network, the following stage consists in studying the performances of our simulator. The stage of evaluation is based on the influence of the increase in the mobile stations number and their mobilities on a set of criteria. As a preliminary phase, we must fix a list of parameters in order to carry out the various scenarios on the simulator as well as to extract precise values reflecting the state of network.

4.2.1 Parameters of Simulation

An example of Mesh network architecture is defined by its geographical coverage (Size with X and Y coordinates) and its population (MPs and STAs). The simulated network is characterized by a cover which is spread out over hundreds of meters, comprises a preset number of MPs nodes and a variable number of stations that varies according to the scenarios of simulation.

Each MP node is defined by:

- Its position in the network: defined by the node coordinates following the axes X and Y,
- Its channel number: allotted in order to avoid the problem of interference,
- Its range of transmission: defined by the coverage of its signal which decreases while moving away from the site of the concerned MP,
- A unique identifier.

Each station is defined by:

- Its position in the network: defined by the node coordinates following the axes X and Y. In order to adjust our simulation with reality, it is necessary that nodes have different positions and carry out random movements in the network,
- The mobility model: the speed $v \in [0, 20 \, m/s]$ and the movement direction follows the angle $\theta \in [0, 360°]$,
- A unique identifier,
- The traffic model:
  - ➢ The arrival of a packet follows the "Poisson process" with average $\lambda = 0,2$ call/hour ,
  - ➢ The duration of the communication follows the "Exponential process" with average $\mu=10$ minutes.

4.2.2 Scenarios of Simulation

After the configuration of the network parameters and nodes and the implementation of the various models, it is essential to simulate the behavior of this simulated network to evaluate its performances.

We will consider two types of exchanged traffic between the nodes:

- Voice communication,
- Web communication

While basing on these types of communications as well as the parameters of simulation, we evaluate the performance of implemented protocols by extracting the simulation results according to three criteria:

> Handoff latency: the time passed between the change of point of attachment request and the association with the new MAP,
> Blocking rate: represents the number of blocked stations at handoff for the total number of stations which requests handoff,
> Loss rate: represents the number of lost packets for the total number of emitted packets.

5   Example of Use

Due to its importance and many the assets brought by this technology, wireless Mesh networks capture the attention of the industrialists and the researchers that allows the appearance of a variety of products and protocols. Indeed, several interesting works were undertaken on this type of networks of which some propose solutions to manage users' mobility, others study the problem of security, and a third category which are interested in the re-authentication after a change of location in the wireless Mesh network.

The endeavor in [12] is to introduce a secure re-authentication mechanism named SWMM (Secure Wireless Mobility Management), which is carried out during the cross of mobile stations by different nodes to allow users fulfilling an effective and reliable handoff as well as a secure access to services offered in Mesh network. In this study, the authors have improved previous methods which deal with the problem of mobility as well as security. Indeed, they have applied the Wireless Mobility Management (WMM) mechanism [13] to support an environment which manages handoff in effective way. In addition, they have also slightly modified this previous method and the EAP-TTLS scheme [14] to provide the network's security.

In order to prove that the proposed protocol SWMM outperforms other existing method, this solution and another suggested in literature, EAP Independent Handover Authentication method (EAP-IHA) [15], have been implemented in our simulator.

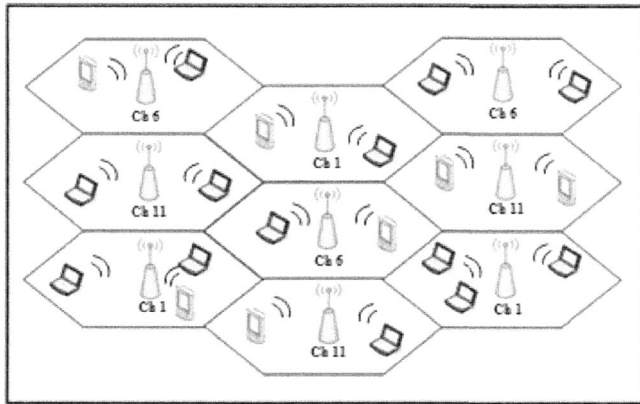

Fig. 3. WMR architecture generated by the simulator for SWMM and EAP-IHA

Figure 3 shows the Mesh network architecture generated by the simulator for the evaluation of the authentication protocols. It is made up of 9 MPs and a variable number of stations. Each MP is characterized by its coverage area and its number of channel. This last parameter is granted to different MPs by the network in order to eliminate the problem of interference.

The obtained results by the simulator following the implementation of these two protocols make it possible to plot the curves of Figure 4.

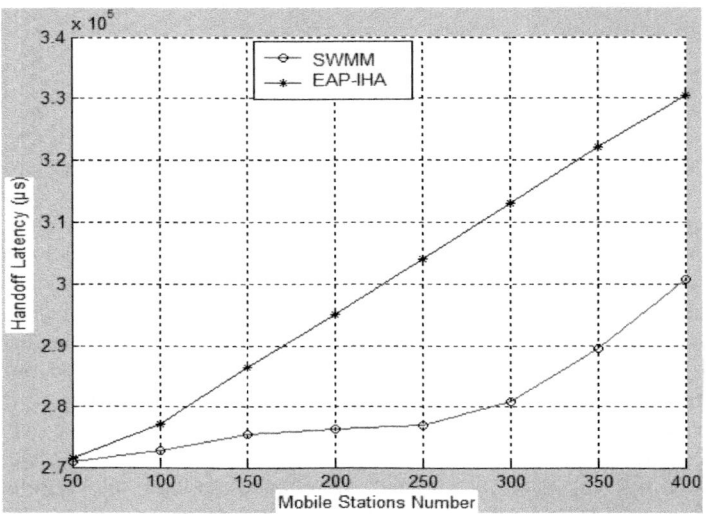

(a) Handoff Latency vs. Mobile stations number

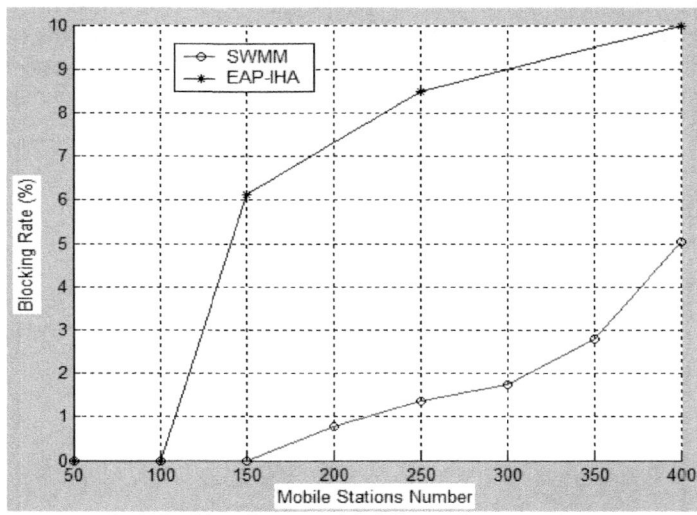

(b) Blocking Rate vs. Mobile stations number

Fig. 4. Simulation results with SWMM and EAP-IHA protocols

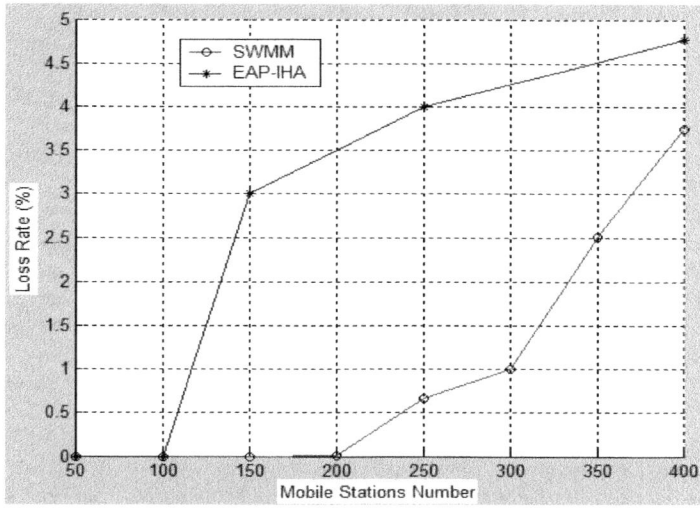

(c) Loss Rate vs. Mobile stations number

Fig. 4. (*Continued*)

The interpretation of these curves makes it possible to evaluate the influence of the increase of network population on the value of handoff latency (a), blocking rate (b) and loss rate (c) on SWMM and EAP-IHA.

6   Conclusion

In this paper, we have presented our Java network simulator which is dedicated to the IEEE 802.11s networks. This tool is classified among the first simulators which treat this type of network since wireless Mesh network is still in the course of standardization by the IEEE organization. Among the advantages of this environment of simulation, we first of all quote the simplicity of the handling of various offered functionalities. Moreover, this simulator allows the easy development and the integration of new modules and protocols. Thereafter, it provides the required measurements to evaluate the performances of these modifications with more reduced time compared to the other existing simulators. Our current simulator presents the first stage towards the implementation of a more advanced tool which allows the treatment of Mesh networks. Indeed, we aim to add a set of supplementary modules which ensure a better handling of network. As future work, we claim to integrate a graphic interface to facilitate the use of the supported modules and the implementation and the validity test of the added protocols.

References

[1] NS-2, The Network Simulator, http://www.isi.edu/nsnam/ns
[2] The ns-3 network simulator, http://www.nsnam.org/

[3]   Andreev, K., Boyko, P.: IEEE 802.11s Mesh Networking NS-3 Model
[4]   Catoio, G., Cibot, L., Laloux, M.: Etendre dynamiquement la couverture d'un réseau Mesh (2006/2007)
[5]   Faccin, S.M., Wijting, C., Kneckt, J., Damle, A.: Mesh wlan networks: concept and system design. IEEE Wireless Communications (April 2006)
[6]   Camp, J.D., Knightly, E.W.: The IEEE 802.11s Extended Service Set Mesh Networking Standard
[7]   IEEE P802.11s/D0.01 Draft Amendment to Standard for Information Technology-Telecommunications and Information Exchange Between Systems - LAN/MAN Specific Requirements-Part 11: Wireless Medium Access Control (MAC) and physical layer (PHY) specifications: Amendment: ESS Mesh Networking (March 2008)
[8]   Qiu, L., Bahl, P., Rao, A., Zhou, L.: Troubleshooting Wireless Mesh Networks
[9]   Holzner, S.: Total Java, Editions Eyrolles (2001)
[10] Ajib, W.: Gestion de transmission d'un flux temporaire de donnés dans un réseau radio mobile d'accès TDMA. Thesis, ENST Paris (2000)
[11] Wai, F.H., Ye, Y.N., James, N.H.: Intrusion Detection in wireless Ad Hoc networks. Introduction to wmobile Computing Technical Report CS4274
[12] Daly, I., Zarai, F., Kamoun, L.: Secure Wireless Mobility Management. In: IEEE 17th International Conference on Telecommunications, ICT 2010 (2010)
[13] Huang, D., Lin, P., Gan, C., Jeng, J.: A Mobility Management Mechanism using Location Cache for Wireless Mesh Network. In: QShine 2006, Waterloo, Ontario, Canada, August 1979. ACM (2006)
[14] Khan, K., Akbar, M.: Authentication in Multi-Hop Wireless Mesh Networks. Proceedings of World Academy of Science, Engineering and Technology 16 (November 2006) ISSN 1307-6884
[15] Izquierdo, A., Golmie, N., Hoeper, K., Chen, L.: Using the EAP Framework for Fast Media Independent Handover Authentication. National Institute of Standards and Technology, USA (2008)

Re-authentication Protocol for Vertical Handoff in Heterogeneous Wireless Networks

Ikbel Daly, Faouzi Zarai, and Lotfi Kamoun

LETI Laboratory, University of Sfax, Tunisia
{ikbel.daly,faouzi.zarai,lotfi.kamoun}@isecs.rnu.tn

Abstract. The Heterogeneous Wireless Networks (HWN) is a type of framework which includes different varieties of wireless technologies. To ensure a robust management and a reliable behavior of HWN, it is necessary to well manage the interworking between its technologies and in particular to secure interworking and roaming between the 3rd Generation Partnership Project (3GPP)/Long Term Evolution (LTE) and Wireless Local Mesh Networks. A whole of solutions has been proposed to solve this problem; we quote mainly Extensible Authentication Protocol-Authentication and Key Agreement (EAP-AKA) which still suffers from some vulnerabilities such as, man-in-the-middle attack, Sequence Number (SQN) synchronization, and disclosure of user identity. In this paper, we propose new re-authentication protocol. The suggested solution proves its effectiveness following the studies of simulation which carried out according to different criteria handoff latency, loss and blocking rate.

Keywords: Re-authentication protocol, Heterogeneous Wireless Networks, 3GPP LTE, Mesh networks, Handoff, Security.

1   Introduction

A heterogeneous network is an association between several types of networks which belong to various generations and technologies. These varieties of equipments manage between them to provide a new range of functionalities with a better quality of services and more security. One of the most important kind of this heterogeneity in wireless networks, we find 3G-WLAN interworking which is the future generation of mobile and wireless communication systems. Indeed, this specific environment integrates two various types of networks; the WLANs networks and the cellular networks of the third generation.

Each category of these networks brings a whole of assets with an aim of guaranteeing the best interworking of various procedures and mechanisms between its components and of fulfilling the user's requirements and needs. Consequently, the integration of these different architectures makes it possible to benefit from the diversity of the advantages brought by each category and thereafter this complementarity improves the effectiveness of network as it guarantees the resolution of some problems such as the security, the interference and the quality of services.

J. Del Ser et al. (Eds.): MOBILIGHT 2011, LNICST 81, pp. 219–230, 2012.

In our study, we will extend our research to illuminate the interworking between SAE/LTE (System Architecture Evolution / Long Term Evolution) network [1], [2] (or simply LTE network) and the Wireless Mesh Network (WMN) [3]. On the one hand, the first type of network, developed by the 3rd Generation Partnership Project (3GPP), presents an improved and more secure version of the system UMTS (Universal Mobile Telecommunication System). In addition, WMN presents one of the promising technologies in the world of wireless communications. Indeed, Mesh network makes it possible to provide a free mobility and the self-configuration of the various equipments of network, extensible zone of cover by the addition of routers, as well as a better quality of services.

In spite of the diversity of the benefit brought by each one of these technologies as well as the multiplicity of the research carried out in this field in order to improve the performances of these two types of networks, the security remains an enormous challenge which needs to be studied. Indeed, the open medium of these technologies increases their vulnerabilities and the risks of attacks. Moreover, the integration of the networks, 3G-WLANs worsens the situation and the environment of association becomes increasingly vulnerable and less protected.

The architecture of LTE-Mesh interworking is illustrated by Figure 1. This environment is composed of the set of components belonging to the Mesh and LTE networks. First of all, in Wireless Mesh Network, we distinguish the following equipments; WMR, Mesh AG and AAA server. The nodes WMRs (Wireless Mesh Router) support the services of Mesh network and make it possible to establish connections with the close nodes in order to ensure the property of the transmissions multi-hops.

The communication with the external networks is carried out by the entity Mesh AG (Mesh Access Gateway). And with an aim of controlling the access to Mesh network, a procedure of authentication must arise in this environment. Indeed, WMN has recourse to an authentication server called AAA (Authentication, Authorization and Accounting). This entity is the responsible for mobility management, the registration, the authentication and the re-authentication of the various equipments of Mesh network.

On the other hand, the LTE network has its own architecture, which is composed by an access network and a core network. The first block, called EUTRAN (Evolved UMTS Terrestrial Radio Access Network), is made of a whole of nodes, noted eNodeBs, which ensure the transmission of the radio signals [4]. Moreover, the AAA server is used to guarantee the authentication between the networks 3G and Non-3GPP while using the ePDG entity (evolved Packet Data Gateway) and to register the users by allotting a whole of parameters such as IMSI (International Mobile Subscriber Identity). This confidential information will be stored in a data base called HSS (Home Subscriber Server).

In addition, the LTE network is made up of other set of entities such as MME (Mobility Management Entity) for the management of mobility and session, PDN GW (Packet Data Network Gateway) for the establishment of external communications and the allowance of the addresses as well as Serving GW (Serving Gateway) for the packets routing.

The remainder of this paper is organized as follows: In Section 2, we give some related works. In Section 3, we detail a new solution to secure interworking and roaming between 3GPP LTE and Wireless Local Mesh Networks. The proposed authentication protocol uses new equipment, named hybrid unit. In Section 4, we describe a simulation method of our scheme and analyze the numerical results derived from simulation and highlight the contribution developed in the previous sections. Finally, we conclude the paper in Section 5.

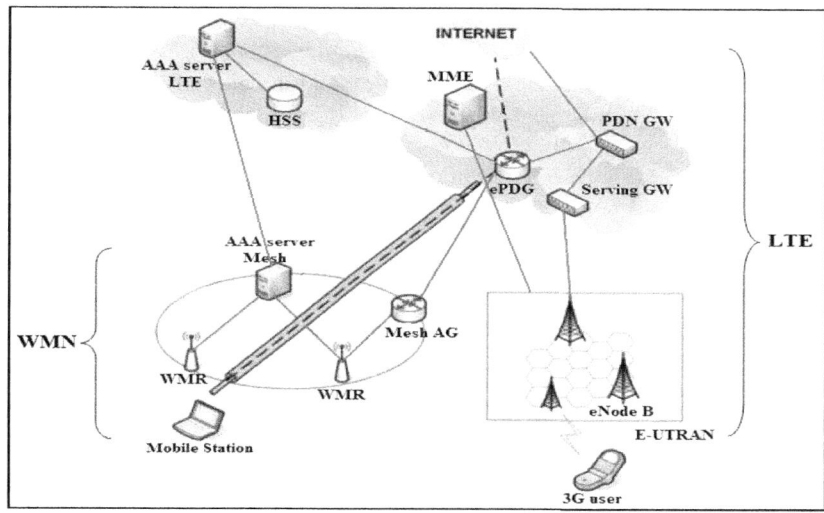

Fig. 1. Architecture of the LTE-Mesh interworking

2   Related Work

Because of the diversity of mechanisms, equipments and technologies within such an LTE network, the constraints and the challenges become increasingly critical and demanding. Indeed, the users, who are by definition mobile, seek to communicate during their moves across different domains without any constraint of connectivity, and having a completely transparent network change.

Moreover, securing the network access is a task whose complexity grows progressively, parallel to the increase in the number of applications and with the degree of opening towards the outside which they imply. Consequently, to provide a robust system of security should solve this problem and maintain the safety of data. Due to the importance of the security aspect in any network and in particular in LTE network, this subject catch sight of the organizations what allows the standardization of some new authentication protocols and the keys management such as UMTS AKA (Authentication and Key Agreement) [5] and EAP-AKA (Extensible Authentication Protocol-AKA) [6]. Besides, the whole of the assets provided by LTE network encouraged the community of researchers to propose new solutions to confront and resolve these problems.

The first study [7] reveals with a whole of mechanisms which make it possible to optimize security following the execution of the vertical handoff in a heterogeneous environment (Heterogeneous Wireless Networks - HWN). First of all, the authors have proposed a new model for optimizing network access authentication procedure. Then, they have detailed a mechanism ensuring the security context transfer.

Our field of research is limited to the first procedure which accentuates the integration of a new intermediate entity between WLAN network and LTE network, called "Interworking Unit". This component ensures a protected communication and safe data exchanges between the various types of networks in the heterogeneous environment.

In spite of the profit illustrated within the execution delay of the re-authentication procedure, the proposed mechanism did not well profit by the addition of Interworking Unit which only plays the role of a protected and safe bridge between the two networks. Moreover, the confidential parameters allotted for each user in a given network should not be circulated towards the other external networks because that makes it possible to increase the risks of attacks.

With an aim of ensuring the security in 3G-WLAN interworking, we have recourse to a protocol called EAP-AKA. But, the latter still suffers from some vulnerability such as Sequence Number synchronization, disclosure of user identity, additional bandwidth consumption and man in the middle. Consequently, a variety of work treated this subject in order to improve this authentication protocol.

Among these studies, we quote [8] which proposes a new protocol of authentication and key agreement based on EAP-AKA, quoted previously. With the intention of overcoming, the vulnerabilities of EAP-AKA, the suggested method exploits the ECDH mechanism (Elliptic Curve Diffie-Hellman) with the encoding technique "symmetric key cryptosystem".

The proposed protocol is composed of four principal phases namely; Initialization, Registration and generation of TK (Temporary Key), Authentication and key agreement and finally Transmission of MSK (Master Session Key). The authors of [8] underline in their model the need for having secure mutual communications between each pair of equipments of the interworking.

Moreover, the present mechanism makes it possible to ensure a "Perfect Forward Secrecy" (PFS) which provides more security in this environment while making it possible to keep from the replay attack. On the other hand, the suggested protocol handles a great quantity of parameters, which are greedy in memory capacities as well as several functions of encoding and of keys generation that carry out to weigh down the mechanism side time of execution and of calculation, memory capacity for data storage as well as the signaling overhead. Thereafter, this solution does not take into account the mutual authentication between the two types of networks 3G-WLAN and supposes the existence of a confidence relation between these two parts. What contradicts the existing.

In this same context, a second solution [9] which treats the protocol EAP-AKA in order to improve the aspect security in the field of 3G-WLAN. The contribution of this work appears by the proposal of more effective, robust and new authentication procedure as well as a more reliable and protected keys management. To solve the problems of security in an environment that includes a 3G-WLAN integrated networks, the project 3GPP (3rd Generation Partnership Project) introduced an architecture which is based on the protocol two-pass EAP-AKA authentication [10]. As it is indicated through its name, the mechanism contains two phases of authentication. The first one is used for the user registration in WLAN network. For the second authentication, the EAP-AKA technique is encapsulated in the protocol IKEv2 (Internet Key Exchange version 2) which allows the registration of client in the field of 3G Public Land Mobile Network (3G PLMN).

One-pass EAP-AKA authentication procedure presents the innovation brought by this work. This mechanism eliminates the duplication of the EAP-AKA protocol execution. What causes a reduction of the authentication overhead and the minimization of messages flow exchanged between the various equipments of the heterogeneous network. In addition, the suggested model does not supervise the mutual authentication between the different equipments intervening in the procedure of authentication and it supposes the existence of confidential relations without specifying the means which ensure the security of confidential data transmission.

According to the study of some existing solutions, we could extract an optimal solution by the proposal of a new authentication protocol. During the establishment of this mechanism, we held in consideration the legitimacy of the connections between the various equipments to ensure more security and the minimization of authentication delay in order to preserve a better quality of services.

3   Proposed Solution

This present work deals with the problem of the lack of security at the time of handoff in a heterogeneous framework in particular the interworking between a WLAN network (Mesh) and a 3G network (LTE). The suggested solution is based on a re-authentication protocol by exchanging a whole of parameters between Mesh and LTE. This interworking between these two various types of technologies is ensured by a new entity called "Hybrid Unit" (HU).

On the one hand, HU is connected by a secure channel to the Mesh network while benefitting from the advantages of the VPN (Virtual Private Network) network which allows the data transmission in safety. On the other hand, this entity is binding on the various components of LTE network. Consequently, it profits from a set of parameters and keys such as an identity and the keys CK, IK and K which make it possible to check its legitimacy beside LTE network as well as to ensure the integrity and the data confidentiality exchanged between these components.

The innovation brought by the proposed re-authentication protocol is the preparative phase which precedes the execution of the point of attachment change. In which we arrange a base of identities and keys that facilitates thereafter the generation of the new identity as well as the new key for the mobile station. Indeed, this stage makes it possible on the one hand to minimize the time of the re-authentication and consequently more quality of services and on the other hand to ensure more safety for the exchanged confidential data and the access to the network. This new protocol is composed of three great phases: Initialization, Pre-handoff and EAP procedure, as it is shown in Figure 2. In the first phase, we prepare a data base formed of the couple identity and key (id, k). This information is shared between AAA server of WLAN network (AAA_{WLAN}) and the hybrid unit (HU) (step 1).

Before carrying out the execution of the point of attachment change mechanism for the mobile station (STA) which migrates from LTE network towards Mesh network, we have recourse to a new procedure, the pre-Handoff phase. This procedure starts by sending a message from the mobile user towards the new point of attachment (WMR). This element contains the temporary identity (TMSI - Temporary Mobile Subscriber Identity) of the station, which identifies in more the identity of its network mother, the first TMSI is transmitted in clear and the second is associated with a Sequence Number (SN) and enciphered by the key CK of STA with an aim of ensuring the confidentiality and the safe and protected data transmission. This association between TMSI and SN makes it possible to minimize the risk of attack especially the Reply Attack (step 2).

After the reception of these data, WMR adds its identity (ID_{WMR}) to the received message and sends it to server AAA_{WLAN} (step 3). Immediately, the Mesh server checks the legitimacy of WMR entity by testing its identity. If step 4 does not cause any doubt on the topic of the wireless Mesh router entity, AAA_{WLAN} transfers, in its turn, the data coming from the station after the addition of its identity (ID_{AAA}) towards the hybrid unit (step 5). In order to ensure more security and in spite of the presence of the secure tunnel, HU checks in addition the identity of AAA server (step 6). Then, it identifies the original LTE network of the mobile STA by referring to its temporary identity TMSI. Afterward, HU sends to MME entity the data which relate to, on the one hand, the station made up of its identity (TMSI) and a Sequence Number (SN) enciphered by the key CK of STA and on the other hand to the Mesh network that contains its identity (ID_{AAA}), which is encrypted with the key CK of HU (step 7).

Following obtaining this information, the MME entity, from LTE network, starts the step 8 that concerns the checking of the station validity and if it is authorized to reach the selected Mesh network while being based on its profile. If the two conditions are valid, MME sends some parameters about STA; its identity IMSI and its key K. The confidentiality of this information exchanged between MME and HU is ensured by the application of encoding with the key CK of the hybrid unit (step 9).

The reception of this last message launches the step 10 for the reservation of an index from the base (identity, key), noted "ind". HU associates this index to the received parameters of STA from its home network (IMSI, K) then it sends the totality of the data to the server AAA_{WLAN} while using the secure channel to ensure the safety of the transmissions (step 11). At this stage, all the necessary elements for the generation of the new parameters are present on the level of the entities; HU and AAA_{WLAN}. Consequently, on the one hand these two components launch the mechanism of calculation of new identity (ID_{STA}) by taking advantage of a preset function "f" and the old identity of STA (IMSI). On the other hand, they generate the new key (K_{STA}) while referring to old key (k) and a function "g" (step 12).

After the generation of the user new parameters, we send them towards the MME entity encoded by the keys CK and IK of HU in order to ensure the confidentiality and the integrity of the transmitted data (step 13). Then, this information (ID_{STA}, K_{STA}) reached the station STA through its home network at step 14, enciphered by the keys CK and IK of STA. This last message encloses the second phase, the pre-handoff procedure.

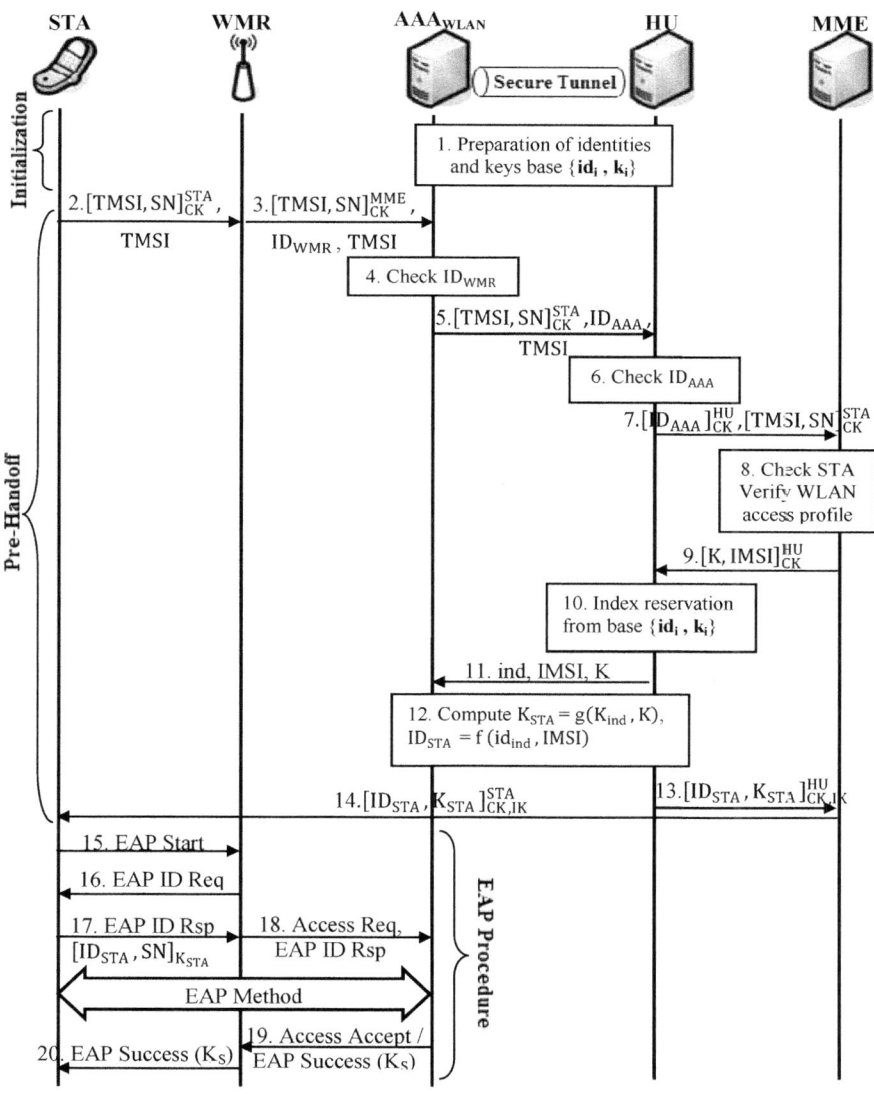

Fig. 2. Re-authentication procedure following the handoff in a heterogeneous network

Following the preparation for the handoff execution, which allows the checking of the legitimacy of various equipments as well as the generation of new identities and keys for the mobile stations, we achieve our proposed re-authentication protocol by the application of EAP procedure. This mechanism is established by sending a message "EAP Start" from STA towards the wireless Mesh router (step 15). The latter answers by a request for identity "EAP ID Req" at step 16. Afterward, STA sends its answer "EAP ID Rsp" containing its identity (ID_{STA}) and a Sequence Number encoded by its new key (K_{STA}) (step 17).

At step 18, WMR transmits this network access request "Access Req" and the received information from message "EAP ID Rsp" to the server AAA_{WLAN}. Thereafter, the application of the EAP method allows the checking of the validity of various equipments and in particular that of STA and the generation of the session key, noted K_S, which is a temporary key shared between the entities STA and WMR. In step 19, the decision of the Mesh server for acceptance or refusal of the access request is mentioned. In case of acceptance, AAA_{WLAN} sends a message "Access Accept / EAP Success" containing the session key (K_S) towards WMR. In its turn, the entity WMR transmits this decision in a message "EAP Success" containing the temporary key (K_S) to station STA (step 20).

4   Performances Evaluation

In order to test and evaluate the performances of the re-authentication protocol for a mobile station which migrates from LTE network towards a Mesh network, we have recourse to apply the method of simulation which presents a software tool that reveals with a fast, economic and effective solution for the test of networks. Indeed, there is a variety of existing simulators which allow the implementation and the analysis of the new integrated protocols performances in different networks topologies. On the other hand, a whole of new technologies are not yet well adjusted in the majority of the existing simulators in particular that of the Mesh and LTE networks. Following the need to appreciate the performances of the suggested re-authentication protocol, we have developed a simulator which allows the integration of the aspect of interworking LTE-Mesh.

This section is devoted to analyze the results of the protocol implementation in our simulator. First, we have defined some details to put into practice our architecture of Mesh network in which we will integrate the Mobile users (STAs) coming from LTE network. Indeed, this simulator specifies various parameters of this type of network which allows simulating its features to study the performances of the new authentication protocol suggested in 3G-WLAN environment. The selected network covers 300m×300m comprising 9 WMRs and a variable number of Mesh clients and also some visited nodes, which migrated from LTE network. To evaluate the performances of our solution, we will consider two types of traffic: voice and Web communication and many scenarios.

While referring on these types of communications as well as the parameters of simulation, we evaluate the simulation's results according many criteria:

- Handoff latency: the time passed between the change of point of attachment request and the association with the new WMR,
- Blocking rate: represents the number of blocked STA at handoff for the total number of STA which requests handoff,
- Loss rate: represents the number of lost packets for the total number of the emitted packets.

For the remainder of scenarios, we have concentrated on the comparison between Mesh stations behavior on the one hand and the visited stations behavior coming from LTE network on the other hand while basing on three criterion; handoff latency, loss rate and blocking rate. In fact, we have fixed the number of Mesh stations at 200 users and we have progressively increased the number of external users.

4.1   Handoff Latency

Figure 3 shows the results of simulation following the implementation of our new protocol of re-authentication in a heterogeneous wireless environment. First of all, the obtained curves illustrate an increase in the values of handoff latency which accompanies the multiplication in the number of external stations which migrate from the LTE network towards the Mesh network. This rise can be justified by the growth of handoff requests carried out on the level of WMN. Consequently, it may cause the heaviness and the overload of the WMRs nodes.

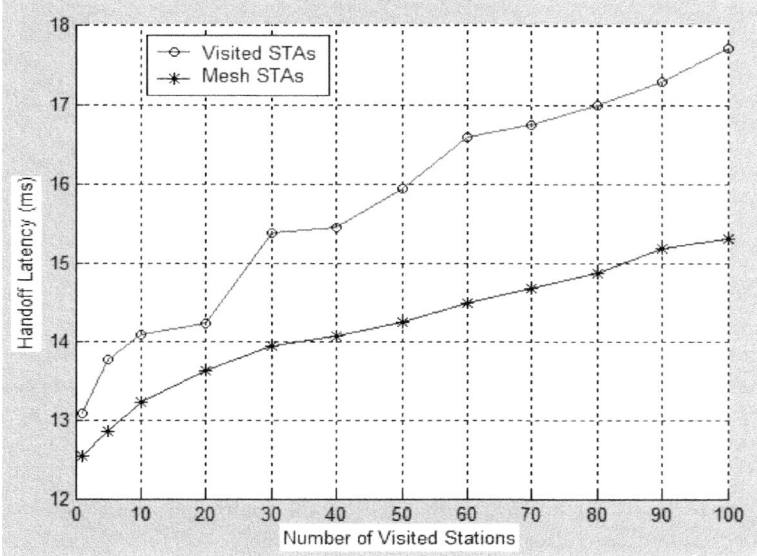

Fig. 3. Handoff latency according to the number of visited Stations for External and Mesh STAs

According to Figure 3, the values of handoff latency for visited STAs visitors are higher than those of Mesh STAs. Indeed, the procedure of legitimacy checking of the external stations which precedes the handoff execution costs more time than the local checking of the Mesh network stations. On the other hand, the obtained values for the visited stations are optimal in the case of LTE-Mesh interworking and with the constraints of quality of service.

4.2   Loss Rate

A second parameter, which allows the evaluation of the new protocols performances and the test of their network behavior, is the loss rate of the packets in communications. Following the increase in the values of handoff latency, mentioned in the preceding paragraph and in Figure 3, the risk the packets loss grows simultaneously.   Indeed, if the values of handoff delay exceed a well defined threshold, which depends on the specific application (example voice or Web application), the data in the course of transmission will be announced as lost.

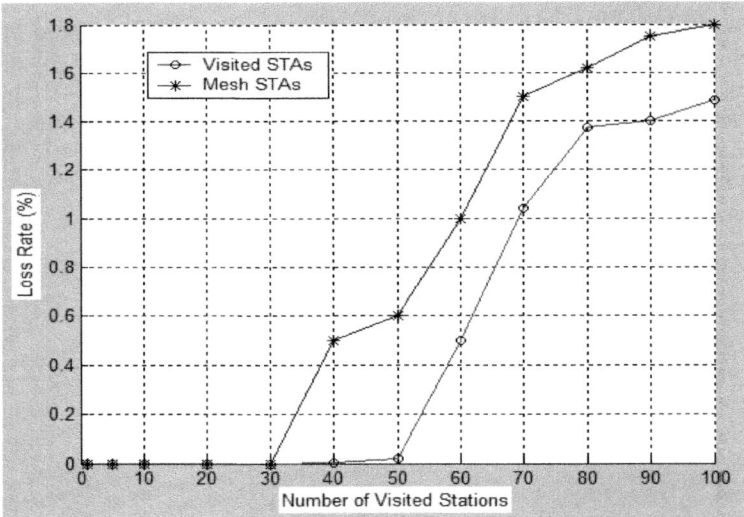

Fig. 4. Loss Rate according to the number of visited Stations for External and Mesh STAs

Figure 4 illustrates the obtained loss rates for the studied protocol with the progressive increase in the number of visited stations to Mesh network. The comparison between these two curves shows a light difference between the loss rates of the two categories of stations. For Mesh STAs, the obtained values by simulation are higher than those of visited STAs. This difference can be justified by the priority granted to the calls for handoff with those of the internal transmissions since these requests present more vulnerabilities and risks of attacks.

The obtained results demonstrate that for the first values of external stations (between 30 and 50), the loss rates are almost negligible. What reflects the effectiveness and the robustness of the implemented re-authentication protocol.

4.3   Blocking Rate

The loss of two successive packets of the same communication results in the blocking of station. This condition becomes increasingly demanding with real time applications. Consequently, the values of blocking rate are related to the results of the obtained loss rates. Figure 5 illustrates the rates of blocking gained with the application of the proposed re-authentication protocol in the environment of simulation for the heterogeneous network.

As in the case of loss rate, the curves begin with negligible values. Then, these rates increase gradually with the increase in the population of STAs, which come from the LTE network. Moreover, a comparison between the curves in Figure 5 shows that the blocking rates for the visited stations are less than those of the local stations that belong to the Mesh network. This result can be explained by the notion of the priority for the handoff service.

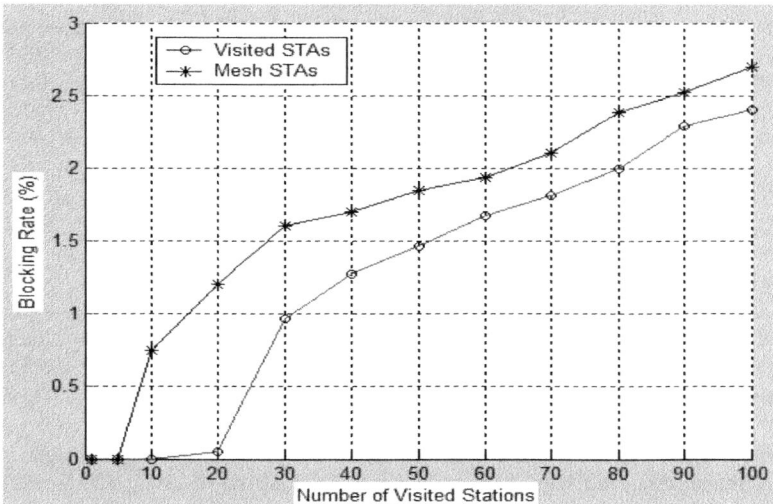

Fig. 5. Blocking Rate according to the number of visited Stations for External and Mesh STAs

5   Conclusion

A heterogeneous wireless network cannot prove its effectiveness and robustness only if it suitably treats its first challenge, which is the interworking between its various technologies. Indeed, this aspect makes it possible to better handle the management of mobility and security. In this paper, we were interested in the study of the LTE-Mesh interworking and in particular ensuring the security during handoff of the stations, which migrate from LTE network towards Mesh network. Therefore, we have proposed a re-authentication protocol for securing interworking and roaming 3GPP LTE and wireless local Mesh network. Moreover, the suggested solution is based on the use of a hybrid unit. This entity makes it possible to establish a preparative phase which precedes the execution of the handoff mechanism. Consequently, that makes it possible to ensure a better security and a mutual authentication between the various components of the heterogeneous network. These findings can be appreciated by the optimal simulation results and the resistance against replay attack. Compared with the existing protocols, our re-authentication protocol can reduce computational overhead, loss rate of transmitted packets and handoff blocking probability. As a future work, we can extend our study to resolve the problem of lack of security in various types of clients' handoff process between others different technologies in heterogeneous wireless network.

References

[1] 3GPP TS 23.402 v1.2.0, 3GPP System Architecture and Evolution (SAE): Architecture Enhancements for non-3GPP accesses, Release 8
[2] Holma, H., Toskala, A.: WCDMA for UMTS: Radio access for third generation mobile communication, pp. 2927–2932. John Wiley and Sons (2004), 978-1-4244-1997-5
[3] Zhang, Y., Luo, J., Hu, H.: Wireless Mesh Networking: Architectures, Protocols and Standards. CRC Press, Taylor & Francis LLC, USA (2006) ISBN: 0849373999
[4] 3GPP Technical Report 25.814, version 7.1.0, Physical Layer Aspect for Evolved Universal Terrestrial Radio Access (UTRA) (September 2006)
[5] Technical Specification 3rd Generation Partnership Project; Technical Specification Group Services and System Aspects; Security of Home Node B (HNB) / Home evolved Node B (HeNB) (3G TS 33.320 version 9.0.0 Release (2010),
http://www.3gpp.org/
[6] Arkko, J., Haverinen, H.: EAP-AKA authentication. RFC 4187 (January 2006)
[7] Rajavelsamy, R., Jeedigunta, V., Osok, S.: A Novel Method for Authentication Optimization during Handover in Heterogeneous Wireless Networks. In: Communication Systems Software and Middleware (COMSWARE 2007), pp. 1–5. IEEE (2007), 9719237, 1-4244-0613-7/09/
[8] Mun, H., Han, K., Kim, K.: 3G-WLAN Interworking: Security Analysis and New Authentication and Key Agreement based on EAP-AKA. IEEE (2009), 978-1-4244-2588-4/09/
[9] Ntantogian, C., Xenakis, C.: One-Pass EAP-AKA Authentication in 3G-WLAN Integrated Networks 48(4), 569–584 (2008)
[10] 3GPP TS 33.234 (v7.2.0) (2006) 3G security; WLAN interworking security; system description. Release 7 (September 2006)

A Distributed Framework
for Organizing an Internet of Things

Jamie Walters[1], Theo Kanter[1], and Enrico Savioli[2]

[1] Mid Sweden University - Sundsvall 85170, Sweden
{jamie.walters,theo.kanter}@miun.se
[2] University of Bologna - Bologna, Italy
enrico.savioli@studio.unibo.it

Abstract. Applications on a future *Internet of Things* require the provisioning of current, relevant and accurate context information to endpoints. Context information existing globally require organization into object-oriented models available locally in APIs as current, relevant and accurate views. Moreover, such applications require support for the highly dynamic interactions influencing continual changes in global context information. Existing approaches, such as the web services, are unable to provide this support partly due to the presupposed existence of a network service brokering context information, relying on DNS; or adopting a presence model for context which does not adequately scale. To this end, we propose a distributed framework for the interconnection of end-points and co-located agent entities, whereby agents are provided with local views of a relevant subset of global context information. We show how to achieve relevant local current views of global context information via ranking in an object-oriented context model. The distributed approach realizes the provisioning of context information in real-time, i.e., with predictable time bounds. Finally, we demonstrate the feasibility of the approach in a prototype based on P-Grid.

Keywords: mediasense, dcxp, p-grid, context proximity, sensor ranking, context metrics, context distance, sensor ranking.

1 Introduction

The increasing interest in the provisioning of applications and services that deliver experiences based on context mandates the continual research into methodologies, architectures and support for delivering the context information required. Constraints on service delivery with respects to real-time availability underpins any such solution. A future connected things infrastructure with an installed device base exceeding billions [1], requires support for a wide range of context centric experiences ranging from personalized and seamless media access, to intelligent commuting and environmental monitoring. This incorporate devices such as mobile phones, personal computers or IPTV boxes; all merging towards the paradigm of *everywhere computing* [2]; the seamlessly connected new world. As users navigate a vast and seemingly endless connected things infrastructure, it

J. Del Ser et al. (Eds.): MOBILIGHT 2011, LNICST 81, pp. 231–247, 2012.

becomes increasingly important to be provisioned with the relevant subsets of information required in order to be enabled with an experience relative to the users' current situation.

1.1 Scenario

John is constantly on the move both for business or pleasure. Within a future cityscape, he encounters multiple information points which maybe used to inform him of the state of his surroundings. Embedded into a digital ecosystem, he is capable of deriving enough information in support of the services wishing to effect changes or deliver him a unique context-based user experience. Such information include temperature, humidity and location as well more complex sources such as audiovisual devices, network connections or traffic conditions. With his smartphone, John is able to connect to and derive representations of context from these points in order to support his applications.

1.2 Analysis

Applications and services wishing to respond relevant to John's current context require this information to be organized and made available in globally accessible end-points. Approaches such as IMS [3] or Senseweb [4] enables the required support, brokering context information via web service portals on the Internet They are, however dependent on DNS as a means of locating service portals, users and applications. Issues with DNS availability due to DoS attacks and configuration errors raises questions about its continued suitability and prompting research into Distributed Hash Table (DHT) based overlays such as Chord [5], Pastry [6] and Tapestry [7] as possible replacements [8].

To this end, early work surrounding the MediaSense architecture implemented the DCXP protocol [9], a Chord based approach capable of provisioning John's context information in support of his dependent applications and services. The DCXP approach produced the ranges in response times deemed adequate enough to support real-time context dependent services. Furthermore, it proved that distributed systems were more capable of achieving this than approaches building on mobile or web services. Other approaches such as [10] explore the option of building context provisioning solutions using DHTs. However, while a DHT provides for a more scalable and resilient approach, it relies on deterministic hashing algorithms for achieving the distribution, indexing and locating of information.

Consequently, this places a limit on their ability to utilize self-organization towards realising a more homogeneous distribution of information located on the overlay. With regards to the persisting of context information, an additional disadvantage of DHTs is their inability to support queries of a range of values, critical in scenarios where John might be trying to locate a service in some approximate area or over a series of context values. While solutions such as [11] have sought to address this problem, this is not done natively, mandating the implementation of additional layers of complexity on the existing overlay. DHT-based implementations such as Chord are limited to a searching complexity of

$O(logN)$ [12]. However solutions seeking to provision John with current and relevant information mandates investigations into alternatives capable of realising improved response times.

We are further mandated to organize John's context information in subsets representative of the dynamic state of the sensor information to which he has access. As John changes state in real-time, such as entering a building or vehicle, he will encounter new sensors or sensor information. This requires that a schema of available context information be maintained and kept current; an evolving meta model as suggested in [13].

In approaching such a possible solution, the use of distributed relational databases such as in [14] and making use of the advanced research in database distribution would not be applicable. This, as such a distribution assumes communication reliability in order to maintain database integrity across wide area networks which cannot be guaranteed in heterogeneous mobile scenarios [15], [16]. This is also undermined by the fact that relational databases are highly inefficient for supporting real-time data manipulation, evolution and querying.

Current metric approaches such as Internet search engines consider the theory of connected things, however relative to static document content. A document's connectivity determines its relevance. This concept of ranking has been explored and used both in a centralized [17] as well as distributed [18] solutions. However, centralized solutions such as Google index only a tiny portion, less than 10 billion of the estimated 550 billion pages, on the relatively static Internet [19]. Any attempt to apply such a centralized solution to the ranking of sensors in an *Internet of Things* would be undermined by its inability to scale well. Distributed solutions based on the PageRank [20] concept would not scale well to accommodate highly dynamic document sets. Current *real-time* searches are realized by targeting known content providers, an approach that could not scale to accommodate the vast and mostly ad-hoc nature of a connect things infrastructure.

In this paper, we present an alternative approach that permits users to browse and locate relevant information in a vast and dynamic *Internet of Things*. Solutions capable of providing broad access to context information and enabling the derivation of context-based metrics. Such metrics include a sensor ranking and context proximity metric detailed further in Section 3. We therefore revisit the MediaSense Framework in an attempt to provide the approach required to support such user activities within real-time. Key to this is our new approach to the overlay structure, substituting Chord with a more resilient and robust P-Grid overlay.

For the remainder of this paper, Section 2 details the revised architecture; Section 3 presents an overview of the metrics while Section 4 summarizes our conclusion and future work.

2   The MediaSense Framework

In response to the shortfalls discussed in Section 1, the MediaSense framework seeks to create a solution towards supporting an *Internet of Things*. A solution

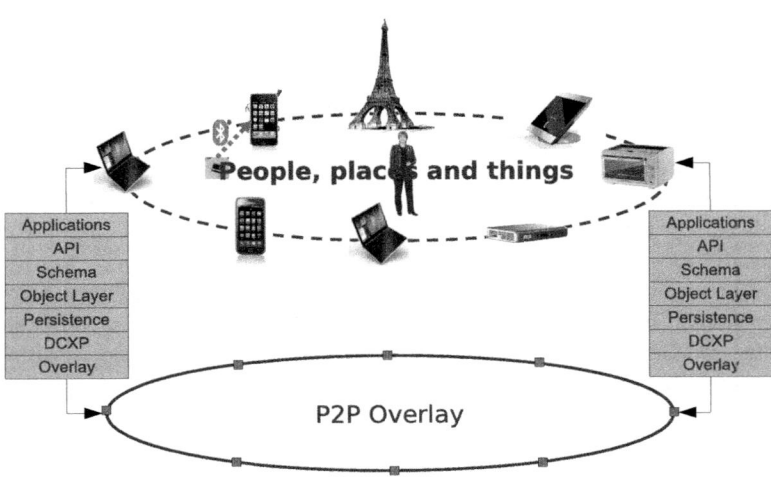

Fig. 1. The MediaSense Framework

in which presentities [21] are regarded as the focal point, enabling support for their dependent applications and services. This from the information gleaned from their interactions and associations within such a digital ecosystem.

Early work on the MediaSense framework realized an architecture for the distributed provisioning of user sensor information within real time constraints, providing the foundation for further work towards supporting the browsing of the dynamic data and interactions existing on an *Internet of Things*.

Our revised solution entails multiple layers of abstraction, enforcing layer logic independence. As a completely decentralized solution, nodes are permitted to freely participate, and realize the components required to supports its functions. Information Points, such as sensors, actuators or even an audio stream, can be registered by a node and be made available for usage at any layer across the solution. This is used to support an application layer exposed to applications and service providers for accessing the framework's functionalities. This masks the complexity of lower layers and their interactions, enabling users to focus on developing context objects, applications or services; having them transparently shared across the network with relative ease.

This components are illustrated in Figure 1 and are detailed in the remainder of this chapter.

2.1   The Overlay

The ability to provision context-centric user experiences from distributed information mandates an underpinning distributed overlay. Our previous work was supported by a Chord based [5] implementation used for maintaining the backbone communications as well as providing an indexing mechanism for information that must be persisted amongst participating nodes, also called

Context User Agents (CUAs). As with typical peer-to-peer protocol implementations, the nodes participating within the overlay act as entry points for applications and services wishing partake in the provisioning of sensor information across the overlay.

Citing issues with DHT based overlays as discussed in Section 1, we have substituted Chord with P-Grid [22] as the overlay of choice.

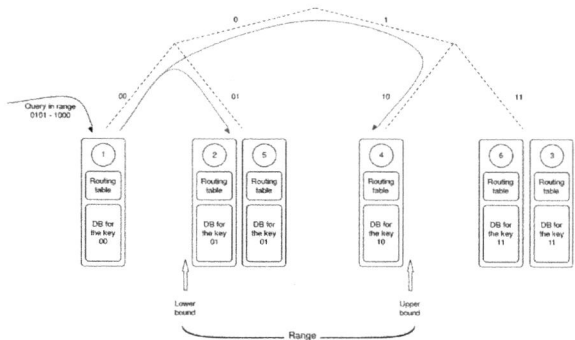

Fig. 2. The P-Grid Distributed Tree Structure

P-Grid. In an effort to increase the functionality of the architecture, we saw the need to move away from a DHT based implementation to overlay structures offering improved resilience, distribution and self organization. P-Grid, as the overlay of choice, shares a common behavior to DHT based implementation with respect to being able to index and locate information. P-Grid, however realizes a distributed binary tree, illustrated in Figure 2.

The key space is partitioned among all the nodes and organized into a tree structure with each node's location determined by the binary bit string representing the set of values for which the node is responsible. With this, it preserves the ordering on data and natively enables the resolution of both specific key, substring and range queries without any pre-existing knowledge. This is achievable with at most the same message complexity of most DHTs and has proven performance of $0.5logN$ versus $logN$ for a Chord based implementation.

The non-deterministic distribution of keys, offers improved resilience in the dynamic environments which are expected to exist in a future *Internet of Things* as it permits a more flexible self-organization in response to a very dynamic set of information. This is complemented with redundancy for fault tolerance; multiple nodes are assigned to the same key partition and nodes hold references to multiple partition holders. As a relatively future proof overlay, it readily permits future extensions and modifications.

While the overlay permits users to make sensor distributed information available his is not sufficient to support the interconnected and evolving *Internet of Things*. It is instead necessary to exploit the overlay as a building block and to rely on a protocol that focuses on data dissemination across interested nodes.

2.2   Distributed Content Exchange Protocol(DCXP) Layer

The DCXP layer primarily deals with the realization of such a protocol. One that permits users to publish and access context information in a structured manner, enabling the enforcement of some access controls. Residing immediately on top of the overlay, DCXP is an application-level protocol designed with the goal of enabling nodes to share and move context information between peers. As with the early implementations of the MediaSense architecture, this layer implements the core protocol employed in the provisioning of context information. These are summarized in Table 1.

One key departure is that the protocol is no longer used to maintain the overlay structure. With this task managed solely by the P-Grid overlay, network composition and state is abstracted from the protocol layer. The functionality of the protocol is now resigned to realising a distributed *publish/subscribe* interface to the resources available on the overlay. To this end, we introduce two new primitives: *TRANSFER* and *SET*.

The *TRANSFER* primitive provides for the ability to relocate context resources in support of applications and services.This is used by the object layer discussed later in Section 2.4. When a node requires a sensor resource that is not locally available, it makes a *TRANSFER* request, the object is then copied and used locally, reducing network messaging overhead, and the considerable demands that can be placed on nodes responsible for a context resource. Such an action could be achieved autonomously or in response to the application requirements.

The *SET* primitive enables interaction with actuators in end points completing the sensor/actuator pair, allowing applications to influence context in response to user preferences and context information.

The *publish/subscribe* functionality of the DCXP protocol is realized through a group of components namely, the Context User Agent, the Context Storage and the use of a Universal Context Identifier. These are discussed further in Section 2.2.

The resulting overlay, built using P-Grid, is used solely for maintaining the connection between nodes, and persisting the data registered by applications and services residing at the nodes. Such nodes, may not need to actively participate in the overlay, as would be the case with mobile devices over heterogeneous and sometimes unreliable connections.

Context User Agent - CUA. A computer wishing to participate within the context provisioning overlay is only required to implement an instance of the CUA. The CUA permits the seamless exchange of context information among sources and sinks, as well as the interaction with the sensors and actuators. Each CUA corresponds to a node on the virtual distributed tree described in section 2. Each CUA further contains some persistence in the form of object oriented databases (OODB) along with an API for creating applications and services consuming and responding to sensor information. It further provides the entry point for registering and resolving a UCI or a query across the CS.

Table 1. The Primitives of the Distributed eXchange Protocol

REGISTER_UCI	Registers a UCI along with the node which is responsible for it.
RESOLVE_UCI	Resolves a UCI to the node which is responsible for it.
GET	Fetches the current context value from the node responsible for a UCI. The reply is sent using a *NOTIFY*.
SET	Changes the current status of an actuator in an end point.
SUBSCRIBE	Makes a subscription request to the node responsible for a UCI, The node then sends a *NOTIFY* message containing the current context value, either at regular intervals or when the value changes.
NOTIFY	Notifies an interested node of the current context value associated with a specified UCI.
TRANSFER	Requests the manager of a resource to transfer responsibility to another node. This might be full responsibility or partial, where the requester re-creates a copy of the resource permitting improved real time performance.

Universal Context Identifier - UCI. All resources on the overlay is persisted using the universal context identifier (UCI) naming scheme. The UCI naming schema provides a URI inspired naming schema with the following syntax:

$$dcxp://user[:password]@domain[/path[?options]]$$

where *dcxp* is the new URI scheme name and *domain* is a Fully Qualified Domain Name (FQDN) relating to where the CI is located. The *user* and *password* arguments are optionally used as a means of authorization. The *path* adheres to the context information namespace hierarchy, permitting the organization and sorting of the items in a logical sense while *options* facilitates further modifiers in the form of *parameter=value* pairs.

An example of a fully qualified UCI adhering to this would be:

$$dcxp://andeen.mccarthy@miun.se/weather/temp?unit=celsius$$

Resources are registered with the Context Storage component residing in the overlay.

Context Storage - CS. Previous implementations, enabled a Context Storage mechanism residing on top of the overlay. With this approach, enabling more useful searches such as range queries involved additional layers of complexity as discussed in Section 1. In order to exploit these native characteristics of the overlay, the CS is now built into the overlay. The key role of the CS remains that of resolving UCIs to physical end point addresses of the responsible nodes, behaving similarly to a dynamic DNS service. Here the substring and range queries enable the locating of resources without needing to know the fully qualified UCI.

Additionally, the CS stores the current value associated with a resource. Where a resource consists of multiple dimensions such as GPS coordinates, each attribute dimension is stored separately. We separate dimensions to further enable independent queries over any value constituting a context information source. An end point is then free to reconstitute and compare the entire n-dimensional value or any valid subset. We further benefit from the overlay's order preservation and range query properties, permitting the acquisition of useful data in support of context metrics or similarity functions.

2.3   Persistence

The CS along with the DCXP protocol enables distributed access to only current context information. Any new values reported in connection to a UCI or context dimension supersedes the current value. This prevents applications from using expired or stale data. Historical information was previously persisted to a centralized relational database by the nodes. The dynamic nature of context information mandates the storage of historical information both on values and the actual context objects. However storing versions of information as different entries on the overlay, would undermine its performance as a direct result of the vast number of attributes persisted at each node; this would be increased by a factor equal to the number of versions.

We therefore solve this problem by using an additional layer; the persistence layer. This consists of a collection of localized Object Oriented Databases residing at each CUA. Each node stores the set of objects that are created locally by the application and services co-located with the CUA. They further store a collection of objects that are being used by these applications and services but originate at a remote node within the overlay. We also persist all observed values for each context dimension permitting temporal views of data evolution, trend and pattern discovery.

Before an object is persisted locally, an attempt is made to persist it on the overlay. If this is successful, then the object being stored is guaranteed to be globally unique and is then stored locally with its UCI on the overlay. If this is not successful, the UCI is being stored already exists and the persistence operation fails. This ensures that locally persisted objects are always globally unique. All objects that are being used by applications and services local to the CUA are stored locally, these contribute to some local schema (Section 2.5) being used at the node.

2.4   Object Layer

The lower layers of the framework utilize only context attributes in the realization of the required functionalities. These are persisted as primitive values on the CS and made accessible through the publish/subscribe interface. Applications residing above the API, however require context objects with which to realize support for users and services.

The Object Layer resolves this gap by permitting the composition of context attributes into context objects. Such as a the latitude and longitude of a GPS sensor being exposed as a 2D location object attached to a presentity. There is no requirement for an object to recompose all the underlying values of a sensor or for all the values to originate from the same sensor. Such composition is opened to implementation on the object layer. Application developers further work only with objects without needing to consider the UCI or location of the primitive sensor attributes and values.

An application requiring use of an object, makes a *TRANSFER* request. This retrieves a description of the object and constructs it locally and made available to the application or service. When an object is created or retrieved, the object layer constructs the object from the attributes it describes and realizes a subscription to all the sensors contributing to its composition. An attempt to retrieve a value would result in a request for the most up-to-date value stored in the CI, for both local or remote objects. An attempt to modify a value would be forwarded to the CI in order to have the value updated. It the object does not reside at the CUA attempting to modify its value, this fails as the modification must be done local to the owner of the object.

Additionally, an object initialized on a remote node, realizes a subscription to the object's home node. If the object is modifies, all nodes with an instance of the object are notified and they may update the object as required. This provides for an always up to date copy of a sensor object on the object layer.

Objects relevant to a presentity are grouped together and presented as a schema made available to applications through the API.

2.5 Schema Layer

In support of localized object views, we introduce the concept of a *Context Schema* [23], defined as:

The collection of information points associated with and contributing to a presentity's current context

where an *Information Point* is defined as:

Any source providing information about the context of an entity or any sink capable of accepting an input effecting changes to an entity's context

Within the schema layer, such a schema is attached to a presentity and encapsulates all the information points and the relationships related to a presentity. An application or service with a requirement to deliver some user-context centric experience subscribes to the current schema description; it realizes a collection of information objects underpinned by a *publish/subscribe* interface to the end points described by the schema. As a presentity traverses a connected things infrastructure it discovers new entities and consequently updates its schema to reflect this. As a result, all subscribing end points receive an updated schema and can adjust their services to accommodate this.

This addresses the scenario where a context network being highly dynamic, rapidly evolves, mandating the need for an historical representation of interactions. This permits us to examine the behavior of entities on an *Internet of Things*, deriving metrics representative of the interactions among these entities.

3   Context Metrics

The support of applications on an *Internet of Things*, mandates the provisioning of current, relevant and accurate context information to end-points. Such information must be derived and represented as a local subset of the global context information domain. With this, users and applications residing at endpoints are capable of having access to relevant information within some predictable window, explore and build dynamic context centric relationships in response to changes in state and context. In this chapter we discuss two context metrics with respects to the implementation detailed in Section 2. However, while these metrics may be implemented on any solutions capable of providing the required data, our architecture provides the most optimal support in a distributed environment.

Firstly, we need to identify similarities among entities providing a base for discovering new entities; and secondly, the need to be able to identify important and useful sources of context information, providing entities with the information needed to evaluate the reliability of context information sources as well as the resulting relationships established over this information.

3.1   Context Proximity

One metric we consider desirable in browsing a network of information, is a context proximity metric, a measure of the *distance* between presentities considering all expressions of context as illustrated in Figure 3. With this, we can create dynamic user-based context-centric clusters of information points and presentities that are capable of enabling applications and services to provide user experiences based on current context.

In Figure 3, P_1 while connected to S_2, derives an implicit but existing relation to P_2 via S_1. The implication being that their connection suggests that P_1 and P_2 share, to some extent a similar context. If S_1 and S_2 are expressing the same context indicator type, i.e. they are two information points of the same type such as a temperature sensor, then P_1 shares a context similar to that of P_2 by a function of the difference between S_1 and S_2; their sensor value proximity.

With this assumption, we explore our context architecture for context information sources that lay within X_S1; the context proximity limit of P_1.

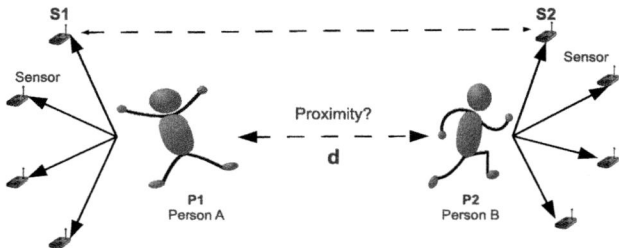

Fig. 3. Context Proximity

We envision that applications will be able to define limits of X_S1 such as: *find all people within 3km with a temperature less than* 5 °C *difference* permitting us to obtain parameters needed to derive the entity sets. Since our solution must remain fully distributed, we located initial nodes by issuing a search using the the range querying function of the underlying P-Grid overlay. This returns a list of entities with respect to the query and constructs a running query at each peer with the following constraints:

1. The peer is responsible for a sensor fitting the criteria of the search
2. The peer is responsible for a sensor S_i with a range such that the set of sensors fulfilling the query from S_1 would be a subset of a query from S_i.

Each peer is then required to:

1. Forward the sensors matching the standing query to S_1
2. Forward the query from S_1 to any node it encounters that matches 1 & 2 above.

This results in a group sensors being returned where for each group G:

$$G = \{S : S \in D : (|V_{S_1} - V_{S_i}| \leq X)\} \tag{1}$$

here, V_{S1} is the current value of S_1 and V_{Si} is the current value of S_i within a domain D.

This is a dynamic set of information points with respect to P_1 its context dimension S_1, that continually evolves to reflect the addition or removal of sensors with respect to their current values as nodes respond to the query.

We consider the fact that not all instances of S_i lie within the same proximity to S_1. This implies that S_1 shares a closer context with some members of G and subsequently those members must be given a higher preference with regards to any context dependent application or services wishing to find context information points in support of delivering some optimal user experience. Noting the varying scales for each sensor, we normalize the euclidean distances with respect to their scales and the distance from S_1 such that:

$$R_{S_i} = f(S_i) = (1 - |V_{S_i} - V_{S_1}| \cdot X_{S_1}{}^{-1}) \; \textbf{\textit{where}} \;\; 0 \leq R_{S_i} \leq 1 \tag{2}$$

A value of 0 being at the edge and 1 being identical to X_{S_1}. This value is useful for applying weighting to the edges connecting S_1 to S_i and subsequently the edges connecting to P_1.

Figure 4 illustrates a possible resulting set of such implicit connections with some degree of context similarity owing to the fact that their underlying sensors are within close proximity. By deriving the degree of this closeness, we can obtain a set of presentities within proximity. Consider P_1 and P_2 connected to sets of sensors such that:

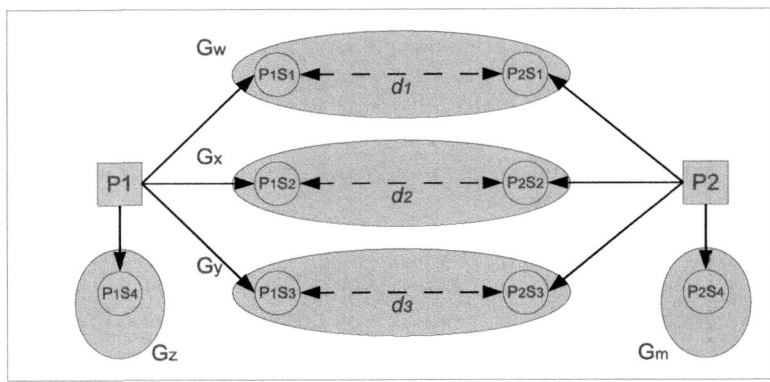

Fig. 4. Determining Presentity Proximity

$$P_1 = \{G_w, G_x, G_y, G_z\} \text{ and } P_2 = \{G_w, G_x, G_y, G_m\}$$

Based on this, we determine PS, the presentity similarity as Jaccard similarity of the set of sensors shared by P_1 and P_2:

$$PS(P_1, P_2) = \frac{|P_1 \cap P_2|}{|P_1 \cup P_2|} = \frac{|G_w, G_x, G_y|}{|G_w, G_x, G_y, G_m, G_z|} \tag{3}$$

This permits the comparison of values that cannot easily be measured discretely such as favorite color, mood, etc. In these expressions of context, we are unable to perform discrete distance measurements, however we can provide mechanisms for grouping together similar values which might equate to a user saying: *I like red, but pink, and purple are also acceptable alternatives.* Therefore if a presentity was comprised entirely of non-discrete values, we could still derive a measurement of distance based on the grouping of these values and finding the degree of similarity between the presentities. An end point could define the dimensions of context considered when calculating similarity, such that for an application only interested distance, temperature and humidity, where P_2 had only two dimensions:

$$PS(P_1, P_2) = \frac{|G_w, G_x|}{|G_w, G_x, G_y|} \tag{4}$$

Further to this, we consider the equation in 2 and adjust the value derived in equation 4 to reflect the distance between the underlying expressions of context supporting the presentities. This is adjusted by a factor of the average of the rank of all the connections between P_1 and P_2. Therefore, we state the distance between two presentities, PR to be:

$$PR = \begin{cases} PS \cdot \dfrac{\sum_{n=0}^{k} R_{Sn}}{k} & , i=0 \\[3mm] PS \cdot \dfrac{\sum_{n=0}^{k} R_{Sn} \cdot P_{Sn}}{\sum_{n=0}^{k} P_{Sn}} & , i>0 \end{cases} \tag{5}$$

where i is the number of dimension restrictions, P, indicated by the application or service. When applying such restrictions, all dimensions must be accounted for. Each unaccounted for dimension will be ignored, effectively given a priority of 0. We provide for this as we consider that an application or service will be able to indicate context dimension priorities, eg. *find all persons within a context proximity of 0.7, prioritize by distance, then temperature* or *find all persons within 5km, prioritize by distance then temperature*. The resulting is a more relevant subset of presentities and information sources accumulated in an end point close to a presentity, application or service.

3.2   Sensor Ranking

While traversing an *Internet of Things*, users will be excepted to encounter masses of information sources such as sensors. A supporting solution should be able to provide application developers and users with as much information as possible in order to select the most relevant and recommended sources available.

Our ranking algorithm consists of two main components, illustrated in Figure 5. Firstly we need to determine the local ranking value for s with respects to P_i. We then need to aggregate the global ranking value for s.

Localized Ranking. In approaching this problem, we adapt a modified version of the *Inverse Document Frequency* algorithm [24]. This is shown in Equation (6) with a sensor s, and a presentity P. SR is the sensor ranking of the sensor s, R is the corpus, the total collection of schemata relative to presentity P with r being all the schemata relevant to P containing a reference to sensor s.

$$SR_s{}^{(P)} = log \frac{|R|}{\{r : s \in r\}} \tag{6}$$

This provides us with a representative metric as to the importance of sensor s relative to P. We consider further, that there exists scenarios where some presentities will be less dynamic or mobile with respect to s. Such an example might be a sensor located in a store; the employees working in the store will by default almost always utilize the sensors that are local to the store accounting for a disproportionately higher value for SR: In such scenarios, taking:

$$log \frac{|R|}{\{r : s_i \in r\}} \tag{7}$$

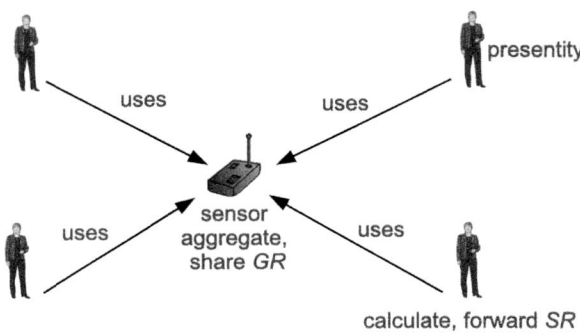

Fig. 5. Determining Sensor Ranking

considers more dynamic presentities traversing an *Internet of Things*. Such as a person that travels around the city interacting with more sensors and subsequently creating more context schemata in fulfillment of service delivery. This is represented by larger ratio of R to $\{r : s_i \in r\}$. Such presentities we argue, indicate a more accurate ranking for s, relative to the wider sensor ecosystem.

We could also calculate a ranking for sensors with respect to some time duration of interest, t, by limiting the schemata used in calculation to those created within t.

Global Aggregation. The second component of our approach is a global aggregation of all the local ranking values SR assigned to s. To achieve this, we calculate the Global Ranking GR by finding the sum of all SR of s such that:

$$GR_s = \sum_{k=0}^{n} SR_k^{(P)} \tag{8}$$

This value is continually calculated as new schemata referencing s are created. We further take into consideration the owner of s, the presentity or domain where it resides or to which it belongs. This we regard as the domain D and assign it a value equal to the average ranking of all the sensors belonging to D. This we call DR, the ranking of D. This value is important to us as it would permit us to identify more connected and important spaces such as domains, buildings or just a collection of deployed sensors. We calculate DR as:

$$DR = \frac{\sum_{k=0}^{n} SR_k^{(P)}}{k} \tag{9}$$

This ranking value can be made available on the overlay, supporting range queries across ranking values and readily accessible. No centralization is required for ranking to occur and new values are updated in real time. The resulting values can be used as indicators of relevancy or importance of sensors on an *Internet of Things*.

4   Conclusion

Users, applications and services on an *Internet of Things* demand sensor information in support of realising context-centric user experiences. Such information must be dynamically organized and provisioned as accurate and relevant subsets of global information in end-points. Additionally, such provisioning must be liberated from the assumption network services enabling the brokering of information or presence models that do no scale well.

In response to this, we presented further work on the MediaSense framework, detailing its re-implementation towards a more robust support for the dynamic properties inherent in provisioning context information among heterogeneous end-points. As the new overlay of choice, P-Grid realizes a more unstructured behavior over DHT solutions. It achieves this by implementing a non-deterministic key-space organized as a distributed binary tree. This, while preserving the ordering of values and allowing us the ability to perform more complex range queries than DHT based solutions.

The ability to perform such queries along with the self-organization behavior of the overlay permits us implement algorithms for deriving metrics ranking entities both personally and globally. Unlike cloud solutions such as Google [17] which are updated at regular intervals, this permit the ranking of resources as they become available. As nodes are added, they respond to any relevant standing queries created by the existing querying nodes. A node recalculates the proximity and ranking of sensors from this continually updated and relevant subset. A We further exploit this overlay to persist ranking values for entities providing a global access to an entity's reputation.

Future work on towards this includes deriving a large sample set capable of generating values for testing and bench marking the performance of the solution. The creation of a simulator for our solution is still required in order to verify scalablity and performance within a large scale deployment. The solution further requires the implementation of mobile nodes enabling performance measurements reflective of its usage in a real world scenario. This would include creating applications on mobile phones or computers. Other work includes the ability to conduct more knowledge discovery by exploiting the properties of the overlay and further work on extending and improving both algorithms.

References

1. Sundmaeker, H., Guillemin, P., Friess, P., Woelfflé, S.: Vision and Challenges for Realising the Internet of Things. In: Cluster of European Research Projects on the Internet of Things (CERP-IoT) (March 2010)
2. Lee, J., Song, J., Kim, H., Choi, J., Yun, M.: A User-Centered Approach for Ubiquitous Service Evaluation: An Evaluation Metrics Focused on Human-System Interaction Capability. In: Lee, S., Choo, H., Ha, S., Shin, I.C. (eds.) APCHI 2008. LNCS, vol. 5068, pp. 21–29. Springer, Heidelberg (2008)
3. Gonzalo, C.: 3G IP Multimedia Subsystem (IMS): Merging the Internet and the Cellular World, p. 381 (2005)

4. Kansal, A., Nath, S., Liu, J., Zhao, F.: SenseWeb: An Infrastructure for Shared Sensing. IEEE Multimedia 14(4), 8–13 (2007)
5. Stoica, I., Morris, R., Karger, D., Kaashoek, M.F., Balakrishnan, H.: Chord: A scalable peer-to-peer lookup service for internet applications. In: Proceedings of the 2001 Conference on Applications, Technologies, Architectures, and Protocols for Computer Communications, vol. 31, pp. 149–160. ACM (2001)
6. Rowstron, A., Druschel, P.: Pastry: Scalable, distributed object location and routing for large-scale peer-to-peer systems. Design (2001)
7. Zhao, B.Y., Huang, L., Stribling, J., Rhea, S.C., Joseph, A.D., Kubiatowicz, J.D.: Tapestry: A resilient global-scale overlay for service deployment. IEEE Journal on Selected Areas in Communications 22(1), 41–53 (2004)
8. Pappas, V., Massey, D., Terzis, A.: A comparative study of the DNS design with DHT-based alternatives. The Proceedings of IEEE, 1–13 (April 2006)
9. Kanter, T., Pettersson, S., Forsstrom, S., Kardeby, V., Norling, R., Walters, J., Osterberg, P.: Distributed context support for ubiquitous mobile awareness services. In: 2009 Fourth International Conference on Communications and Networking in China, pp. 1–5. IEEE (August 2009)
10. Baloch, R.A., Crespi, N.: Addressing context dependency using profile context in overlay networks. In: 2010 7th IEEE Consumer Communications and Networking Conference (CCNC), pp. 1–5. IEEE (January 2010)
11. Ratnasamy, S., Hellerstein, J.M., Shenker, S.: Range Queries over DHTs. IRB-TR-03-009 Intel Corporation (2003)
12. Ding, G., Bhargava, B.: Peer-to-peer file-sharing over mobile ad hoc networks. In: Proceedings of the Second IEEE Annual Conference on Pervasive Computing and Communications Workshops, 2004, pp. 104–108. IEEE (2004)
13. Kanter, T.G.: Going wireless, enabling an adaptive and extensible environment. Mobile Networks and Applications 8(1), 37 (2003)
14. Stonebraker, M., Aoki, P.M., Litwin, W., Pfeffer, A., Sah, A., Sidell, J., Staelin, C., Yu, A.: Mariposa: a wide-area distributed database system. The VLDB Journal The International Journal on Very Large Data Bases 5(1), 48–63 (1996)
15. Barbara, D.: Mobile computing and databases-a survey. IEEE Transactions on Knowledge and Data Engineering 11(1), 108–117 (1999)
16. Ulusoy, O.: Transaction processing in distributed active real-time database systems. Journal of Systems and Software 42(3), 247–262 (1998)
17. Google (2010), http://www.google.com
18. Zhu, Y., Ye, S., Li, X.: Distributed PageRank computation based on iterative aggregation-disaggregation methods. In: Proceedings of the 14th ACM International Conference on Information and Knowledge Management, pp. 578–585. ACM, New York (2005)
19. Li, J., Loo, B., Hellerstein, J., Kaashoek, M., Karger, D., Morris, R.: On the Feasibility of Peer-to-Peer Web Indexing and Search. In: Kaashoek, M.F., Stoica, I. (eds.) IPTPS 2003. LNCS, vol. 2735, pp. 207–215. Springer, Heidelberg (2003)
20. Sankaralingam, K., Sethumadhavan, S., Browne, J.C.: Distributed pagerank for p2p systems. In: Proceedings of the 12th IEEE International Symposium on High Performance Distributed Computing, pp. 58–68. IEEE (2003)
21. Plaice, J., Kropf, P.G., Schulthess, P., Slonim, J.: DCW 2002. LNCS, vol. 2468, pp. 345–392. Springer, Heidelberg (2002)

22. Aberer, K., Cudré-Mauroux, P., Datta, A., Despotovic, Z., Hauswirth, M., Punceva, M., Schmidt, R.: P-Grid. ACM SIGMOD Record 32(3), 29 (2003)
23. Walters, J., Kanter, T., Norling, R.: Distributed Context Models in Support of Ubiquitous Mobile Awareness Services. In: Par, G., Morrow, P. (eds.) S-CUBE 2010. LNICST, vol. 57, pp. 121–134. Springer, Heidelberg (2011)
24. Robertson, S.: Understanding inverse document frequency: on theoretical arguments for IDF. Journal of Documentation 60(5), 503–520 (2004)

An Analysis of Smart Antenna Usage
for WiMAX Vehicular Communications

Mihai Constantinescu, Eugen Borcoci, Tinku Rasheed, and David Hayes

Polytechnic University of Bucharest, Romania
Create-Net, Italy
Plasma Antennas, U.K.
{mihai.constantinescu,eugen.borcoci}@elcom.pub.ro,
tinku.rasheed@create-net.org,
dh@plasmaantennas.com

Abstract. WiMAX communications for vehicular use is a topic of significant interest in the research and industry communities, for both V2V (vehicle-to-vehicle) and V2I (vehicle-to-infrastructure) scenarios. This paper presents results of an experimental, simulation-based study, for mobile WiMAX V2I communications for different mobile station (MS) speeds. The scenario results are applicable to both omni-directional and beam forming smart antenna use. This study describes a database consolidation process in order to determine multi-dimensional regions where different lower layer parameters have influence on the overall performance of WiMAX V2I communications. Based on the multi-dimensional graphs, optimal parameter sets and network topology information can be provided by the network operator to vehicular MS, (e.g. a car or a train), providing essential support for the mobile station's smart antenna tracking systems and allow adaptation of the major WiMAX parameters to its speed and network topology.

Keywords: WiMAX vehicular, smart antenna, cross-layer optimization.

1 Introduction

WiMAX communications for vehicular use has gained continuous attention from the research community, for both V2V (vehicle-to-vehicle) and V2I (vehicle-to-infrastructure) applications. Given the high number of WiMAX physical and MAC layer parameters, that influence the overall performance in mobility scenarios, experimental simulation studies of complex scenarios are very helpful in determining the combined effect of such parameters.

In Ref [4], the authors propose a study of the feasibility of using WiMAX for V2I communication on a static setting in urban environment and perform a comparison with use of WiFi. Pegasus, a system providing wireless connection roaming at high rates over multiple interfaces, uses network information for user locations and used paths for effective and balanced utilization of the available bandwidth [5]. Ref [6] evaluates an architecture based on IEEE 802.21 framework, integrating both mobility and Quality of Service (QoS) mechanisms, through an advanced mobility scenario using a real

J. Del Ser et al. (Eds.): MOBILIGHT 2011, LNICST 81, pp. 248–257, 2012.
© Institute for Computer Sciences, Social Informatics and Telecommunications Engineering 2012

WiMAX testbed. In Ref [7], mobile WiMAX trials are analyzed to investigate the vehicular downlink performance for a number of on-car antenna configurations.

This paper presents a detailed experimental study related to WiMAX V2I communications using an OPNET™ v14.5 [1] simulation environment. Our aim is to determine multi-dimensional regions where different lower layers parameters have influence on the overall handover performance in mobility scenarios related to WiMAX V2I communications. The OPNET WiMAX simulations results in our work are consolidated in multidimensional graphs, named as: decision spaces. These decision spaces present, in aggregated form, the performance obtained in a WiMAX V2I mobility scenario, related on a specific trajectory. The results can be used as a method of optimizing vehicular communications by guiding the tracking and scanning algorithm and controlling the hand-over (HO) decisions between the base-stations.

The research leading to these results are supported partially by the European Community's Seventh Framework Programme (FP7/2007-2013) within the framework of the SMART-Net project (grant number 223937). The SMART-Net (SMART-antenna multimode wireless mesh Network) project [13], based on which the architectural requirements are drawn for this work, has among its objectives, studies and experimental analysis of hybrid mesh networks, including mobility issues. SMART-Net project is developing a heterogeneous access network solution incorporating multi-radio access technologies (RAT) and smart antennas to offer advanced wireless broadband solutions [2][3]. Terminal/user mobility is an integral part of the architecture and it is supported by a mobility management and control framework. The networking technologies considered in SMART-Net are IEEE 802.11x, 802.16x and 802.15x. Specifically micro and macro – mobility are studied and innovative solutions are targeted both for horizontal and vertical handover types. The IEEE 802.16e/WiMAX mobility solutions constitute a major area of work and investigation of the project.

The paper is organized as follows: Section 2 details the scanning, tracking and the HO algorithm for smart antenna use, Section 3 describes the simulations and the decision spaces in the context of smart antenna use. Section 4 outlines the conclusions, identifies a number of open issues and suggests some future work.

2   Scanning, Tracking and the HO Algorithm for Smart Antenna Use

The work presented in this paper is a continuation of a set of detailed studies on WiMAX mobility. The initial results have been shown in a study of HO performance for WiMAX mobility in [8], continued with an WiMAX HO conditions evaluation towards enhancement through cross-layer interaction proposed in [9], together with a SIP-based cross-layer optimization for WiMAX Hard HO method, described in [10]. In depth analysis of WiMAX V2I communications are presented in [11].

In this paper, simulation results are consolidated into multi-dimensional regions, where different lower layers parameters have influence on the overall handover performance in mobility scenarios. The BS and MS scanning processes are a combination of omni-directional and smart antenna scan modes designed to adapt the beam selection according to the MS's movement through the network of BSs and to assure the directed beam alignment between the linked MS and BS pairs.

Thus the smart antenna mode, a mapping table termed as scanning table, where the BS's corresponding beam will be used during MS communications with the BS (Table 1). That table will have one entry for each BS: BS ID – corresponding beam – SNR value measured on omni-directional mode.

Table 1. Scanning table

BS_ID	Corresponding beam	SNR_omni-directional mode
1	11	52dB
2	1	47dB

The corresponding beams are found when the MS scan is in its omni-directional mode. During that period the MS discovers all BSs within the current coverage area and identifies the corresponding beam/angle for each BS. The first mapping table is updated on each scan.

A second mapping table, named HO mapping table, based on the first one, will provide essential information for smart HO (Table 2). That table will have 6 entries:

- Serving BS ID – corresponding beam - SNR value measured on smart antenna mode;
- Serving BS ID – predicted beam - SNR value estimated for smart antenna mode;
- Back-up BS ID – corresponding beam - SNR value measured on smart antenna mode;
- Back-up BS ID – predicted beam - SNR value estimated for smart antenna mode;
- Target BS ID – corresponding beam - SNR value measured on smart antenna mode;
- Target BS ID – predicted beam - SNR value estimated for smart antenna mode.

Table 2. HO mapping table

BS_ID		beam		SNR_smart antenna mode
Serving BS_ID	1	Corresponding beam	11	58.2dB
		Predicted beam	10	60.7dB
Back-up BS_ID	2	Corresponding beam	1	55.8dB
		Predicted beam	12	58.3dB
Target BS_ID	3	Corresponding beam	4	46.9dB
		Predicted beam	3	49.4dB

Each corresponding beam value for each BS is taken from the first table. A scanning in smart antenna mode along that beam will provides the related SNR value.

The serving BS is the BS where the MS is currently connected. The back-up BS is the BS where the MS could be connected in the case of the serving BS having unexpected unavailability. That BS is on coverage area, and MS keeps the related data for back-up situations.

The target BS is the BS chosen as the target for the next BS. Due to the added gain obtained during the smart antenna mode, the MS has a larger effective coverage area than a standard omni-directional antenna. It can be noted that in these situations, the number of HOs could be advantageously decreased, based on a smart HO algorithm. The MS will decide to skip some BSs along its trajectory, so the target BS could be different from next BS with maximum SNR.

The BSs locations are available from the network operator using periodical updates. The update mechanism is outside the scope of this paper. However, it is believed the solution proposed in [10] could also be used successfully for the information exchange update routines.

That is, based on GPS information, the MS identifies its geographical position, direction, speed, and computes the angle of the beam for each BS under its coverage. Each current beam assignment is made during the scanning process and recorded into first mapping table, the MS will use the computation results to predict the angle/beam for the next position/scanning time stamp. The scanning time stamps, as mentioned, are determined one at a time, and the MS position will be established and related to the next scanning time stamp reference.

Simulations have demonstrated that the proposed predicted beam selection process is very useful, especially for high speed mobile users, when the corresponding angle for each BS is changing quickly and there is a need for a dense scanning process. In fact, based on predicted simulation results, it is noted that the scanning duration could be decreased, so moderating the application's throughput. The scanning, tracking and smart handover for a mobile user using a smart antenna is depicted in Fig. 1.

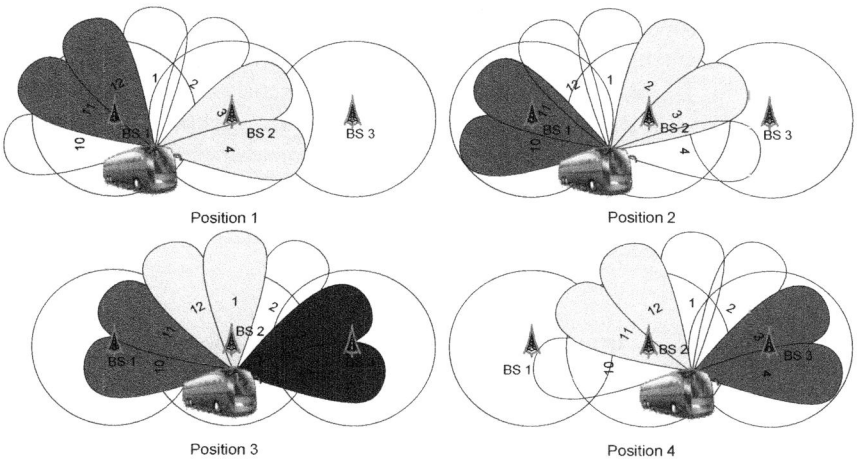

Fig. 1. Serving BS, scanned BS and HO target BS for smart antenna mode

Position 1:

- MS on BS_1 coverage for both smart antenna and omni-directional mode, next to BS_2 coverage in omni-directional mode;

- MS attached to BS_1, keeping BS_2 as back-up;
- Serving BS: BS_1, corresponding beam: 12, predicted beam: 11 (red);
- Back-up BS: BS_2, corresponding beam:4, predicted beam: 3 (yellow);
- If in omni-directional mode, MS should initiate the scanning process. In smart antenna mode, BS_1 still far from HO SNR threshold, so the target BS is not available at that moment;
- Target BS: not used.
- However, in case of forced HO due unexpected serving BS failure, the MS will initiate HO to the back-up BS, currently BS2.

Position 2:

- MS next to BS_1 coverage limit for omni-directional mode, and on BS_2 coverage for omni-directional mode;
- MS attached to BS_1, keeping BS_2 as back-up;
- Serving BS: BS_1, corresponding beam: 11, predicted beam: 10 (red);
- Back-up BS: BS_2, corresponding beam: 3, predicted beam: 2 (yellow);
- If in omni-directional mode, MS should initiate the HO process, serving BS: BS_1, target BS: BS_2.
- In smart antenna mode, BS_1 still far from HO SNR threshold. However, the scanning process should be started soon (position 3);
- Target BS: not used.

Position 3:

- MS on BS_1 coverage for smart antenna mode only, on BS_2 coverage for omni-directional, and on BS_3 coverage for smart antenna mode;
- MS attached to BS_1, keeping BS_2 as back-up;
- BS_3 available on smart antenna mode and selected as possible HO target;
- Serving BS: BS_1, corresponding beam: 11, predicted beam: 10 (red);
- Back-up BS: BS_2 , corresponding beam: 1, predicted beam: 2 (yellow);
- Target BS: BS_3, corresponding beam:4, predicted beam: 3 (blue);
- BS_1 near HO SNR threshold, so MS should initiate the HO process: serving BS: BS_1, target BS: BS_3. the HO is a smart one, due BS_2 was skipped from HO process.

Position 4:

- MS on BS_3 coverage for smart antenna mode and next to coverage limit for omni-directional mode, on BS_2 coverage in omni-directional mode;
- MS attached to BS_3, keeping BS_2 as back-up;
- Serving BS: BS_3, corresponding beam: 4, predicted beam 3 (red);
- Back-up BS: BS_2, corresponding beam: 12, predicted beam: 11 (yellow);
- Target BS: not used.

Moving through the BS network, the MS will discover BS_4, and the conditions from Position 1 will be encountered again, having BS_3 as serving BS and BS_4 as back-up BS (not shown in Fig.1).

Fig.2.A and B present some advantages of smart antenna use. From fig.2.A, it can be seen the extended coverage effect obtained due to smart antenna use. The first graph shows the application throughput when MS is using an omni-directional antenna, 6dBi gain (blue). There are some communication gaps, due to lack of coverage. On the same conditions, but when MS is using a smart antenna with 13.5dBi cross-over gain (red), and 16dBi peak gain (green), the application throughput is completely different, as seen in the second and the third graph.

On the fig.2.B, the MS has an omni-directional antenna, 6dBi, on first graph (blue), and a smart antenna with 13.5dBi cross-over gain (red), and 16dBi peak gain (green). Even though the application throughput seems to be the same for all three graphs, the extended coverage allows to MS a smart HO decision, so two unnecessary HO are avoided (third graph-green).

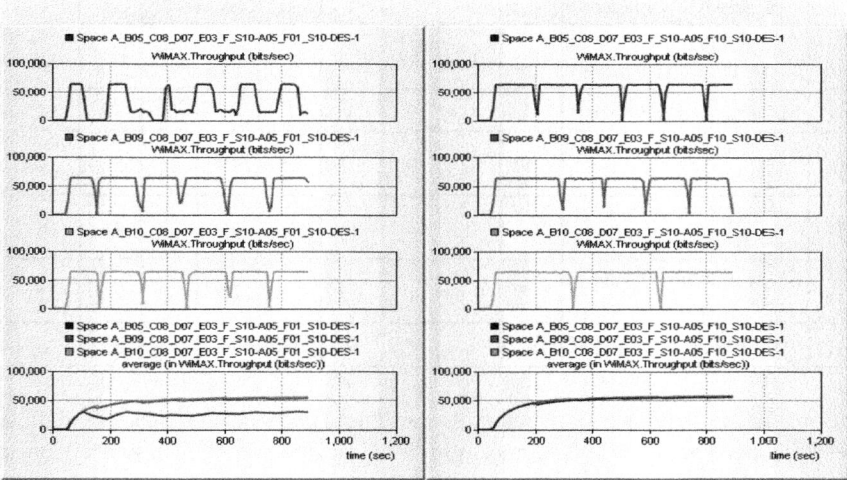

Fig. 2. A and B Advantages of smart antenna use: extended coverage –left (A) and smart HO-right (B)

3   Decision Spaces

A smart antenna, of cylindrical type, 12 beams, was considered, detail beam patterns for which are provided by Plasma Antennas (see [12]). It can operate on both omni-directional and beam selected modes (named smart antenna mode for easy reference).

The smart antenna has the following gains:

- Omni-directional gain, with all 12 beams active, acting as a standard omni-directional antenna: 6dBi;
- Smart antenna cross-over gain, measured between two adjacent beams: 13.5dBi;
- Smart antenna peak gain, measured on the direction of the active beam: 16dBi.

Given the high number of WiMAX physical and MAC layer parameters influencing an inter-dependent mode, the overall performance in mobility scenarios, a detailed experimental study of WiMAX V2I communications has been performed, in order to determine multi-dimensional regions where different lower layer parameters have influence on the overall handover performance.

The simulations results are consolidated in multi-dimensional graphs, named decision spaces. These decision spaces present in aggregated form the communications performance obtained in a WiMAX V2I mobility scenario, related to a specific trajectory. Each multi-dimensional graph - decision space is consolidating the results of 100 simulations, each simulation corresponding to one parameter pair from Table 3.

Table 3. Main WiMAX parameters analyzed on simulation scenarios

	WiMAX Parameter	1	2	3	4	5	6	7	8	9	10
A	MS Maximum Tx Power (W)	0.1	0.2	0.3	0.4	0.5	0.6	0.7	0.8	0.9	1
B	MS Antenna Gain (dBi)	-1	0	2	4	6	8	10	12	14	16
C	MS Scanning Threshold (dB)	3	6	9	12	15	18	21	24	27	54
D	MS HO Threshold Hysteresis (dB)	0.4	2	4	6	8	10	12	14	16	18
E	BS Maximum Tx Power (W)	0.5	0.8	1	1.3	1.5	1.8	2	2.3	2.5	2.8
F	BS Antenna Gain (dBi)	6	8	8	9	10	12	14	15	16	21

Considering only two parameters as variables, one decision space is defined for each MS speed value (10m/s, 20m/s, 30m/s, 40m/s and 50m/s). These decision spaces provide an overall estimation about the MS communication capabilities on the related network conditions (network topology, BS locations, antenna gain, Tx maximum power), combined with mobile constrictions (MS speed, scanning method, antenna gain, Tx maximum power).

Since the network conditions remain unchanged on usual operations, MS could use the decision spaces corresponding to its speed to adapt its own parameters (scanning method/parameters, Tx maximum power). As example, Fig. 3. presents four decision spaces (DS1..DS4), corresponding to the following situations:

- MS maximum Tx power variable 0.1..1W;
- MS antenna gain fixed: 6dBi for DS1 and DS 3, and 13.5dBi for DS2 and DS4;
- MS scanning threshold fixed: 24dB;
- Scanning method fixed: N=4, P=240, T=10, where N=number of scanning frames, P=number of interleaving frames, used for data, T=number of N_P cycles;
- MS HO threshold hysteresis fixed:12dB;
- BS maximum TX power fixed: 1W;

- BS antenna gain variable: 6.....21dBi. That parameter is unchanged for operator, but in can be different from one BS to another, so for each BS is corresponding a vertical section in the decision space, according with BS antenna gain value;
- MS speed fixed for each decision space: 20m/s (72km/h) for DS1 and DS3, and 40m/s (144km/h) for DS2 and DS4.

The DS1 and DS2 could be used for MS with an omnidirectional antenna, moving with 20m/s –DS1, and 40m/s respectively -DS2, and the DS3 and DS4 are for an MS with a smart-antenna as in Ref [12], at the same speeds.

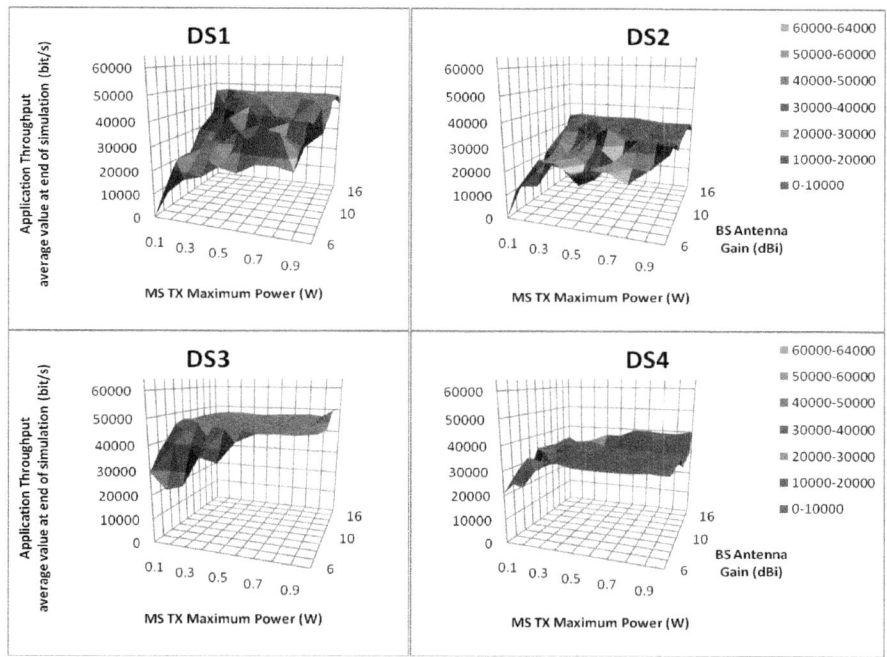

Fig. 3. Decision spaces for MS with omni-directional antenna (6dBi) and smart antenna (cross-over gain 13.5dBi), moving with 20m/s, and 40m/s respectivelly

Accordingly, with the system behavior described in DS1 and DS2, the MS communication capabilities are decreasing with the speed. At lower speeds (20m/s), the increase of the MS Tx maximum power allows a significant increase in application throughput. At higher speeds (40m/s), the increase is not so significant, and after a specific value of MS maximum Tx power, the application throughput reaches a maximum, so bigger values for MS Tx power will have almost no additional effect on the communication capabilities. Knowing that threshold value, the MS could limit its Tx power so reducing the possibility of interference.

Analyzing DS3 and DS4, the same communication capabilities decrease is observed. However, due to smart antenna use, the coverage area is bigger when compared to omni-directional use, and the application throughput reaches maximum for small value of MS

Tx maximum power. Again, knowing the related threshold value, MS could limit its Tx power, and could gain an important reduction in potential interference.

A DS1 against DS3 comparison highlights the importance of smart antenna use. The maximum communications capability is dramatically increased using smart antennas, and that maximum is reached for low value of MS Tx power. Fig.4 compares the vertical section through DS1 and DS3, analyzing the benefits of smart antenna use for a MS moving with 20m/s on the two situations (omni-directional antenna, smart antenna).

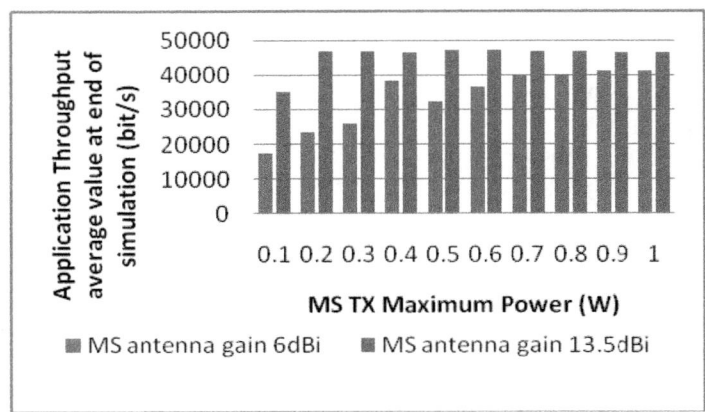

Fig. 4. Vertical sections through DS1 and DS3, showing the benefits of smart antenna use. MS speed=20m/s

According to these sections, a maximum value is obtained for omni-directional antenna use for MS Tx power=0.4W, with a decrease from 0.5W to 0.7W when the initial maximum is reached again. On contrary, using a smart antenna, the maximum is obtained for MS Tx maximum power=0.2W and remain steady. The MS could stop increasing its Tx maximum power, and look for additional fine tuning, as smart HO methods, to improve further its communication capabilities.

4   Conclusions

In conclusion, this research has demonstrated through a set of OPNET simulations that optimal parameter sets, obtained from the decision spaces described above, can be provided by a network operator to mobile stations (MS), helping it to adapt dynamically its behavior and to obtain the maximum throughput possible from the network at different speed, antenna gain, maximum transmission (Tx power), and scanning values.

With regard to future research, it is recommended that the use of smart antennas on both the BSs and MSs be considered for high user densities and more complex journeys using extensions of the current OPNET representations. In this way, the problems of co-channel interference may be assessed and protocols developed to facilitate dynamic beam and power management for both the mobile users (i.e. the MSs) and their WiMAX access points (i.e the BSs), thus offering greater frequency re-use, better overall coverage and the possibility of increased cell sizes in sparsely populated areas. Moreover, the use of smart antennas for BSs, in the context of a highway or other transport infrastructure, will offer greater benefits in terms of coverage and reduced interference.

However a number of open issues remain, such as BS beams synchronization with MSs current positions, (possibly requiring GPS timing information), sub-frames allocation, BS resource allocations in respect of smart handover protocols.

Furthermore, advances in smart antenna technology, such as electronically selecting beams in both azimuth and elevation, especially at BSs, should also be incorporated in the OPNET HO simulations to quantify the resulting gains in performance, quality of service and coverage. The introduction of LOS blockage caused by buildings and natural objects such as trees should also be introduced to access the resilience of the HO process.

References

1. http://www.opnet.com (last accessed 16.03.2011)
2. Wendt, S., Kharrat-Kammoun, F., Borcoci, E., Selva, B., Tonnerre, A., Hamadani, E.: D2.1 – Requirements and Specifications of SMART-Net Target Scenarios. ICT European FP 7 SMART-Net project, https://www.ict-smartnet.eu
3. Wendt, S., Kharrat-Kammoun, F., Borcoci, E., Cacoveanu, R., Lupu, R., Hayes, D.: ID2.4b – Network Architecture and System Specification. ICT European FP 7 SMART-Net project, internal WP2 deliverable, https://www.ict-smartnet.eu
4. Chou, C., Li, C., Chien, W., Lan, K.: A Feasibility Study on Vehicle-to-Infrastructure Communication: WiFi vs. WiMAX. In: 2009 Tenth International Conference on Mobile Data Management: Systems, Services and Middleware, pp. 397–398 (2009)
5. Frangiadakis, N., Roussopoulos, N.: A Pegasus over WiFi and WiMax Demo: Connectivity at high speeds. In: 2009 5th International Conference on Testbeds and Research Infrastructures for the Development of Networks & Communities and Workshops, pp. 1–3 (2009)
6. Matos, R., Sousa, B., Neves, P., Sargento, S., Curado, M.: Advanced Mobility in Broadband Wireless Access Scenarios. In: 2009 IEEE International Conference on Wireless and Mobile Computing, Networking and Communications, pp. 214–220 (2009)
7. Zaggoulos, G., Nix, A., Halls, D.: Novel Antenna Configurations for Wireless Broadband Vehicular Communications. In: 2010 6th International Conference on Wireless and Mobile Communications, pp. 21–26 (2010)
8. Constantinescu, M., Borcoci, E.: Handover Performance Evaluation in WiMAX Mobility Scenarios. In: 7th International ICST Conference on Broadband Communications, Networks, and Systems, BROADNET-MOWAN (2010)
9. Constantinescu, M., Borcoci, E., Rasheed, T.: WiMAX Handover Conditions Evaluation Towards Enhancement Through Cross-layer Interaction. In: 2010 Sixth Advanced International Conference on Telecommunications, pp. 422–427 (2010)
10. Constantinescu, M., Borcoci, E.: A SIP-Based Cross-Layer Optimization for WiMAX Hard Handover. In: The 8th International Conference on Communications, COMM (2010)
11. Constantinescu, M., Borcoci, E.: Handover Optimization in WiMAX Vehicular Communications. Accepted to The Sixth International Conference on Digital Telecommunications, ICDT (2011)
12. http://www.plasmaantennas.com/products/selectabeam-portfolio/selectabeam-sc-1000.html
13. SMART-Net, SMART-Antenna based Multi-mode Wireless Mesh Networks, European Research Project, https://www.ict-smartnet.eu

Real Time Traffic Capabilities Evaluation of a Hybrid Testbed for WiMAX Networks with Smart Antenna Support

Şerban Obreja[1], Irinel Olariu[1], Alexey Baraev[2], and Eugen Borcoci[1]

[1] University Politehnica Bucharest, Bucharest, Romania
[2] CreateNet, Trento, Italy
serban@radio.pub.ro, alexey.baraev@create-net.org

Abstract. The ever-increasing adoption of the Internet and multimedia services rise new challenges for the next generation network solutions, especially on the wireless ones. WiMAX is one of the new technologies developed for broadband wireless networks. It offers high data rate services while providing high flexibility for the radio resource management. Adding smart antenna support and developing new scheduling and routing algorithms for wireless mesh networks based on WiMAX technology will recommend it as a very attractive solution for the next generation wireless networks. In this paper a basic simulation testbed for a WiMAX network with smart antenna support is proposed. It is based on OPNET simulation tool and uses the System-in-the-Loop function to interconnect the simulated system with a real network, for a better functional and performance evaluation. The advantage of using smart antennas is illustrated by the simulation results.

Keywords: WiMAX networks, smart antenna, OPNET, System-in-the-Loop.

1 Introduction

In the last years wireless communications had an exponential growth, with a huge social end economical impact. The current services require high data rates at the radio interface which pushed forward the development of high capacity wireless technologies, such as WiFi, WiMAX or LTE. The throughput increase on the radio interface was made possible by the development of new complex techniques in signal processing at the physical level, such as turbo coding, MAQ modulation, OFDMA, MIMO. However, last-mile access network technologies represent a significant congestion spot between the high-capacity core backbone network and the high-capability terminal equipments. In order to reduce these requirements new solutions to exploit these technologies need to be developed. The IEEE 802.16 technology and WiMAX based sytems constitute an attractive solution for metropolitan and rural areas, [10]. Recently, advances made at physical layer introducing directional smart antennas showed a possible increase in performance, while keeping the transmitting power at the same level as for omni-directional antennas case.

J. Del Ser et al. (Eds.): MOBILIGHT 2011, LNICST 81, pp. 258–266, 2012.

In this paper a basic simulation testbed for a WiMAX network with smart antenna support is proposed. It is based on OPNET simulation tool and uses the System-in-the-Loop function to interconnect the simulated system with a real network, for a better functional and performance evaluation [1]. The testbed was developed in the framework of the SMART-Net FP7 project [3]. The SMART-Net project defined an architecture and then designed and developed a Wireless Mesh Network solution mainly based on WiMAX and WiFi technology, by introducing smart antennas support in WiMAX. The project proposed efficient scheduling and routing algorithms, to enhance the capacity and to provide scalability, reliability and robustness for such a system [2],[5]. Inside this project, both performance evaluation based on simulations and real life testbed have been conducted. In order to combine the capabilities of the two techniques, cooperation between the two frameworks has been achieved by coupling them using the technique called System- in-the-loop, [1].

This paper presents a basic simulation testbed for evaluating the smart antennas integration on standard OPNET WiMAX nodes and also for evaluating the smart antennas capabilities [4]. It represents a starting point for the development of a combined simulation model-HW model for the SMART-Net proposed system. Adding smart antennas support in OPNET WiMAX nodes was the first step in starting integrating the already developed components of the SMART-Net OPNET system model.

This paper is organized as follow. The second section is a short description of the SMART-Net system and of the smart antennas. The third section is dedicated to the testbed. It is presented its structure and the simulation scenarios. The fourth section provides the simulation results. The last section presents conclusions and guidelines for future work.

2   Smart-Net System

2.1   Smart-Net Features

Smart-Net solution for Broadband Wireless Access is based on a system architecture which consists on multimode devices with smart antennas support. These devices are interconnected in a partial mesh topology which has a central point, Smart Gateway (SMG), acting as gateway linked to the backhaul networks. The other nodes of the network are either SMART Stations (SMS) or SMART Relays (SMR). A SMR is an operator's equipment which is specifically used to forward data traffic to the users, allowing coverage extension and cooperative diversity, while a SMS is a subscriber station that also enables data transfer for other users based on the service provider policy [3], [10].

System capacities can be significantly improved by introducing multi-hop capabilities and mesh networking in wireless access networks. The use of relay stations results in a signal to noise ratio enhancement, allowing obtaining higher throughput and coverage while reducing the device transmitting power. Indeed the user station does not have to communicate directly with the central station and then, by minimizing the transmit power levels, co-channel interference can be reduced [8], [9].

Using omnidirectional antennas in wireless networks create inherently interference, which decrease the capacity of the system. A significant increase in capacity is obtained by using smart antennas. They feature a directivity that can be controlled by higher levels protocols (Layer 2, Layer 3) in the network node, allowing its orientation towards the destination node and thus reducing interference. In the same time, directional antenna can be used to increase radio coverage.

The Smart-Net project introduced smart-antenna support on WiMAX equipments and developed some algorithms for scheduling and routing in a multihop relay based WiMAX mesh network [2], [6]. To validate the proposed solutions two testbeds were developed during the project. A real life testbed consisting of Wimax equipments with smart antennas installed. The WiMAX equipments used in the testbed are produced by Thales Company and the smart antennas are produced by Plasma Antennas Company. Both are members of the Smart-Net project. In parallel, a simulated testbed was developed using the OPNET network simulator [4]. The smart antennas were modeled in the OPNET and integrated in WiMAX nodes. Experiments were performed to evaluate the smart antennas and the proposed algorithms.

Also, it was proposed by the project to perform experiments by combining the real life and simulated testbeds in order to obtain a more complex experimental system. For the testbeds interconnection the System-in-the-Loop (SITL) function provided by OPNET was used. Such approach is presented in [7]. OPNET and SITL are used to evaluate wireless tactical networks. Two simulated tactical units communicate via simulated wireless link. Two real life tactical units are interconnected with the simulated network via SITL. The capacity of the simulated tactical units to exchange real traffic, generated by a military application installed on the real life unit, while they move on a predefined trajectory over a virtual terrain, is evaluated. In this paper a similar approach is used to evaluate the smart antennas integration on standard WiMAX node and the smart antennas performances related to the standard omnidirectional ones.

2.2   Smart Antennas

Smart antenna systems intelligently combine multiple antenna elements with signal-processing capability to optimize its radiation and reception patterns automatically [3]. They have a certain number of high gain beams with low sidelobes, which minimize interference both on transmit and receive, without using complex adaptive nulling algorithms. Low sidelobe multi-beam antennas have the advantage over adaptive systems in that they suppress a very large number of interferers in a consistent, predictable way. Adaptive systems are limited by their degrees of freedom (e.g. number of radios), their adaptation time and might not work well (i.e consistently) when the signal of interest is at the edge the receiver's sensitivity. However, they potentially have the advantage of allowing the suppression of interfering signals that are close in angle (within a beamwidth) to the source-of-interest.

When receiving, smart antennas can maximize the sensitivity in the direction of the desired signal and minimize the sensitivity towards interfering sources.

For reasons of cost and consistency of performance common smart antennas are switched or selectable multi-beam antennas, requiring only a single radio. These antennas have multiple fixed beams and the system switches very rapidly between these beams. A simultaneous coverage of these beams is also possible, for broadcast purposes, in the form of an omnidirectional or a sectoral beam.

For the SMART-Net project, two types of multi-beam antennas, capable of WiMAX operation, have been designed and implemented [2], [3].

- An active, 12 beam cylindrical array antenna with omni-mode
- A passive 9 beam planar array antenna with sectoral mode.

The active 12-beam cylindrical antenna with it 360° coverage has been selected to be most suitable for mesh and nomadic Point to Multipoint operation, with typical ranges up to 20 km, depending on the modulation rate. The passive 9 beam planar antenna, with its narrow beams has been selected to be most suitable for medium range backhaul and relay operations. A representation of both antennas, suitable for inclusion within OPNET, has also been provided, but as *simulated* data.

Besides the multibeam antennas, a switching algorithm is used to choose the appropriate beam among the available antenna beams. This algorithm is based on a learning interval in which, based on SINR, the best beam is chosen for each destination. Based on the decision took by the selection algorithm, when a smart node (a node equipped with smart antennas) needs to communicate with another smart node, the beam with the best SINR is used. Because the best beam is decided in the learning phase, the switch operation is very fast, a few nanoseconds. Some performance degradation is expected in the mobile nodes case, because the learning interval lasts a few milliseconds. In this paper only evaluation of smart antenna on fixed WiMAX nodes is presented. For mobile nodes the smart antenna integration is not ready.

3   Real Time Simulation of Smart Antenna Using the Smart-Net Simulated Testbed

3.1   Simulated Testbed Infrastructure

The testbed infrastructure consists of a simulated WiMAX network which is interconnected with real devices in order to introduce real time traffic in the simulation (figure 1). The System-in-the-Loop is an OPNET facility which allows real time communication between real and simulated parts of the networks within the simulation loop [1], [7]. By using SITL, OPNET simulation exchanges the packets between simulations and real networks in real-time. The SITL gateway represents an external device through which the simulation exchanges the packets, where the WinPcap library is used to route those packets selected by user defined filter, from an Ethernet network adaptor, to the simulation process.

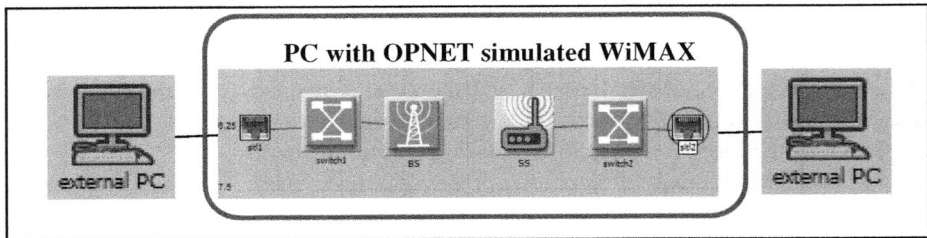

Fig. 1. Testbed infrastructure: a simulated WiMAX network interconnected with external devices via OPNET System-in–the-Loop functionality

The simulation runs in real-time and exchanges packets with the external hardware via an Ethernet link. The requirement of using Ethernet link between the real devices and the SITL gateway introduces limitation in developing joint real and simulated wireless network scenarios. Joint scenarios for evaluating scheduling and routing algorithms for wireless mesh networks are not possible with SITL. The interconnection scenarios were limited at evaluating the effects on real time traffic while crossing the interconnected testbeds.

There are three main simulation topologies for using the SITL module:

- real-to-real (communication between real devices over simulated network)
- sim-to-sim (communication between simulated devices over real network)
- sim-to-real (communication between real and simulated devices).

3.2   Simulated Scenarios

The detailed view of the simulated WiMAX network is presented in figure 2. A real-to-real SITL topology is used for these scenarios. It consists of a single WiMAX link, between a Base Station and a fixed Subscriber Station, which is concatenated with Ethernet links at both ends. These Ethernet links are used to interconnect, via SITL gateways, the WiMAX simulated network with the external stations which are both acting as real time streaming server and player. The smart antennas were installed on the simulated WiMAX nodes. Introducing smart antenna support on standard WiMAX nodes requires modifying the radio transceiver pipeline stages. The beam selection algorithm was introduced in the pipeline stages together with the 12 beam cylindrical array and 9 beam planar array antenna models.

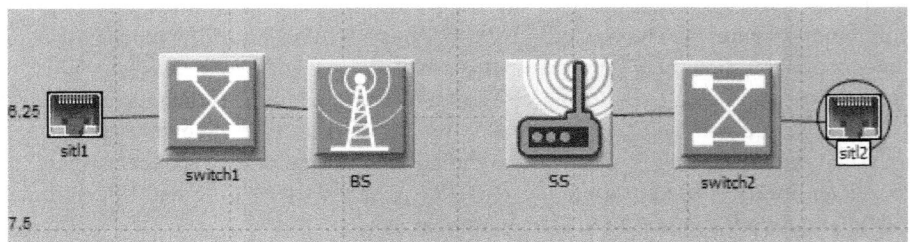

Fig. 2. The OPNET simulated scenario. The Base Station and a fixed Subscriber Station are each connected to separate SITL gateways through Ethernet switches.

Given the complexity of the components, the experiments presented in this paper focused on functional evaluation. Further simulations need to be performed in order obtain precise performance evaluation for the smart antenna in Smart-Net system.

The following two scenarios were created in order to evaluate the smart antennas integration in standard WiMAX nodes using the OPNET SITL tool. The first one uses the topology given in figures 1 and 2 with standard WiMAX nodes (with omnidirectional antennas). It is used to evaluate the SITL behavior and as a reference for the second scenario. The second one uses the same topology but WiMAX nodes with smart antenna support (modified transceiver pipeline stages). This scenario is used to evaluate the smart antennas behavior and their performances related to the omnidirectional case.

4   Simulation Results

As mentioned in the previous section two scenarios were proposed for evaluating simulated smart antennas integration on WiMAX nodes. Both scenarios are based on real time flows which are sent through the simulated WiMAX network using the OPNET SITL functionality. This approach allows the evaluation of the simulated system model to transport real time traffic flows, like video streaming, interactive voice. More realistic evaluation, closer to a real life one is obtained in [9].

In the first scenario, a movie with a rate around 2.5 Mbps is streamed from the streaming server through the simulated WiMAX link. On the same link and in the same direction (downlink) it is transmitted a noise UDP traffic with the rate of 5Mbps. All the flows are transmitted as Best Effort. A total of around 7.5 Mbps throughput is transmitted on the downlink. The WiMAX physical parameters are: 20MHz bandwidth, 2048 subcarriers, 10.94 kHz subcarrier frequency spacing, symbol duration of 102,86 ms, frame duration of 5ms [10]. The antenna gain is set at 15 dB. Adaptive modulation and coding is used on the physical interface.

Both scenarios have been repeated by varying the distance between the BS and SS. Also the SS orientation, related to the BS which is fixed, is changed. This is useful for the second scenario when smart antennas are used. By changing the orientation, the functionality of the beam selection algorithm is evaluated. If it chooses the right beam, this will be reflected in the link quality. The simulation duration is set at 5 minutes for each experiment. The first scenario results are shown in figure 3. It illustrates the throughput on the WiMAX link in each selected experiment. For each experiment standard omnidirectional antenna was used. As it was expected the capacity of the WiMAX link decreases while the distance between nodes is increased. On figure 3, the curve with the smallest throughout correspond to the case when we have the biggest distance between nodes. The WiMAX capacity decrease is illustrated also by the perceptual evaluation of the movie quality.

For the second scenario the same parameters have been used and moved the SS in the same positions as those used in the first scenario. The planar 9 multibeam smart antennas were installed on WiMAX nodes. In all experiments performed the WiMAX link capacity was similar or better than in case of omnidirectional antenna. This assertion is sustained by the results illustrated in figure 4 and figure 5. Figure 4 presents the throughput obtained in the omni-beam and smart antenna case when the BS and SS are at the same distance and in the same positions. The higher throughput curve corresponds to the smart antenna scenario. For small distances between the BS and SS node the throughput is the same. When the distance is increased the difference between the throughputs obtained in each case is increased – the higher throughput being obtained when using smart antennas.

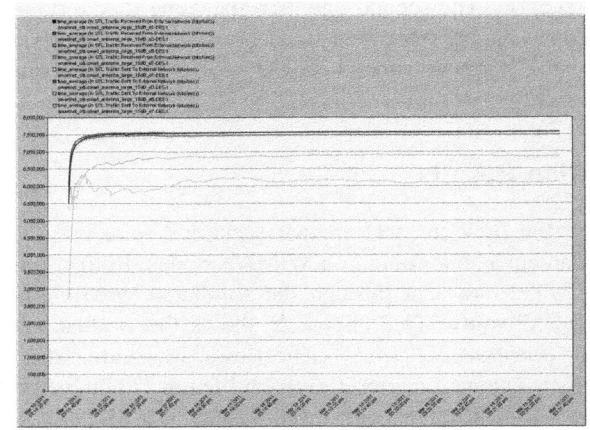

Fig. 3. First scenario results: The capacity of the WiMAX link decreases while the distance is increased

Because the direction of the WiMAX link was modified while moving the SS node and in all cases the link quality was good, it means that the beam selection algorithm is working correctly. To better evaluate the selection algorithm the following experiments were performed in the second scenario: the SS node was moved in several positions situated on a circle centered in the BS node. The link capacity and movie quality remained the same for all experiments. This showed that the smart antennas are able to choose the right beam for each case, offering better link quality due to better signal to interference ratio obtained when a narrow beam is used.

The real video flows transmitted through the simulated network offered us the possibility to perceptually evaluate the smart antenna benefits. Figure 5 presents a capture from the received movie in each of the two mentioned experiments. A frame with a poor perceptual video quality, obtained for the standard omni-antenna, is presented in figure 5a. Figure 5b shows a video frame taken in the case when a similar simulated topology (the same positions for BS and SS) as the one used to obtain figure 5a was used. The perceptual movie quality decreased less when smart antennas were used. From these results we can say that the WiMAX nodes with smart antennas present superior performances comparing with the standard ones. Some performance degradation could be obtained when the nodes are situated at border between two beams. A larger set of experiments must be performed in order to "catch" and illustrate this degradation.

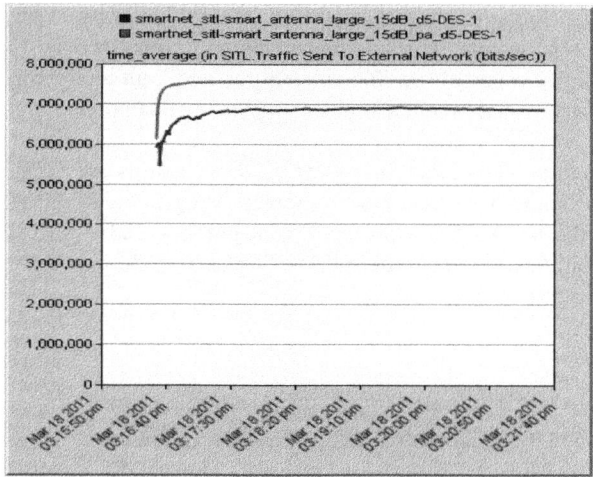

Fig. 4. Wimax throughput for the omni-beam and smart-antenna cases, for the same distance between the BS and SS nodes and the same nodes positions

Fig. 5. Perceptual illustration of the smart antenna benefits. Received movie snapshot for: a) omnidirectional antenna. b) smart antenna.

5   Conclusions

This paper presents a basic simulation testbed used to evaluate beam selection algorithm and switched beam antennas performances for WiMAX technology. The discussed solutions are proposed in the Smart-Net project. The presented testbed was built on a OPNET (version 14.5) Platform and it uses OPNET SITL module to interconnect the simulated network with real life devices. Even if SITL module has some limitations, which restrict its use for wireless mesh networks, it proved to be very useful for performance evaluations, because it allows introducing real time traffic in the simulated network, and then analyzing how it is affected while passing through the simulated devices.

For the future, the simulated testbed will be developed by implementing with the presented smart nodes the wireless mesh network proposed in Smart-Net project. Then, the proposed algorithms, for scheduling and routing in this WiMAX based mesh network with smart antenna support, will be evaluated.

Acknowledgments. The research leading to these results are supported partially by the European Community's Seventh Framework Program (FP7/2007-2013) within the framework of the SMART-Net project (grant number 223937). The research is also supported partially by the Program POSDRU/89/1.5/S/62557 within the framework of EXCEL project.

References

1. OPNET 14.5 PL0 documentation, http://www.opnet.com/support
2. Wendt, S., Abdallah, A., Hayes, D., Foul, T.A., Borcoci, E.: Project Deliverable, D3.3a: Preliminary description of cross layer optimization. ICT European FP 7 SMART-Net project
3. Wendt, S., Kharrat-Kammoun, F., Borcoci, E., Cacoveanu, R., Lupu, R., Hayes, D.: Project Deliverable, ID2.4b: Network Architecture and System Specification. ICT European FP 7 SMART-Net project, internal WP2 deliverable (October 2010)
4. Kammoun, F.K., Meddour, D., Baraev, A., Rasheed, T., Hamadani, E., Tonnerre, A., Borcoci, E., Enescu, A., Ciochina, S.: Project Deliverable, D4.1: Large Scale Simulation Testbed Specifications. ICT European FP7 SMART-Net project Deliverable (January 2009)
5. Baraev, A., Rasheed, T., Selva, B., Tonnerre, A., Wendt, S., Hamadani, E., Borcoci, E., Badoi, C., Cacoveanu, R.: Project Deliverable, D4.3b: Assessments on System Level Implementations and Simulations, ICT European FP 7 SMART-Net project (May 2010)
6. Mostafavi, M., Hamadani, E., Vural, S., Borcoci, E., Constantinescu, M., Niculescu, D., Rasheed, T., Riggio, R., Gomez, K., Baraev, A.: Project Deliverable, D3.2b: Performance analysis of efficient routing protocols for multimode mesh networks (October 2010)
7. Mohorko, J., Fras, M., Čučej, Ž.: Real time "system-in-the-loop" simulation of tactical networks. In: 16th International Conference on Software, Telecommunicati ons and Computer Networks, SoftCOM 2008, September 25-27, pp. 105–108 (2008)
8. Kortebi, R., Meddour, D.-E., Gourhant, Y., Agoulmine, N.: SINR-Based Routing in Multi-Hop Wireless Networks to Improve VoIP Applications Support. In: IEEE Proceeding of CCNC 2007, Las Vegas, USA (2007)
9. François, J., Bertrand, M., Meddour, D.-E.: Video Streaming Experiment on Deployed Ad hoc Network. In: The Proceeding of the IEEE International Conference on Testbeds and Research Infrastructures for the Development of Networks and Communities (TridentCom), Orlando, USA (May 2007)
10. IEEE Standard 802.16-2004, Air Interface for Fixed Broadband Wireless Access Systems

Multipath Network Coding in Wireless Mesh Networks

Velmurugan Ayyadurai, Klaus Moessner, and Rahim Tafazolli

Centre for Communication Systems Research (CCSR), University of Surrey,
Guildford, UK
{v.ayyadurai,k.moessner,r.tafazolli}@surrey.ac.uk

Abstract. A practical wireless network solution for providing community broadband Internet access services are considered to be wireless mesh networks with delay-throughput tradeoff. This important aspect of network design lies in the capability to simultaneously support multiple independent mesh connections at the intermediate mobile stations. The intermediate mobile stations act as routers by combining network packets with forwarding, a scenario usually known as multiple coding unicasts. The problem of efficient network design for such applications based on multipath network coding with delay control on packet servicing is considered. The simulated solution involves a joint consideration of wireless media access control (MAC) and network-layer multipath selection. Rather than considering general wireless mesh networks, here the focus is on a relatively small-scale mesh network with multiple sources and multiple sinks suitable for multihop wireless backhaul applications within WiMAX standard.

Keywords: mesh networks, multipath, multihop, network coding.

1 Introduction

Today's world depends a lot on wireless to provide faster connectivity and more broadly than anyone may have expected. Future wireless systems are expected for very high demands in terms of data rate, latency, reliability and robustness. A wireless communication system relies on wireless links between wired infrastructure devices and end user devices for voice and data transmission.

The mesh network approach is to employ mobile stations as intermediate routers to establish multihop communication paths between mobile stations and their corresponding base stations. Utilizing this, an opportunistic routing will take advantage of the spatial diversity and broadcast nature of wireless networks to combat the time-varying links by involving multiple neighboring nodes as forwarding candidates for each packet relay. This breakthrough has created a vast interest in developing protocols for wireless mesh networks (WMNs).

Currently, WMNs are very diverse and different in several aspects. One aspect of WMN uses mobile station as intermediate access points on fixed locations by network operators or simply be other idle mobile stations that are not transmitting their own data. Although, depending on how radio resources are allocated for routing paths of

J. Del Ser et al. (Eds.): MOBILIGHT 2011, LNICST 81, pp. 267–274, 2012.
© Institute for Computer Sciences, Social Informatics and Telecommunications Engineering 2012

active connections, different protocols at the medium access control and routing layers can be designed. Radio resources for mobile stations at different hops may be allocated in time division duplex (TDD) or frequency-division duplex (FDD) mode. Frequency bands other than the cellular frequency bands can also be used for relaying. The intermediate mobile stations will serve toward various objectives, such as enhancing data rate coverage and enabling range extension in WMNs.

With this motivation, there has been a recent interest in both academia and industry in the concept of multipath network coding in the networking perspective with scheduling and resource allocation algorithms to improve throughput with a cross-layer perspective. Network coding is first proposed in [1] for noiseless wireline communication networks to achieve the multicast capacity of the underlying network graph. The essential idea of network coding is to allow coding capability at network nodes acting as routers in exchange for capacity gain, i.e., an alternative tradeoff between computation and communication.

2   Network Coding Scenario and Related Works

Wireless network properties require a strategy for medium access control (MAC) in order to coordinate reliable packet transmissions between source-sink pairs. The problem of throughput increase and delay decrease strongly depends on the cross-layer interactions between MAC and the network layers. Hence, they need to be jointly designed for efficient wireless network operation. For stable operation of the queues, the possible underflow of relay packets opens up new questions regarding the optimal queue management and the optimal use of network coding on the instantaneous queue contents.

The focus of our discussions is based on relay-based WMNs, which essentially generalizes the examples shown below in various ways depending on the network model. The intermediate mobile station receives multiple transmissions that contain the same data and relaying amounts to mere multi-hopping only when the receiver obtains one copy of the data. For unicast communication in [2], the back pressure algorithms achieve the maximum stable throughput region at the expense of poor delay performance. On the other hand, the capacity analysis of wireless networks has been limited to saturated queues with infinite delay in [3].

In Fig.1a, suppose the source S wants to multicast two bits X_1 and X_2 to two sinks Y_1 and Y_2 simultaneously. Each of the paths in the network is assumed to have a unit capacity of 1 bit per time slot (bps). With traditional routing, each mobile station between S and the two sinks simply forwards a copy of what it receives. It is then impossible to achieve the theoretical multicast capacity of 2 bps for both sinks; sink in the middle can only transmit either X_1 or X_2 at a time. However, with network coding, as shown in Fig.1b, the intermediate mobile station can perform coding, in this case a bitwise exclusive-or operation, upon the two information bits and generate X_1+X_2 to multicast towards its outgoing paths. Sink Y_1 receives X_1 and X_1+X_2, and recovers X_2 as $X_2 = X_1 + (X_1 + X_2)$. Similarly, sink Y_2 receives X_2 and $X_1 + X_2$ and can recover X_1. Both sinks are therefore able to receive at 2 bps, achieving the multicast capacity.

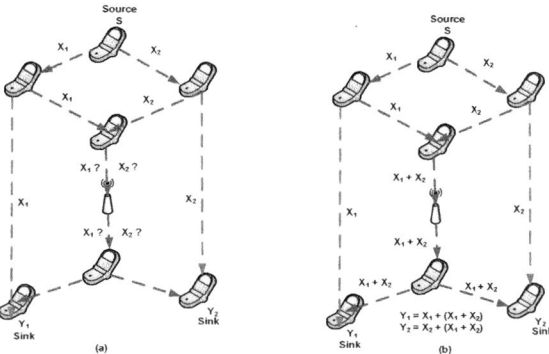

Fig. 1. The butterfly network example of network coding. (a) With traditional routing, the link in the middle can only transmit either X_1 or X_2 at a time. (b) With network coding, the relay node can mix the bits together and transmit $X_1 + X_2$ to achieve the multicast capacity of 2 bps.

With this simple mathematical and potential for practical network coding, the communications and networking communities have devoted a significant research efforts to utilize in a number of wireless applications, ranging from opportunistic routing in mesh networks. In the above example, the network coding operation is bitwise exclusive-or, which can be viewed as linear coding over the finite field i.e. Galois Field GF (2).

Following the seminal work of [1], Li et al. in [4], showed that a linear coding mechanism suffices to achieve the multicast capacity. Ho et al. in [5], further proposed a distributed random linear network coding approach, in which nodes independently and randomly generate linear coefficients from a finite field to apply over input symbols without a priori knowledge of the network topology. They proved that receivers are able to decode with high probability provided that the field size is sufficiently large. These works lay down a solid foundation for the practical use of network coding in a diverse set of applications. After the initial theoretical studies in wireline networks, the applicability and advantage of network coding in wireless networks were soon identified and investigated extensively in [6]. The performance gains have been verified through experimental results in [7].

There is a lot of work for opportunistic routing in wireless mesh networks, with or without network coding. COPE [7], MORE [8] and MC^2 [9] investigate network coding with opportunistic routing in wireless networks with broadcast transmissions, focusing exclusively on the throughput improvements. In this work we use redundant paths to send uncoded packets in order to recover the information using packets from another path, thus decreasing the delay in transmission with a gain in throughput. The goal of this paper is to investigate the performance that can be achieved by exploiting path diversity through multipath forwarding and redundancy through network coding.

3   Opportunistic Mesh Networks - Multipath Network Coding

In the Mesh mode, several mobile stations can constitute a small multipoint to multipoint wireless connection, without specific uplink and downlink sub-frames. It may enable direct communication between mobile stations that can also be used as relays to forward other's data. Two kinds of routing and scheduling are used to coordinate transmissions: centralized and distributed. Here, for the easy operation and high reliability, we focus on the centralized approach. The centralized approach organizes all the nodes of the WiMAX network in a tree structure rooted at a particular node, namely the base station. The ways in which the tree is built and the choice of the links used have a deep impact on the capacity that a WiMAX backbone may offer.

Using multipath routing in mesh networks provide multiple alternative paths through a network, which can yield a variety of benefits such as fault tolerance, increased bandwidth, or improved security. The multiple paths computed might be overlapped, edge-disjointed or node-disjointed with each other. In Fig.2 shows the multipath mesh network transmission, adopting the multihop technology based on packet combining and delay management.

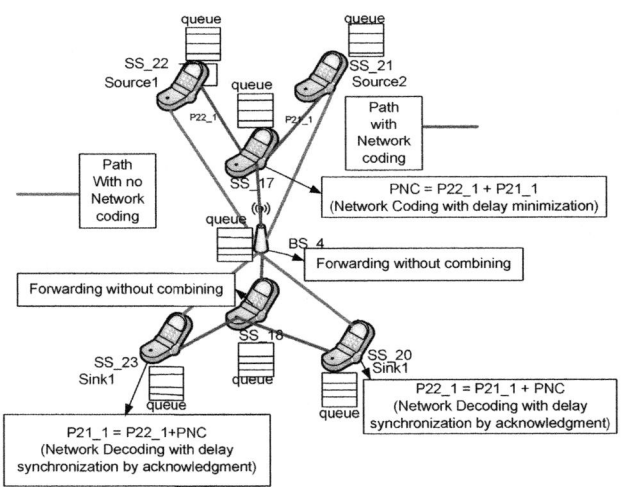

Fig. 2. Multipath mesh routing with network coding and buffer management

3.1   Multipath Routing Tree

In centralized routing, the base station determines the path for all the mobile station in a centralized manner like PMP mode, and traffics from or to the base station can be relayed by other mobile stations through a multihop route which is different from PMP mode. Using multipath routing mechanism, are able to increase spatial reuse rate and next-hop selection. The model assumed is a multiple source unicast mesh network with directional links using smart antennas. Multipath routing makes maximum utilization of network resources by giving the source nodes a choice of next hops for the same destination and to lower delay. The choice of multipath entails that, to synchronise across various paths, buffers need to be established at the network nodes so that the packets can be stored and sequenced appropriately. The nodes inside the network (except the source and sink) act as routers, do not decode the information but simply forward coded packets that have been previously received from the source or the previous node as shown in Fig.2.

The scheduling first established with fixed multipath routing tree and assign traffics to routes according to multipath routing tree. Let base station be level = 0. The node one hop from base station will be level (n) = 1(relay). The source nodes can only choose the nodes that are level (n) = i-1 as their next hop (n_p). Node n must be in transmission range of node n_p and is connected to node n_p. The source nodes transmit copies of packets belonging to a single flow on all the paths. The base station can effectively works on the principle that higher performance can be achieved by utilizing more than one feasible path.

3.2   Network Coding with Delay Management

It has been shown from Fig.2, the intermediate nodes in a network, when performing network coding are able to combine a number of packets received into one or more new outgoing packets. Network coding permits instead of binary field operations, moving to larger field sizes, being able to perform more complex operations when combining incoming packets in intermediate nodes, becoming one of the most successful network coding algorithms as it permits achieving network capacity when multicasting, with relatively low complexity. In network coding each data unit is processed using finite fields F_q with 'q' a prime number or, considering a Galois Field (GF), $q = 2^m$ for some integer m, where F_2m refers to [0, 2^{m-1}].

The intermediate network nodes used to encode the data packets by combining the two packets from sources SS_22 and SS_21 into (PNC = P22_1 + P21_1) with a FIFO queue management system as shown in Fig.2. The FIFO queue management system used to calculate the processing delay based on the service capacity and arrival rate time of the bursty packet sources. The intermediate network nodes need to calculate the data synchronisation time by getting an acknowledgment from sinks. The network coding delay uses the equation (1) and throughput uses the equation (2). Each path is given with equal service capacity (μ) and the arrival rate λ varies depending on the bursty packet sources.

$$D_{NC} = \frac{2}{(\mu - \lambda_1)(\mu - \lambda_2)} \tag{1}$$

$$Thr = \frac{1}{D_{NC}} \tag{2}$$

4   Simulation and Analysis

The simulation scenario is constructed using OPNET 14.5 simulator on the existing SMART Net project using smart antennas. The scenario for data flow in one direction has sources (SS_22, SS_21) and sinks (SS_23, SS_20) with two paths as shown in Fig.3. The packets of two sources have to reach two sinks simultaneously. The first path is through the intermediate network node using packet combining with forwarding. The second path is through the base station (BS_4) with intermediate network node (SS_18) for forwarding to get the path delay for data synchronisation. The traffic from the sources should be of same type for combining the packets of same size. The traffic assumed for source nodes SS_22 and SS_21 is voice packet and traffic for source nodes SS_23 and SS_20 is data. The average delay is measured with the average time that a data packets from source node to destination node. The average throughput is measured with the bit of data packet received.

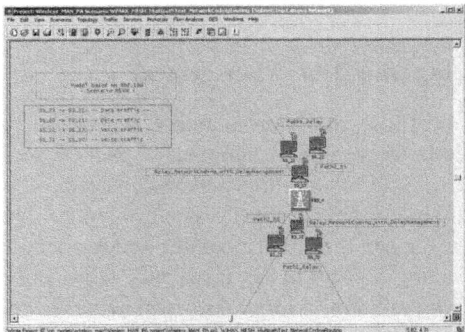

Fig. 3. Multipath Network Coding Scenario in OPNET

The figures Fig.4a and Fig.4b show the output of data traffic in SS_23 for average delay and average throughput. The first path is through SS_23->SS_18->BS-4->SS_17->SS_22 using packet combining and forwarding on the intermediate node SS_18 with delay in buffering. The second path is through SS-23->BS-4-> SS_17->SS-22 with packet forwarding without delay in buffering. The figures Fig.5a and Fig.5b show the output of data traffic in SS_20 for average delay and average throughput. The first path is through SS_20->SS_18->BS-4->SS_17->SS_21 using packet combining and forwarding on the intermediate node SS_18 with delay in buffering. The second path is through SS-20->BS-4-> SS_17->SS-21 with packet forwarding without delay in buffering. The analysis shows with multipath network coding and delay synchronization, there is very high throughput improvement, but only a slight reduction in delay. This gives the required possibility to use multipath network coding feature in mesh networks for high throughput applications.

The figures Fig.6a and Fig.6b show the output of voice traffic in SS_22 for average delay and average throughput. The first path is through SS_22->SS_17->BS-4->SS_18->SS_23 with packet combining and forwarding on the intermediate node SS_17 with delay in buffering. The second path is through SS-22->BS-4->

SS_18->SS-23 with packet forwarding without delay in buffering. The figures Fig.7a and Fig.7b show the output of voice traffic in SS_21 for average delay and average throughput. The first path is through SS_21->SS_17->BS-4->SS_18->SS_20 with packet combining and forwarding on the intermediate node SS_17 with delay in buffering. The second path is through SS-21->BS-4-> SS_18->SS-20 with packet forwarding without delay in buffering. The analysis shows with multipath network coding and delay synchronisation, there is very large reduction in delay, but only a slight improvement with throughput. This gives the required possibility to use multipath network coding feature in mesh networks for delay-sensitive applications.

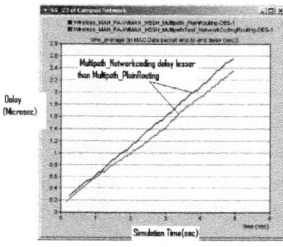

Fig. 4a. SS_23 delay for data traffic

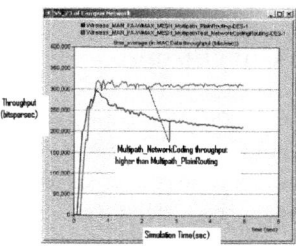

Fig. 4b. SS_23 throughput for data traffic

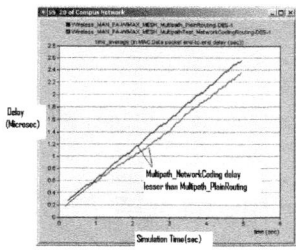

Fig. 5a. SS_20 delay for data traffic

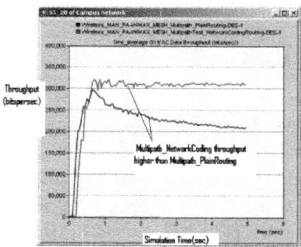

Fig. 5b. SS_20 throughput for data traffic

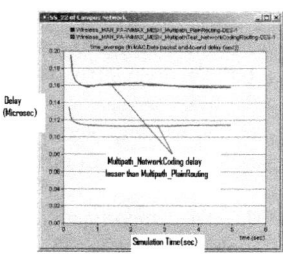

Fig. 6a. SS_22 delay for data traffic

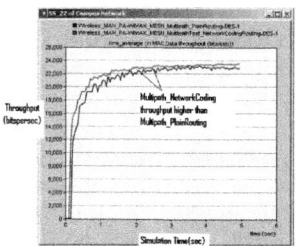

Fig. 6b. SS_23 throughput for data traffic

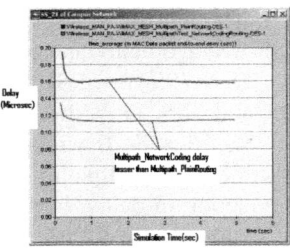

Fig. 7a. SS_21 delay for data traffic

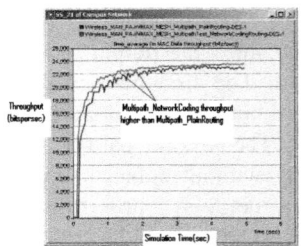

Fig. 7b. SS_21 throughput for data traffic

5   Conclusion

A scenario for multipath routing using network coding for traffics of voice and data are simulated in WiMAX mesh networks. The analysis shows an improvement in average delay and average throughput with a tradeoff. The simulation provides the suitability for delay-sensitive applications. The future wireless networking expectations for very high demands in terms of data rate and latency can be improved with the existing resources but changing the way of forwarding and computational requirements.

References

1. Ahlswede, R., Cai, N., Li, S.-Y.R., Yeung, R.W.: Network Information flow. IEEE Trans. Inf. Theory 46(4), 1204–1216 (2000)
2. Tassiulas, L., Ephremides, A.: Stability properties of constrained queuing systems and scheduling for maximum throughput in multihop radio networks. IEEE Trans. Autom. Control 37(12), 1936–1949 (1992)
3. Gupta, P., Kumar, P.R.: The capacity of wireless networks. IEEE. Trans. Inf. Theory 46(4), 1204–1216 (2000)
4. Li, S.-Y.R., Yeung, R.W., Cai, N.: Linear network coding. IEEE Trans. Inform. Theory 49(2), 371–381 (2003)
5. Ho, T., Medard, M., Koetter, R., Karger, D.R., Effros, M., Shi, J., Leong, B.: A random linear network coding approach to multicast. IEEE Trans. Inf. Theory 52(10), 4413–4430 (2006)
6. Wu, Y., Chou, P.A., Kung, S.-Y.: Information exchange in wireless networks with network coding and physical-layer broadcast. In: Proc. 39th Annual Conference on Information Science and Systems, CISS (2005)
7. Katti, S., Katabi, D., Hu, W., Rahul, H., Medard, M.: The importance of being opportunistic: practical network coding for wireless environments. In: Proc. 43rd Annual Allerton Conf. Communication, Control, and Computing (September 2005)
8. Chachulski, S., Jennings, M., Katti, S., Katabi, D.: Trading structure for randomness in wireless opportunistic routing. In: ACM SIGCOMM (2007)
9. Gkantsidis, C., Hu, W., Peter, K., Radunovic, B., Gheorghiu, S., Rodriguez, P.: Multipath code casting for wireless networks. In: ACM SIGCOMM (2005)

2G/3G-Connect: A Tool for Strategic Techno-Economical Studies on National-Wide Mobile Networks under a Hybrid 2G/3G Architecture

Alberto E. García[1], Laura Rodríguez de Lope[1], Klaus D. Hackbarth[1], Dragan Illich[2], and Werner Neu[2]

[1] University of Cantabria, Santander 39005, Spain
agarcia@tlmat.unican.es
[2] WIK Consult, Bad Honef, Germany

Abstract. The current increment of the service portfolio for mobile communications and the related traffic load require an extension of the capacities in the mobile networks. As a consequence, mobile network operators have to replace part of the traditional equipment based on 2G GSM/GPRS by the 3G equipment based on UMTS and even start the path in direction to the 4G installing HSPA. This leads to techno-economical studies which require a corresponding service network model and its implementation in form of a computer support planning tool. This paper presents such a model and the structure and characteristics of the corresponding tool 2G/3G-Conncet. The paper also indicates corresponding applications.

Keywords: high-level Petri nets, net components, dynamic software architecture, modeling, agents, software development approach.

1 Introduction

The growing penetration of smart phones for mobile communication causes the necessity of increasing the capacities in the corresponding mobile networks which can't satisfy the additional demands by simple 2G technology based on GSM/GPRS and brings also to limits the GPRS improvement by upgrading them with equipment under the enhanced GPRS data rate for GSM evolution (EDGE). Hence, mobile network operators are accelerating the implementation of UMTS, currently envisaged in the 2100 MHz band. In some countries, mobile network operators offer so called broadband mobile access (BMA), mainly applied by end-user with laptop computers. This service by one side, offered in competition to fixed broadband access under corresponding xDSL technology, is based on the mobility paradigm mainly of the young generation. By the other side, it is applied for covering so called white areas where xDSL or future PON architecture can not be implemented by economic reasons, and hence broadband mobile access overcomes the so called digital division in rural area. It can

J. Del Ser et al. (Eds.): MOBILIGHT 2011, LNICST 81, pp. 275–290, 2012.

be shown that for BMA the capacity offered by UMST is not sufficient and hence a corresponding enhancement is required and implemented by additional HSPA or HSPA+, see [7]. HSPA or HPSA + can be seen as a bridge technology for a smooth development in the direction of 4G under the long term evolution paradigm, see [14], similar as EDGE provided an enhancement of 2G technology in the direction of 3G one.

Similar changes happens in the fixed network part of the mobile network were traditional technology based on ATM and SDH equipment or corresponding leased lines for transport and mobile switching centres are replaced by technology either over Ethernet based radio links, or corresponding leased lines. For higher demands even leased dark fibre withEthernet over fibre technology might be applied in the BSS/UTRAN part, while the NSS part will apply full IP/MPLS systems. The classical mobile switching centres are hence substituted by corresponding Soft-switch equipment composed by Media Gateways and MSC call servers. Under the paradigm of Fixed-Mobile Integration (FMI) it can be expected that the BSS, the UTRAN and NNS will be at least partially integrated with infrastructure of the NGN in case that both networks are operated by the same operator, see [5].

As a consequence, current public mobile networks are strongly hybrid in the cell deployment where equipment ranging from simple GSM over GSM/GPRS or EDGE coexists with UMTS and HSPA or HSPA+. In the fixed network part current ATM-SDH transport paradigm is changed to carrier Ethernet and IP/MPLS. The cell deployment and the application of the corresponding technology mix depend strongly on the services offered by the public mobile network operator and the traffic requirement from the end users, see [8]. Hence, entities involved in the market place of mobile services and corresponding networks require strategic planning tools which allow to study the consequences of the increasing bandwidth demand by current and future applications from both, the technical implication but also the economical one.

This paper presents a computer tool for techno-economical studies in hybrid 2G/3G mobile networks based on a LRIC cost model and a scenario concept, which handles the required large parameters set, likethe number and distribution of potential end user in a country, the type of services to be integrated and the corresponding traffic demand, the applied range of technology and other parameters like radio spectrum in different frequency bands.

The paper provides in the second chapter a description of the model for the network design and dimensioning and provides in the third chapter an introduction to the applied TELRIC cost model for the economical analysis of the network configuration resulting form the network planning part. The fourth chapter outlines applications of the 2G/3G-Connect tool and the last chapter provides conclusions and indicates future extensions of the model and the corresponding tool.

2   Description of the Model for 2G/3G-Connect Tool

Telecommunication service and network models for strategic techno-economical studies are divided mainly into three main parts:

- Scenario generation
- Network planning, composed by:
  - Cell deployment
  - Fixed network design and dimensioning
  - System assignment to the network elements.
- Economical analysis by a corresponding cost model.

The model considers that current and mainly future PLMN has to support a strong set of applications and hence must provide a set of services supporting them. The services are differentiated not only by their bandwidth requirements but also by their quality of service (QoS) parameter. 2G/3G-Connect approximates QoS by three parameters: the accessibility to a service, expressed by a maximum value for the blocking probability under a traffic load in the common business hour (BH), the mean delay value for a connection between its source and destination location (end-to-end delay, e2e) and its jitter. The first parameter is considered mainly in the cell deployment applying an extension from the Erlang-loss model for multi-services, see [10]. The second and third parameters are applied mainly in the dimensioning of the output interfaces for the layer-2,3 equipment in the fixed network part and use a corresponding waiting model. For this purpose, the model for 2G/3G-conncet applies the concept of equivalent bandwidth requirement under a value which lies between the mean bandwidth and the maximum bandwidth of a service, see [6]. For this purpose 2G/3G-connect allows that each service gets associated a so called QoS class which describes the required QoS parameter.

Additionally, 2G/3G-Connect includes several options for improving the service availability which are:

- Doubling key network element and providing a traffic distribution on equal terms.
- Assignation of lower level nodes to two higher ones e.g. RNC locations to core network locations and providing a traffic distribution on equal terms.
- Congestion avoidance against unforeseen traffic loads by over-dimensioning of the transmission capacities. For this purpose 2G/3G-connect allows to provide corresponding parameters individually for each network level.

This chapter describes the main functions of the scenario generator and the associated network planning. Figure 1 shows the main functional blocks treated in this chapter implemented in the 2G/3G-connect tool by corresponding program modules.

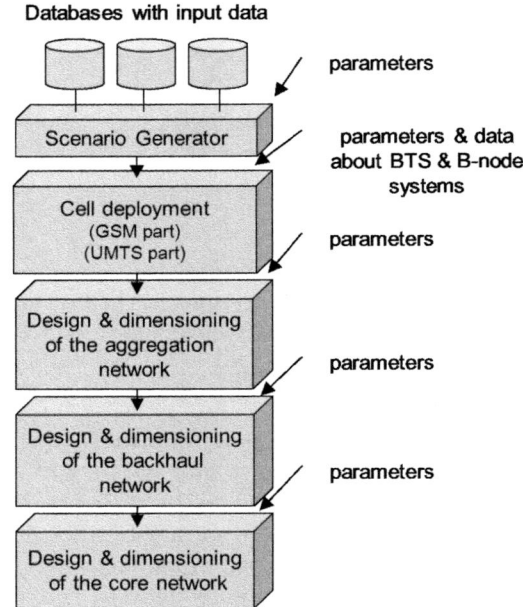

Fig. 1. Modules for the network planning implemented in 2G/3G-connect

2.1   Scenario Generator

The objective of the scenario generator is to provide the input data for a national mobile network planning. The model for 2G/3G-Connect is composed by the following scenarios:

- Geographical scenario, composed by geographical, topographical and population data.
- Service scenario, which defines the services set handled by the 2G/3G network.
- User and traffic scenario which considers different types of user and its traffic-service relation.
- Architecture scenario, defining the technology applied for the cell deployment and the fixed network part and its relation with the user and the geographical scenario.
- Frequency and spectrum scenario, indicating the frequency bands and the amount of spectrum associated to the applied cell technologies.

Geographical scenario. 2G/3G-Connect considers for the geographical scenario a list of small geographical units with their corresponding coordinates, topographical situation defined by three attributes (flat, hilly, mountainous) and the number of inhabitants; often these data are deduced from Postal Area Codes, see [2]. The geographical scenario includes also data about the road and the railway network of the country and data about special areas with high population concentration as airports, railway-stations, shopping malls etc.

Based on these data the scenario generator creates areas and joints adjacent areas to districts. The areas are then classified depending on their population density in rural, suburban and urban.

For considering mobility of the population and traffic estimation for roaming, the geographical scenario considers additional data about economical activities in the areas and classifies the areas as business or residential ones. For roaming, the scenario generator also considers data about hotels and its occupation.

Service scenario. The service scenario describes the characteristics of the different services offered by the considered network. The main parameters are:

- Mean packet length and mean bandwidth requirement separated for the uplink and downlink packet stream.
- Mean duration of the service.
- Parameters for the traffic distribution to different destinations.
- Assignation of the service to a QoS class.

Table 1 shows a corresponding example for the most important services considered by 2G/3G-Connect.

Table 1. Typical service scenario for 2G/3G mobile networks (M2M: relative traffic between mobile user in the same network, M2F: relative traffic from a mobile user of the considered network and a user in an other one, F2M: inverse from M2F, M2ICIP: traffic from a mobile user to an IP network over a corresponding interconnection point, M2 MobSer traffic between a mobile user and a server connected with the considered mobile network)

service characteristics	Parameters for capacity requirement					Traffic distribution parameters					
	Mean Bandwidth		Mean packet length		Mean duration	M2M	M2F	F2M	M2 ICIP	M2 Mob Ser	QoS class
	up link	down link	up link	down link							
Dimension	kbps	kbps	bytes	bytes	min	-	-	-	-	-	-
Real time voice	12.2	12.2	25	25	3	0.4	0.3	0.3	0	0	1
Other real time serv.	16	64	100	100	15	0	0	0	0.8	0.2	1
Streaming to content server	1	80	3.0	256	5	0	0	0	0.7	0.3	2
Guaranteed data	20	80	30	256	1	0.1	0	0	0.7	0.2	3
Best effort	20	80	30	256	3	0.01	0	0	0.6	0.3	4
SMS	9.6	9.6	100	-	0.001	0	0	0	0	1	4
MMS	40	40	1000	-	0.002	0	0	0	0	1	4
Mobile broadband access	40	160	256	256	5	0	0	0	0.4	0.6	4

User and traffic scenario. The user and traffic scenario classifies the total number of users into three different categories: business, premium and standard, and associates them a traffic matrix which indicates the BH traffic of each user type to the corresponding services. As 2G/3G-connect considers hybrid sites, the corresponding scenario must provide for each service input data about the traffic distributions over different cell technologies installed in the same hybrid site; table 2 shows an example related with the service scenario shown in table 1.

Table 2. BH traffic values and their distribution over the different cell technologies in case of hybrid sites

Service and traffic per user	GSM traffic per user in hybrid sites	UMTS traffic replaced by HSPA	BH traffic per user (Erlang or messages		
			Business	Premium	Standard
real time voice	0.8	0	0.05	0.005	0.006
other real time services	0.2	0	0.01	0.0025	0
streaming to content services	0.0	0.5	0	0.005	0
guaranteed data with business server	0	0.5	0.002	0	0
best effort to general server	0	0.5	0.001	0.01	0.002
SMS	0.2	0.1	0.1	0.05	0.01
MMS	0.2	0.2	0.01	0.02	0
Mobile Broadband Access	0	1	0.01	0.005	0

Architecture scenario. 2G/3G-connect considers as network architecture the complete spectrum of technologies ranging from second generation GSM mobile networks, including GPRS and EDGE, up to third generation by UMTS including its extension to HSPA. For this purpose the scenario generator associates to each area a cell type depending on its type and user density. Table 3 shows the cell type and indicates the criteria for its selection.

Concerning the fixed network part 2G/3G-connect considers for bandwidth aggregation in the UTRAN/GERAN part layer-2 equipment based on carrier Ethernet and for the routing function concerning the traffic to be routed among the core network location layer-3 equipment based on IP/MPSL. For the transmission, 2/3G-Connect differentiate mainly between operators which provide a proper physical infrastructure, mainly based on radio links, and operators which use digital or optical leased lines from a different operator. The selection can be provided individually for each network level, e.g. an operator decides to implement proper radio links in the UTRAN/GERAN network part and use electrical or optical leased line for the core network part.

Table 3. Cell types consider by 2/3G-connect

Cell type	Identifier	Related parameters
GSM/GPRS	1	GSM "up to" user threshold for each area type
GSM/EDGE	2	As type 1
UMTS	3	GSM and GSM/UMTS "up to" user thresholds
UMTS/HSPA	4	As type 3
GSM/UMTS	5	GSM/UMTS "up to" user threshold for each area type
GSM/UMTS/HSPA	6	As type 5

Frequency and spectrum scenario. The frequency and spectrum scenario gives the frequencies associated to the different cell technology and the corresponding spectrum. 2G/3G-Connect allows a flexible frequency and spectrum assignment; table 4 shows an example. As a consequence, 2G/3G-Connect provides the possibility for studying the influence of different combinations of frequency and spectrum assignment to the network design and dimensioning and the corresponding economical implications.

Table 4. Example for a frequency and spectrum assignment

Frequency Band	GSM Spectrum (MHz)	UMTS Spectrum (MHz)
800	Not Applicable	0
900	8	0
1800	18	0
2100	Not Applicable	15
2600	Not Applicable	0

2.2   Network Planning

Strategic Telecom Network Planning studies the development of the corresponding telecommunication network under a medium up to long term development. For this purpose different scenarios, as shown in the last section are considered. Network planning is provided by two main steps, network design and network dimensioning. 2G/3G-Connect bases the network design on the set of districts and their corresponding areas which compose them, as shown in section 2.1, and determines the locations for higher level functions and for interconnection points with other networks.

Based on these locations, the network design determines the structure for the logical level and the topology of the physical one. 2G/3G-Connect considers the architecture as determined by the 3GPP under Release 4, see [1] and some extensions for HSPA considering the corresponding extensions from 3GPP in Release 5 and 6. Figure 2 shows the main functional blocks considered in the model for 2G/3G-connect.

The dimensioning corresponds to the traffic routing over the network structure and topology and provides the required capacity in each network element, mainly based on locations (nodes) and links connecting them. Finally, 2G/3G-Connect assigns corresponding equipment to each network element providing the bandwidth required for the capacities resulting from the network dimensioning.

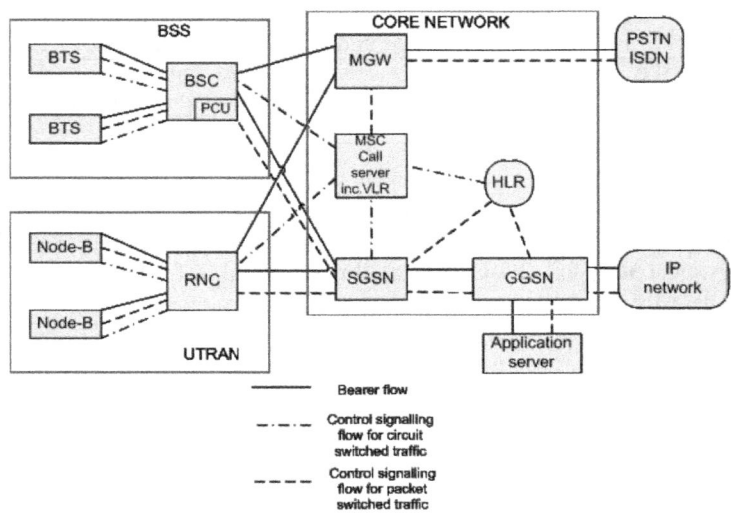

Fig. 2. Functional blocks of a 2G/3G mobile network

Cell deployment. 2G/3G-Connect receives from the scenario generator a list of districts with their corresponding areas, and for each area the required cell types. Based on these data the cell deployment determines for each area the number of sites, the required capacities and the corresponding radio equipment (BTS/node B/HSPA).

The cell deployment calculates for each area the amount of traffic, separately for each service, handled by the radio equipments under its corresponding QoS parameter determined by the QoS class associated to the service. These traffic values are then the main input data for the fixed network design and dimensioning.

Fixed network design and dimensioning. The fixed network part of a mobile network is composed by the following parts:

- Aggregation network part, covering the location of the base station sites up to the locations where the controller equipments (BSC, RNC) are installed.
- Backhaul network part, covering the controller locations up to the core network locations.
- Core network part, connecting the different core network locations and providing interconnection to other networks.

In both 2G as 3G mobile networks the traffic between the user equipment to other users, a server or a corresponding interconnection point with other networks must always be switched hierarchically up to the core network part. Only the core network part provides corresponding routing functions to distribute the traffic to other users inside the network, a proper server or to an interconnection point for users or servers connected in other networks. Hence, 2G/3G-Connect provides the corresponding network design starting from the site locations and the corresponding traffic demand up to the core network part.

Aggregation network part. The aggregation network connects the site locations in each area to a corresponding aggregation point (cell hub). 2G/3G-Connect considers the centre of each district as cell hub location and a pure star structure for these connections. Note that the cell hub location does not provide any traffic aggregation but only one for physical capacity requirements. Hence the layer-2 equipment in the cell hub location provides grooming of physical groups of lower capacity connections from the sites (E1 or 10Mb Ethernet) to higher ones (E3, STM-1, FE or 1GE).

Once the capacities requirements of the cell hub location are determined 2G/3G-Connect detremines the controller node locations taken the number of locations as an input parameter and assign each cell hub location to one controller location. For this purpose 2G/3G-Connect applies a heuristic algorithm based on p-median model resulting from graph theory and applied already in former network design and costing model, see [3]. Figure 3 shows the logical structure for this network part.

Concerning the topology for the physical layer between the cell-hub locations and the controller ones, 2G/3G-Connect considers a tree structure and applies an algorithm, already used in former network design models which allows calculating

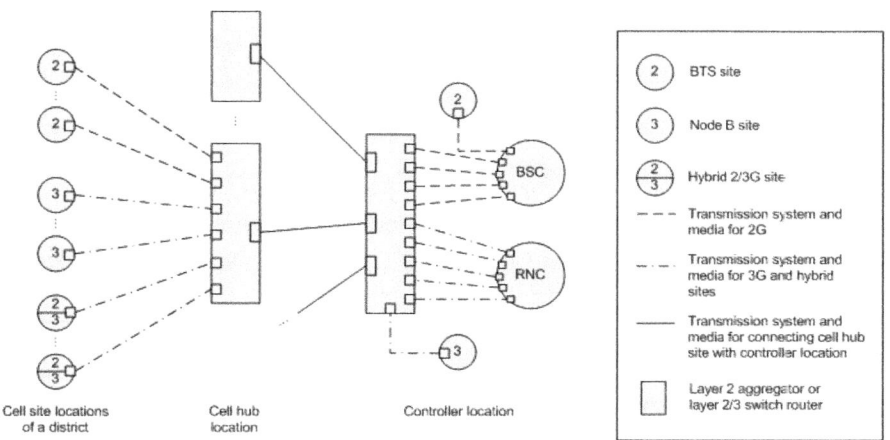

Fig. 3. Logical structure of the aggregation network for 2G/3G mobile networks

different tree structures, ranging from a pure star structure where the amount
of traffic flow is minimised up to a pure tree structure which minimises the total
geographical lengths (minimal spanning tree), see [2]. The optimal tree structure
depends on the applied equipment and its length and capacity depending cost.
From a practical point of view, a star structure is applied under leased line
connections including dark fibre while radio link systems require a tree structure,
minimising the number of radio link systems considering the distance limitations
of the RF system (typically 50km). Fig. 4 shows an example for a tree structure.

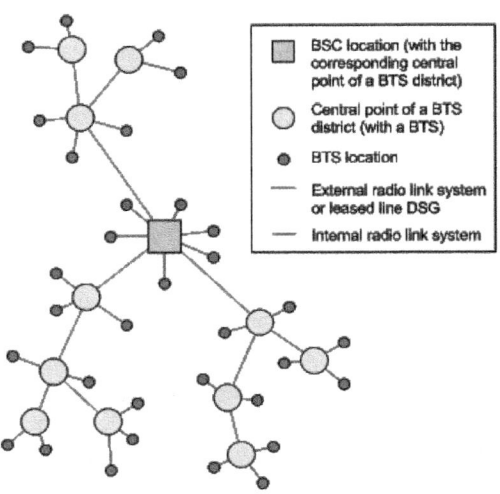

Fig. 4. Example for a tree topology in an aggregation network

Once the logical structure and the physical topology is calculated, 2G/3G-
Connect provides the traffic routing up the controller node locations and the
required capacities for the system assignment for the star links between the sites
and the cell hub locations and the tree links between the cell hub locations and
the controller node ones.

Backhaul network part. From the aggregation network design and dimensioning results the backhaul (controller) node locations and the corresponding
traffics from the different services handled by the controller nodes. 2G/3G determines in the backhaul network part the location for the core node locations as
a subset from the backhaul ones and assign each controller node location to one
core node location; 2G/3G-Connect provides an option to provide an assignment
to two core node locations for reasons of network availability. For this purpose
2G/3G applies a similar algorithm as in the design of the aggregation network
part. Figure 5 shows the logical structure of the backhaul network.

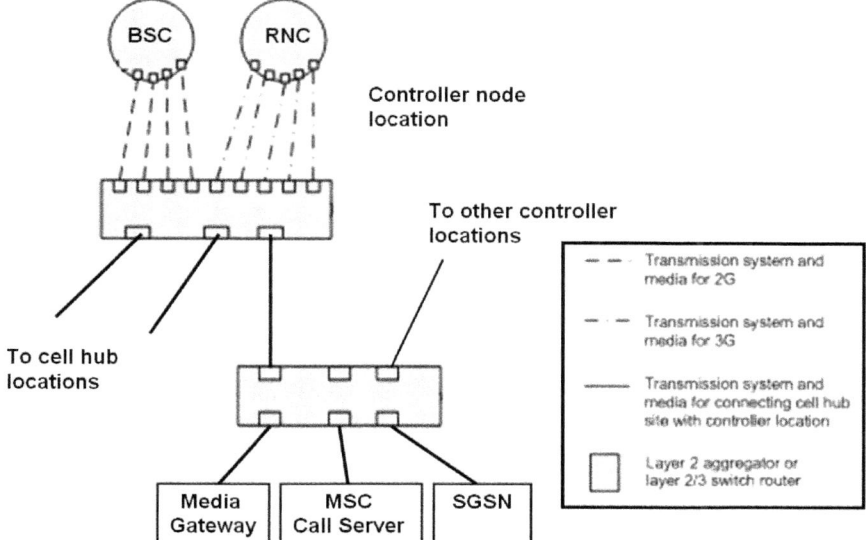

Fig. 5. Logical structure of the backhaul network for 2G/3G mobile networks

2G/3G-Connect considers for the physical topology again a star topology, and in case of applying double assignation a corresponding double star. Star- or double star structures are mainly applied for leased line including dark fibre. Due to the high traffic concentration in this network part 2G/3G-Connect considers in case that an operator implements a proper physical infrastructure a mapping of the logical star or double star structure into a set of ring topologies. The corresponding algorithm allows considering geographical obstacles and a limitation of the number of nodes which can be included into a ring; figure 6 shows an example. The calculation of the ring topologies is based on a heuristic algorithm for the travelling salesman problem; see [9].

Once the logical structure and the physical topology are calculated, 2G/3G provides the traffic routing from the controller node locations to the core nodes ones and determines the required capacities for the system assignment on the star or ring links.

Core network part. From the backhaul network design results the core network locations and the aggregated traffic from each service. Based on the total aggregated traffic in the core nodes, 2G/3G-Conncet selects the location of interconnection points to PSTN/ISDN&PLMN but also to IP based network and determines the locations where different types of servers and other central elements are installed. 2G/3G-Connect provides this selection based on the traffic weights of the core nodes, while the number of core nodes over which these elements should be distributed is given as an input parameter. The dimensioning of these elements is based on corresponding capacity drivers. 2G/3G-Connect

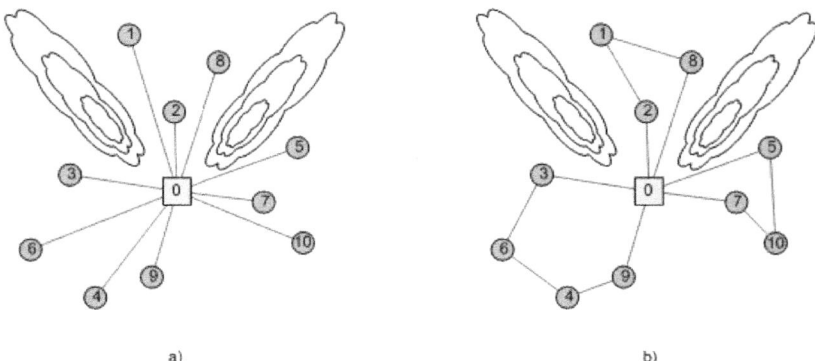

Fig. 6. Example of mapping a star structure into multiple ring topologies

considers that the capacities should not be fully used but an overhead is provided to anticipate a no estimated traffic increase. Table 5 shows the core network elements considered by 2G/3G-connect and an example for the corresponding input parameters considered in the dimensioning.

After the determination of the locations for the core network elements, 2G/3G-Connect distributes the traffic resulting from the different services to the corresponding elements. In case that this element is installed in a different core location 2G/3G-Connect routes the traffic to the corresponding one. For this purpose 2G/3G-Connect calculates for each service a traffic matrix between the core nodes and applies a distribution based on the core nodes weight related with the corresponding service e.g. the number of aggregated user. Figure 7 shows an example for this traffic distribution.

The traffic between different core node locations is routed over the logical structure of the core network which determines the required capacity for the physical layer. From practical studies results that the number of core node location in national mobile networks is small and hence the traffic concentration is high. Hence, 2G/3G-Connect considers for the logical structure a full meshed one and for the physical topology again a full meshed or optionally a ring topology. The last is mainly applied when the operator provides a proper transmission infrastructure while the first one is mainly appliedy under the application of leased lines.

3   Cost Model

2G/3G-Connect contains an economical evaluation based on a long run incremental cost (LRIC) model. In the LRIC model the cost evaluation contains two main cost types: Investment cost (CAPEX) and cost for the operations (OPEX). The first are calculated based on the network configuration resulting from the network planning and corresponding annual cost are deduced bt distributing the total investment cost over the life time of the network elements. In general,

Table 5. Core node location

Position	Driver for dimensioning and costing	Utilization ratio	N of SwRo locations
Media Gateway	N of E1/STM1 ports, BH traffic	70%	All SwRo nodes
MSC call server including VLR	BHCA	67%	Input
HLR	Number of subscribers, BHCA	80%	Input
EIR	Number of subscribers	60%	Input
SMSC	Number of SMS per s	80%	Input
SGSN	a. BHCA b. n of Attached Subscribers c. Throughput, in Mbps	78%	Input
GGSN	a. Throughput, in Mbps b. PDP context	77%	All SwRo with IP Interface
OAM	Considered in OPEX	n.a.	N.a.
Billing	Considered in indirect cost	n.a.	N.a.
IN	a. BHCA b. traffic over all services	80%	Input
Network management system (AAA, DNS functions)	BHCA over all services	80%	Input
MGW Interface card to the PSTN/ISDN/PLMN for circuit switched voice traffic	N of E1 ports	80%	At all SwRo with PSTN/ISDN interconnection
MGW Interface card to the PSTN/ISDN/PLMN for packet switched voice traffic	N of E1 ports	80%	At all SwRo with PSTN/ISDN interconnection

OPEX is considered by a mark up factor to the CAPEX. Figure 8 shows the different steps of the LRIC. Its application to 2G mobile network can be find in [6] and an overview about LRIC models is given in [11].

2G/3G-Connect applies a variant of LRIC referred to as total element long-run increment cost (TELRIC) which considers for the cost determination for the different services the routing path in the network. It calculates the cost for a unit of service as the weighted sum of the costs of all network elements used for the service, see [12]. The weights for the network elements are the so-called routing factors which reflect the way a particular service is routed over the various network elements and, in particular, how often a network element is used for the service.

The (cost) routing factors are determined from the network planning and hence are closely correlated with the traffic routing. In mobile networks, the traffic is always routed to the next core node location where a distribution is provided. From this follows that for a traffic unit for example for on-net voice traffic, there is always a routing factor of two used for all network elements from the cell site to the corresponding core network location. TELRIC provides a

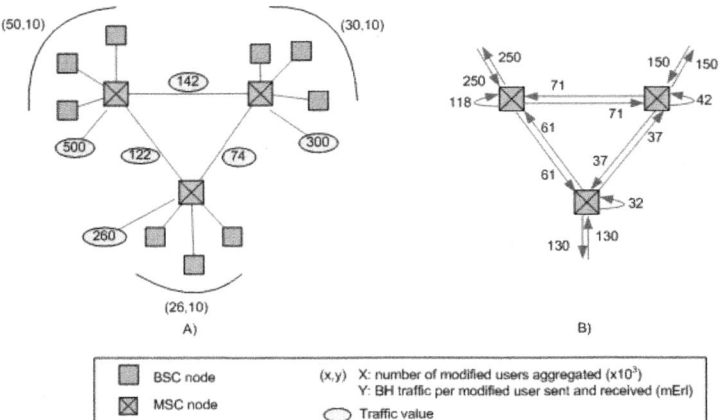

Fig. 7. Example for the traffic distribution of the aggregated traffic in the core nodes based on the weights resulting from the number of aggregated users

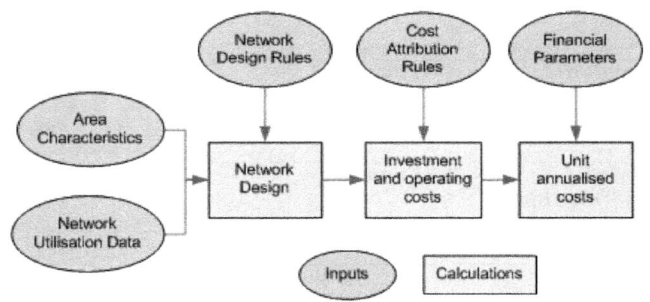

Fig. 8. Scheme for the LRIC model

distribution of network cost to the services that corresponds closely the services' use of the different network elements.

A special cost evaluation results from the so called marginal LRIC (also named pure LRIC) which considers only the cost increment provided by the traffic for an additional service. The next chapter provides an example.

4   Applications

The 2G/3G-Connect tool allows a strong set of applications which can be divided into three types:

- Practical techno-economical studies
- Applications for applied research studies
- Applications in higher level education.

2G/3G-Connect provides an instrument for practical techno-economical studies as strategic studies for mobile network operators who want to estimate the medium-long term tendency in its network evaluation and corresponding investment under different assumption of service and traffic evolution. These studies might be required also from equipment providers due to the fact that a number of mobile network providers require a complete solution "key in hand". But even financial institutions might require results for corresponding studies for risk estimation of credits given to network operators for the modernisation or a new implementation of a corresponding network. Last but not least 2G/3G will have a wide range of application in telecom regulation for national or international regulation entities to fix cost limits and parameters for different types of service e.g. for studying the influence of different parameters as QoS, service availability etc. to the corresponding service cost. For this purpose 2G/3G-Connect supports both TELRIC and pure LRIC as required from the current EU-regulation for mobile networks, see [4].

2G/3G-Connect can support applied research studies for network simulation under different technology scenarios for future qualitative and quantities equipment development or for the introduction of a new type of technology. An example is the study of the implication of changing the classical ATM layer equipment by new carrier Ethernet equipment and its evolution into the direction of equipment under MPLS-TP.

It is planned that 2G/3G-Connect will be applied in master courses for mobile network planning as currently provided by a consortium of five Spanish Universities; see [13].

5   Conclusions and Future Extensions

The contribution shows that the 2G/3G-Connect tool, based on a corresponding LRIC-cost model, allows a wide range of techno-economical studies for national wide mobile networks. Due to the incorporation of mobile broadband access using HSPA technology 2G/3G-Connect provides also a step in the direction of 4G networks based on LTE.

The team which developed the model and the corresponding tool is working on an extension for the future service and network evolution which considers in the fixed network part "fixed mobile integration (FMI)" under a common core network and an extension in the cell technology deployment by LTE based on OFDM as defined in release 8 of the 3GPP; see [5], [14].

References

1. Overview of 3GPP Release 4, 3GPP doc. Nr. TP-040053,
   http://www.3gpp.org/ftp/tsg_t/TSG_T/TSGT_23/Docs/PDF/TP-040053.pdf
2. Mobile Termination Cost Model for Australia, published by the Australian Competition & Consumer Commission (2007),
   http://www.accc.gov.au/content/item.phtml?itemId=783055&nodeId=1a2eee9
   394ef3123590dbf874692a13b&fn=Mobile+termination+cost+model+for+
   Australia+(WIK+report).pdf

3. A analytic cost model for broadband networks, Bundesnetz-agentur Bonn (2005), http://www.bundesnetzagentur.de/media/archive/2078.pdf

4. European Commission, Recommendation on the Regulatory Treatment of Fixed and Mobile Termination Rates in the EU C(2009) 3359 final, Brussels (2009), http://www.ictregulationtoolkit.org/en/Publication.3837.html

5. Esteve Rothenberg, Fixed-mobile Convergence in TISPAN/3GPP IMS. VDM Verlag (2008)

6. Garca, A.E., de Lope, L.R., Hackbarth, K.: Application of cost models over traffic dimensioning with QoS restrictions. Annales of Telecommunication 65(3/4), 3–14 (2010)

7. Holma, H., Toskala, A. (eds.): WCDMA for UMTS- HSPA evolution and LTE, 5th edn. John Wiley and Sons (2010)

8. Holma, H., et al.: UMTS services in Holma. In: Toskala, A. (ed.) WCDMA for UMTS- HSPA evolution and LTE, 5th edn. John Wiley and Sons (2010)

9. Lin, S., Kernighan, B.: An effective heuristic algorithm for the travelling salesman problem. Operation Research 21, 498–516

10. Lindberger, K.: Blocking for multi-slot heterogeneous traffic streams offered to a trunk group with reservation in Traffic Engineering for ISDN, Design and Planning. In: Bonatti, M., Decina, M. (eds.) Proc. 5th ITC Seminar, pp. 151–160 (1988)

11. Noumba, P., et al.: A Model for Calculating Interconnection Costs in Telecommunications, Mixed Media edited for The World Bank (2004)

12. The use of Long Run Incremental Cost (LRIC) as a costing methodology in regulation, OFTEL (2004), http://www.ofcom.org.uk/static/archive/oftel/publications/mobile/ctm_2002/lric120202.pdf

13. Master y Doctorado en Tecnologias de la Informacion y Comunicaciones en Redes Mviles, http://www.ticrm.es/vigente/index.php

14. Toskala, A., Holma, H.: UTRAN long term evolution. In: Holma, H. (ed.) WCDMA for UMTS, 5th edn. John Wiley and Sons (2010)

Strategic Mobile Network Planning Tool for 2G/3G Regulatory Studies

Juan Eulogio Sánchez-García,
Amir M. Ahmadzadeh, Beatriz Saavedra-Moreno,
Sancho Salcedo-Sanz, and J. Antonio Portilla-Figueras

Universidad de Alcalá, Spain
antonio.portilla@uah.es

Abstract. Techno-economical studies in telecommunication networks are important and quite complicated topics. One of the major challenges is to estimate the investment in new or additional network equipment to consider a new service, or to compare two technologies in the same or different frequency band. In order to carry out this kind of analysis, a nationwide network design and dimensioning study has to be done, what requires powerful and robust algorithms with a high degree of flexibility to obtain accurate results avoiding huge amount of input data. This paper presents a strategic network planing tool which implements algorithms for 2G and 3G network dimensioning. This tool can be used by mobile network operators or regulatory authorities to develop or support techno-economic analysis in communication networks.

Keywords: GSM and UMTS design algorithms, strategic studies, network investment.

1 Introduction

Mobile telecommunication market has become one of the most relevant contributors to the nation economies in developed countries. The fast evolution from 2G Global System for Mobile Telecommunications (GSM) to 3G Universal Mobile Telecommunication System (UMTS), and the increasing demand for new data services, makes that the mobile operators have to move fast to deploy network infrastructure to cover this demand. However, an incorrect network deployment (due to an incorrect technology or frequency band selection) may derive into huge investments with non-desired results, i.e lack of coverage or capacity. To avoid this nightmare scenario, mobile network operators must carry out exhaustive strategic studies to determine which technology have to be deployed, which frequencies must be used and, of course, where the deployment should be deployed (cities, towns) and even in which area of each specific city (urban, suburban and rural) [8].

Strategic studies are hard to be carried out because they have to consider nationwide multi-service scenarios, with important differences in demography, topography and user and service profile in the different cities considered [1]. Thus,

J. Del Ser et al. (Eds.): MOBILIGHT 2011, LNICST 81, pp. 291–302, 2012.

in order to do this kind of studies, it is mandatory to use software tools that implement algorithms for the network design and dimensioning. In this paper we present one tool and the algorithms for 2G and 3G network dimensioning, specifically implemented to support strategic and techno-economic studies in telecommunication networks. The algorithm for 2G GSM is based on the independent calculation of the cell range by propagation, using an empirical propagation model, and by traffic load, using a classical Erlang formulation based on the slotted feature of the GSM system. The algorithm for 3G is quite more evolved due to the features of the 3G UMTS access technique which requires a joint optimization of the capacity and propagation. After calculating the cell range for a single site, a procedure to estimate the number of sites (BTS or Node B) required to provide coverage to a specific area is presented, considering several types of BTS and Node B.

The rest of the paper is structured as follows. Section 2 and section 3 describes the models and methods for designing the network. Section 4 presents a study for the metropolitan area of Bilbao (Spain) and a nationwide example for the comparison among 2G GSM and 3G UMTS in the same and different frequency bands. Finally, in the last section we present some conclusions about the tool and the studies carried out with it.

2   Models and Methods for 2G-GSM Dimensioning

The objective of a network dimensioning problem is to calculate the number of network elements (Base Stations (BTS)) to be deployed in each area of a specific city (urban/suburban/rural). To calculate the number of BTS, the area covered by each one has to be obtained finding the value of the cell range of the BTS under study. To do this, we consider the following input parameters:

- A set of services S. Each service is defined by several parameters: user density in the area, individual traffic per user and binary rate of the service.
- A set of 2G-BTS, BTS. The main parameters of each BTS are: Transmission Power, Number of Sectors, Number of transceiver (TRX) per sector and cost of the BTS (C_i).
- Radio Propagation parameters P, where the fading margins, frequency bands (900 and 1800 MHz), and different propagation losses are specified.
- General parameters of the operator G, such as market share, market penetration and amount of spectrum allocated in each frequency band.

The cell range of a 2G BTS is the minimum value between the cell range by radio propagation R_P and the cell range by capacity R_T. Next we will show how these two values can be estimated for 2G networks.

2.1   Calculation of the Cell Range by Radio Propagation R_P in 2G Networks

Propagation coverage studies mainly imply two steps. The first one is to calculate the maximum allowed propagation loss in the cell, defined here as $L_{pathloss}$, and the second is to use an empirical propagation method to calculate the cell radius for this pathloss. Typical methods are the Okumura Hata COST 231 model, [2], or the Walfish Ikegami model [3].

The value of $L_{pathloss}$ is calculated using a classical link budget equation:

$$P_{Tx} + \sum G - \sum L - \sum M - L_{pathloss} = R_{Sens} \tag{1}$$

where P_{Tx} is the transmitter power, $\sum G$ is the sum of all gains, $\sum L$ is the sum of all the losses, R_{Sens} is the receiver sensitivity and finally. Note that all the parameters in Equation (1) are inputs of the system, defined in the set BTS and P, and therefore $L_{pathloss}$ can be obtained from this equation.

As it was mentioned before, the cell radius by propagation is obtained applying the $L_{pathloss}$ into an empirical propagation method. In our work we have used the 231-Okumura Hata model because it is broadly considered as the most general one in mobile networks applications [4]:

$$\begin{aligned} L_b = {} & 46.3 + 33.9 \cdot Log(f) - 13.82 \cdot Log(h_{BTS}) - a(h_{Mobile}) \\ & + (44.95 - 6.55 \cdot Log(h_{BTS})) \cdot Log(R_P) + C_m \end{aligned} \tag{2}$$

where f is the frequency in MHz, h_{BTS} is the height of the BTS in meters, h_{mobile} is the height of the mobile user in meters and R^P is the cell radius by propagation in Km. Note that $a(h_{Mobile})$ and $C(m)$ are parameters defined in the COST 231 specification. They provide the influence of the height of mobile terminal and the type of city, respectively, and they are defined as follows:

$$a(h_{Mobile}) = (1.1 \cdot Log(f) - 0.7) \cdot h_{Mobile} - (1.56 \cdot Log(f) - 0.8) \tag{3}$$

$$C_m = \begin{cases} 0dB \ for \ medium \ sized \ cities \ and \ suburban \ centres \\ 3dB \ for \ metropolitan \ centres \end{cases} \tag{4}$$

Note that if we are considering different services, the receiver sensibility may change, the propagation coverage study has to be done specifically for each service, and of course for the uplink and the downlink channels. Therefore the formulation explained above, and the value R_P, has to be applied for each service $i \in S$ and for each direction (Uplink (UL) and Downlink (DL)), obtaining a set of two vectors, containing, for each service, the cell radius by propagation, ($R_{P_{UL}}$ and $R_{P_{DL}}$):

$$\mathbf{R_{P_{DL}}} = \{R_{P_{DL\_i}}; i = 1, \dots, S\}$$
$$\mathbf{R_{P_{UL}}} = \{R_{P_{UL\_i}}; i = 1, \dots, S\} \tag{5}$$

Cell range by propagation in the frequency band considered is the minimum value among them R_P, which is the most restrictive one.

2.2 Calculation of the Cell Range by Traffic R_T in 2G Networks

A 2G GSM network is a hard blocking system. This means that the capacity, i.e., the maximum number of users that a given BTS can support, depends directly on the amount of hardware in the BTS, [4]. Furthermore, GSM is a time division multiple access (TDMA) synchronous slotted system. Therefore the capacity required by any service can be expressed as multiple of the basic capacity unit, the slot of the TDMA frame. Following this reasoning, the process for calculating the cell range by traffic is the following:

The maximum number of available traffic channels per sector in the BTS is calculated using Equation (6):

$$N_{TCH_{Sector}} = N_{TRX_{Sector}} \cdot N_{TCH_{TRX}} \tag{6}$$

where $N_{TRX_{Sector}}$ is the number of TRX per sector and $N_{TRX_{Sector}}$ the number of traffic channels available for the cell per TRX.

With this value and using the most restrictive blocking probability Pb_i of the services $i \in S$, the maximum offered traffic A^O_{Sector} in the sector is calculated applying the inversion of the Erlang B formulation:

$$A^O_{Sector} = E_B^{-1} \left(N_{TRX_{Sector}}, Min(Pb_i) \right) \tag{7}$$

From the set of services S we can calculate the amount of traffic demanded by a single user a_{User} by means of Equation 8:

$$ad_{User} = \sum_{i=1}^{S} a_i \dot{S}lots_i \tag{8}$$

The maximum number of potential customers in the sector is calculated by the division of the total traffic supported by the sector by the individual user traffic:

$$M_{User} = \frac{A^O_{Sector}}{ad_{User}} \tag{9}$$

And finally the cell range by capacity can be estimated considering the number of sectors in the BTS and the user density:

$$R_T = \sqrt{\frac{M_{Users} \cdot N_{Sectors}}{\pi \cdot \rho_{User}}} \tag{10}$$

2.3   Estimation of the Cell Range of a Single Site

The algorithm for a single site dimensioning works as follows: for the lowest frequency band available, i.e., the best in terms of radio propagation, the values of R_T and R_P are calculated as described before. If $R_P < R_T$ the site (BTS) is considered as *propagation driven* and the cell range $R_C = R_P$. Otherwise the site is considered as *traffic driven*. In this case we need to check out whether the use of a second frequency band (if available) makes the cell range R_C to increase. In the second band, the same BTS type or another BTS type can be installed, therefore a new value of the cell range by traffic R_T^2 and by propagation R_P^2 have to be calculated. With the value $R_C^2 = Min(R_T^2, R_P^2)$ the number of users served in the second band are obtained. These users are already served, and therefore we do not to consider them any more in the cell range calculation in the first band. Now, the R_T^1 is re-calculated. Note that $R_T^1 > R_T$ because we have removed the users in the second band. Finally we compare again R_P with R_T^- and select the minimum one as the final cell range of the site, $R_C = Min\left(R_P, R_T^1\right)$. The complete process is depicted in Figure 1.

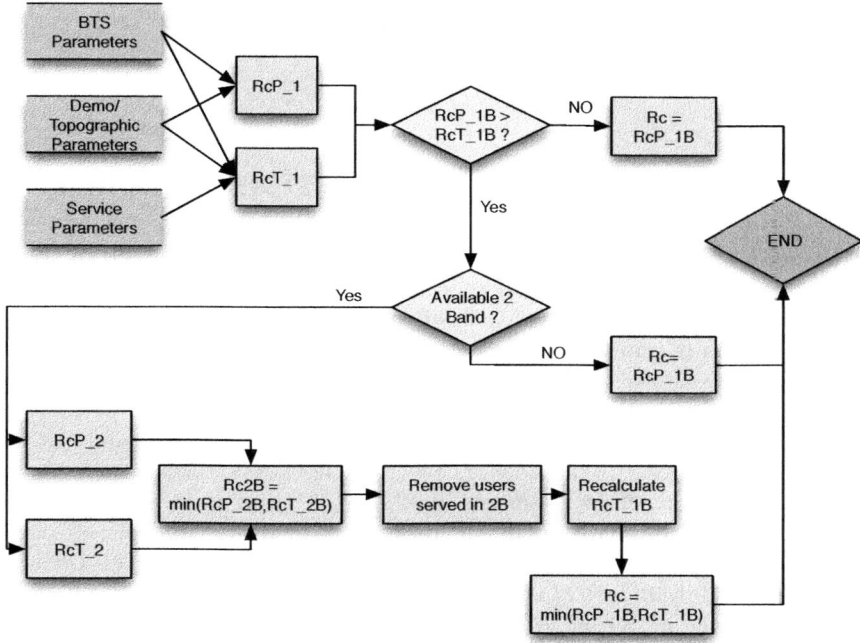

Fig. 1. GSM Single Site Dimensioning Process

2.4   Estimation of the 2G Network Deployment in a Specific Area

The algorithm for single site dimensioning is executed for each type of BTS specified in the scenario, and, in case of availability of second frequency bands, it is used for any allowed combination of BTSs in the same site. Each BTS has associated a cost factor, C_i. Let $\phi = \{\phi_1, \phi_2, \cdots, \phi_k\}$ be the K feasible solutions of BTS combinations in the area, and $C(\phi_k)$ the cost of k the solution.

The combination ϕ_k will have associated the corresponding value of the cell/site radius $R_{C_{\phi_k}}$. Therefore the area covered by the k combination, Sup_{ϕ_k} is calculated by means of the classical circle area formula. The number of sites required for providing coverage in a specific area of the district i (urban/suburban/rural) is calculated by means of Equation (11).

$$N_{Sites\ Area} = \frac{Sup_{Area}}{Sup_{\phi_l}} \tag{11}$$

If we consider that the k solution have the BTS i for the first band and the BTS j for the second band, the cost of the solution is as follows:

$$C(\phi_k) = N_{Sites\ Area} \cdot (C_i + C_j) \tag{12}$$

The algorithm selects the combination ϕ_l, which fulfils:

$$\phi_l \rightarrow C(\phi_l) = Min\,[C(\phi_k)]_{k=1}^{K} \tag{13}$$

3   Models and Methods for 3G-UMTS Dimensioning

Let us consider a 3G mobile network based on WCDMA technology, where the mobile operator provides a set of S services with similar parameters than the set specified in 2G-GSM. As it was defined in section 2.2, propagation coverage studies mainly imply two steps. The first one is to calculate the maximum propagation loss allowed in the cell, and the second is to use an empirical propagation method to calculate the cell radius for this path loss. In this case the difference in the receiver sensibility comes from different Eb/No values of the S services. Another point to consider is that in 3G networks, that are soft-blocking systems, the amount of users are not limited by the amount of hardware in the 3G BTS (named Node B), but by the amount of interference. The interference is measured by means of the interference margin, IM, that has to be taken into account in the link budget of the propagation studies [7].

As it was done in 2G, we obtain for 3G a set of two vectors containing, for each service, the cell radius by propagation, $(R_{P_{PUL}} and R_{P_{DL}})$.

Let us focus now on capacity studies. As it is done in propagation studies, cell radius must be calculated independently for the uplink and the downlink. The equations that determine the radius in both directions are quite similar, so we will focus on the calculation of the cell radius for the downlink case, since this is the most restrictive direction [6]. The interference margin IM determines the maximum load of the cell, η_{DL} , by means of the following relation:

$$\eta_{DL} = \frac{1}{10^{IM/10}} - 1 \tag{14}$$

The load factor of the cell, that is in fact the capacity of the cell, must be allocated to the different services, yielding the load factors of the each service $L_{Total\_DL\_i}$:

$$\eta_{DL} = \sum_{i=1}^{S} L_{Total\_DL\_i} < 1 \tag{15}$$

The number of active connections of each service Nac_{DL_i} is calculated by dividing the total load factor of each service type i over the average individual downlink load factor of the connections of the service:

$$Nac_{DL\_i} = \frac{L_{Total\_DL\_i}}{L_{DL\_i}} \tag{16}$$

where the donwlink load factor is defined by the following equation, [5]:

$$L_{DL\_i} = \frac{(Eb/No)_{DL\_i} \cdot \sigma_i}{(W/Rb_i)} \cdot [(1 - \phi) + f] \tag{17}$$

where

- ϕ is the so called downlink orthogonality factor,
- Rb_i is the binary rate,
- σ_i is the activity factor of the service i,
- f is the average inter-cell interference factor, and
- W is the bandwidth of the WCDMA system.

The total offered traffic demand, A_{DL_i} in Erlang, is obtained by using the inversion of the Erlang B Loss formula. The inputs for this algorithm are the maximum number of active connections in the cell Nac_{DL_i} and the Quality of Service (QoS) of the service expressed by the blocking probability Pb_i:

$$\frac{A_{DL\_i}}{(1 + f)} = E_B^{-1} \left(Pb_i, Nac_{DL\_i} \cdot (1 + f) \right) \tag{18}$$

The number of users in the cell ($M_{DL\_i}^{users}$) is obtained from the division of the total offered traffic demand for service i, (A_{DL_i} in Erlang), by the individual traffic of a single user of this service, a_i:

$$M_{DL\_i}^{users} = \frac{A_{DL\_i}}{a_i} \tag{19}$$

The cell radius for each individual service is calculated as a function of the number of sectors in the node B, $N_{Sectors}$, the number of users of service i per sector, M_i^{users} and the user density, ρ_i, as follows:

$$R_{T_D L_i} = \sqrt{\frac{M_{DL\_i}^{Users} \cdot N_{Sectors}}{\pi \cdot \rho_i}} \tag{20}$$

Note that this process has to be done also for the uplink direction (UL). Therefore, at the end we will have obtained another set of two vectors (one for the uplink and one for the downlink), with the cell radius by capacity of each service. Note also that at the end of this entire process we will have obtained a set of four vectors, two coming from the propagation studies and two coming from traffic studies. The final cell radius R_C, will be the minimum value of all of them:

$$R_C = MIN \left[R_{P\_DL\_i}, R_{P\_UL\_i}, R_{T\_DL\_i}, R_{T\_UL\_i} \right] \tag{21}$$

This process can be usually divided into two problems:

- The *outer problem* is to find the optimum value for the interference margin
  IM, for balancing the cell range between propagation and capacity.
- The *inner problem* is to find the best possible allocation, given a value of the
  IM over the complete set of services S.

The outer problem is solved by making an iterative process to equilibrate the
value of the cell radius between the resulting value calculated by propagation
studies and the resulting one calculated by capacity studies. This is done by
means of increasing the value of the interference margin, IM, when the cell
radius by propagation is higher than by capacity or vice-versa.

For the inner problem the model applies a novel heuristic that is explained in
[9] and it is summarized in the scheme of Figure 2.

The procedure of calculating the cell range is repeated for all possible Node B
specified in the input parameters. Next, following a similar procedure as specified
in subsection2.4, the optimum Node B configuration, in terms of investment cost,
is selected for the deployment in the area.

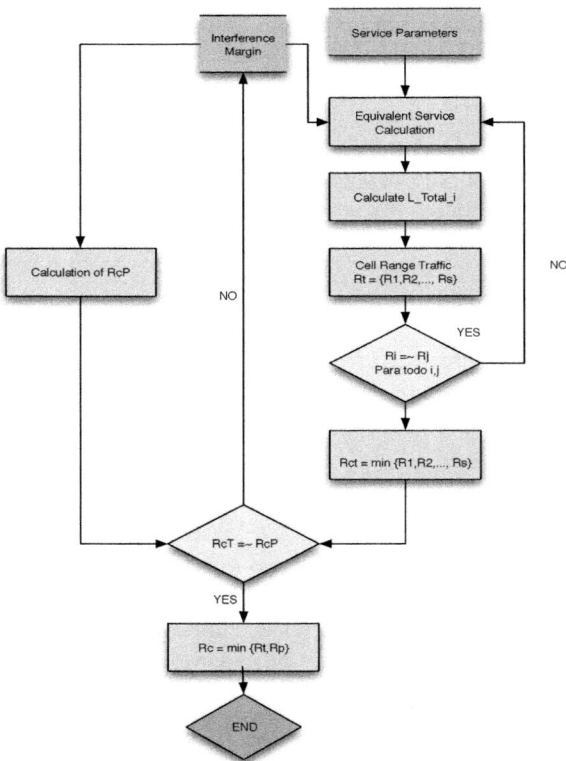

Fig. 2. UMTS Cell Range calculation algorithm

4   Experiments and Results

In order to test the performance of the algorithms and the validity of the results we propose in this article two different set of experiments will be carry out. The first set consists on calculating the network deployment in a specific city of Spain, in this case Bilbao, for both technologies, 2G-GSM and 3G-UMTS. This study allows us analyzing the results in a limited scenario. The second set consists on running the algorithms for a Spanish nationwide scenario, that is composed of 5400 districts representing 100 % of the Spanish population and 100% of the Spanish territory.

In order to make a fair comparison, we have defined the same set of services and traffic per user for 2G and 3G. Specific values are shown in Table 1, where a_U, a_S, a_R stand for the individual user traffic in urban suburban and rural respectively, and Sp is the service penetration in %. We consider that the mobile network operator has a 20% of market share.

Table 1. User service briefcase for the study

Service	Rb_i Kbps	a_U	a_S	a_R	Sp %	DR_i %
Voice	12.2	6.5	6.13	6.04	100	0
RTS	64	0.925	0.96	0.97	100	50
Streaming	64	0.465	0.26	0.22	100	80
Guaranteed Data	144	1.215	0.42	0.24	100	80
Best Effort	144	0.16	0.13	0.12	100	80
SMS	64	0.25	0.25	0.25	100	20
MMS	64	0.25	0.25	0.25	100	20

4.1   Experiments for a Single City (Bilbao)

In this experiment we consider only the metropolitan part of the city of Bilbao. It has a population of 126072 inhabitants with 31 percentage of urban area, 69 of suburban area and without rural area, because it is an important metropolis. We have performed the deployment for GSM mono-band in 900 MHz, and for GSM multi-band in 900 MHz and 1800 MHz, and UMTS 2100 MHz. The results are shown in Table 2.

Table 2. User service briefcase for the study carried out in Bilbao

Technology	Sites/Units Urban	Sites/Units Suburban	Sites/Units Rural	Cost M€
GSM 900 MHz	14/14	7/7	0	2714
GSM 900/ 1800MHz	14/14	7/7	0	2714
UMTS 2100 MHz	5/5	6/6	0	1090

Note that the results for GSM 900 MHz and GSM 900/1800 MHz are similar. This is due to the individual traffic is very low, and therefore, as the second band is only used when there is an excess of traffic in the first band, there is no need to use the second band, so in this example, the operator has free spectrum in 1800 MHz to use it or to receive incoming by hiring it to virtual mobile network operators. Note also that, despite the UMTS system is working in 2100 MHz, as Bilbao only has urban and suburban areas that are densely populated (and therefore concentrates the traffic in small extensions) this 3G system obtains better performance than 2G-GSM, in terms of number of sites and costs.

4.2 Nationwide Experiments

In this subsection we tackle a nationwide scenario. We consider the same service briefcase and frequency distribution as in the single city experiments. The results are shown in Table 3. Note that in this experiment we show global figures and we do not particularize for each area.

Table 3. Results from the nationwide scenario (Spain)

Technology	Sites/Units	Cost M€
GSM 900 MHz	15801	2921.5
GSM 900/ 1800MHz	15667/16041	2914.8
UMTS 2100 MHz	22444	3669.2

Note that in this case the GSM multi-band solution is the best one in terms of cost, since, although it has more BTS units, the reduction in the number of sites and the corresponding saving in infrastructure costs, compensates the effect. An interesting result to be analyzed is the performance of UMTS, providing a solution worse than GSM. The explanation is quite easy to understand when we analyze the results related to each area type and depending on the frequency, see Table 4.

Table 4. Comparison between UMTS 2100 MHz, and UMTS 900 MHz, for different area types

Technology	Urban	Suburban	Rural	Total
UMTS 2100	1205	958	**20281**	22444
UMTS 900	1205	425	**8191**	9821

In urban areas, where the Node B is *traffic driven*, the results in UMTS 900 and UMTS 2100 are identical. In suburban areas where sometimes the Node B is *propagation driven*, the number of Node B in UMTS 2100 is almost twice than in UMTS 900. However the main difference, and also the reason that UMTS 2100 is more expensive than GSM 900/1800, is in rural areas. Rural areas are characterized by very large low populated terrain extensions. In this areas, all Node B are *propagation driven*, and the propagation in 2100 MHz is much worse than in 900 MHz. A fair comparison between GSM and UMTS has to be done in the same frequency band. The results for this comparison are shown in Table 5.

Table 5. Comparison between the investment cost (in M€) in GSM 900 MHz, and UMTS 900 MHz, for different area types

Technology	Urban	Suburban	Rural	Total
GSM 900	317.9	135.6	2467.9	2921.5
UMTS 900	103.4	49.3	1402.7	1555.5

Note that the investment in GSM 900/1800 MHz is almost twice the investment in UMTS 900, considering only the radio access network. This comparison gives an idea of the saving in cost of using a most evolved technology, as 3G in the provision of mobile services.

5   Conclusions

This paper presents a strategic network planning tool for the estimation of the investment in mobile access networks deployments. The tool implements advanced algorithms for the dimensioning of 2G-GSM and 3G networks, under multi-service environments. This kind of tools are very valuable for mobile network operators to carry out techno-economic studies, in order to test the viability of new network investments, or to decide about the technology and frequency bands for the network deployment.

We have presented a nationwide study in Spain, where we compare 2G and 3G technologies in the same and different frequency bands. The results demonstrates that 3G-UMTS is much better than 2G in urban and suburban areas despite of the frequency band, and that only in case of rural propagation driven areas, 2G 900 MHz is better than 3G 2100 MHz. The results also show that 3G UMTS in 900 MHz is the best option despite the area. This is a very relevant result, since in Europe we are currently involved in a rearranging process to reallocate the 900 MHz spectrum.

References

1. Um, P., Gille, L., Simon, L., Rudelle, C.: A Model for Calculating Interconnection Costs in Telecommunications. PPIAF and The World Bank (2004)
2. European Cooperation in the Field of Scientific and Technical Research EURO-COST 231, Urban Transmission Loss Models for Mobile Radio in the 900 and 1800 MHz Bands, Revision 2, The Hague (1991)
3. Walfisch, J., Bertoni, H.L.: A Theoretical Model of UHF Propagation in Urban Environments. IEEE Transactions on Antennas and Propagation AP-36, 1788–1796 (1988)
4. Boucher, N.J.: The cellular radio handbook. J. Wiley & Sons (2001)
5. Holma, H., Toskala, A.: WCDMA for UMTS. J. Wiley & Sons (2001)
6. Laihoo, J., Wacker, A., Novosad, T.: Radio Network Planning and Optimisation for UMTS. Wiley (2002)

7. Ruzicka, Z., Hanus, S.: Radio Network Dimensioning in UMTS Network Planning Process. In: IEEE 18th International Conference on Applied Electromagnetics and Communications, ICECom 2005 (2005)
8. Neu, W., Hackbarth, K., Portilla-Figueras, A.: Analysis of Cost Studies presented by Mobile Network Operators, http://www.osiptel.gob.pe
9. Portilla-Figueras, A., Salcedo-Sanz, S., Hackbarth, K.D., López-Ferreras, F., Esteve-Asensio, G.: Novel Heuristics for Cell Radius Determination in WCDMA Systems and their Application to Strategic Planning Studies. EURASIP Journal on Wireless Communications and Networking 2009, Article ID 314814 (2009)

Impact of the HSPDA-Based Mobile Broadband Access on the Investment of the 3G Access Network

Juan Eulogio Sánchez-García, Amir M. Ahmadzadeh,
Silvia Jiménez-Fernández,
Sancho Salcedo-Sanz, and J. Antonio Portilla-Figueras

Universidad de Alcalá, Spain
antonio.portilla@uah.es

Abstract. There are several technologies for providing broadband services over wireless and cellular networks. The fundamental one in the evolution from 3G to 4G is probably the High Speed Downlink Packet Access (HSPA) technology. There are many works in the literature tackling the problem of HSPA performance and capacity. Most of the developed techniques involving HSPA capacity are related to the system operation. This approach is specially useful when the network planner tries to evaluate how the system works, however, it is not the case when the mobile network operator is doing the business plan and whise to evaluate the return of investment. This paper provides a simple and novel methodology for estimating the additional investment required to provide High Speed Down-link Packet Access (HSDPA) in a 3G mobile network given a user service profile. This method is useful for techno-economic studies for mobile operators, consulting firms and national regulatory agencies.

Keywords: HSPA, Dimensioning, Service Profile, Regulatory Studies.

1 Introduction

In the evolution from 3G to 4G, High Speed Packet Access (HSPA) (and its evolved version, HSPA+) technologies have become a milestone due to they allow the provision of high binary rate data services to mobile users [1]. Furthermore, they have produced a collateral effect that will become an important market opportunity for mobile network operators: the convergence, and even sometimes substitution, of the fixed broadband access (typically based on xDSL or HFC) by the mobile broadband access based on HSPA and HSPA+ [2].

Network design based on HSPA and HSPA+ technologies has been deeply studied in the literature. However most of the works are oriented to the study of the performance of HSPA and HSPA+ in the daily operation of the network. Furthermore, most of these studies are based on simulations of a particular environment [3,4]. To our knowledge, there are not any previous work related with the study of the network investment required to provide HSPDA based services on a wide area.

J. Del Ser et al. (Eds.): MOBILIGHT 2011, LNICST 81, pp. 303–311, 2012.

In this paper we explain a simple methodology to estimate the number of Nodes B required to provide mobile broadband access in a specific area under a multi-service user profile. This method is applied to a nationwide scenario, to obtain an estimation of the network investment required by a mobile network operator to provide this service. This kind of strategic studies are a powerful tool for mobile network operators when analyzing the profitability of new services.

The rest of the paper will be structured as follows: Section 2 provides a summary description of the HSPA technology. In Section 3 we will describe the methodology that we apply to calculate the HSPA Node B cell range and hence to estimate the number of Nodes B required for providing coverage to an specific area. Section 4 provides the study at a nationwide level. Finally we show some conclusions on the study carried out and some lines of future work.

2   HSDPA Technology Background

HSDPA is a 3G mobile access technic specified firstly in the Release 5 of the 3GPP [5]. The main difference with the physical layer of the normal Wide-Band Code Multiple Access (WCDMA) is a new downlink (DL) time shared channel: the High Speed Physical Downlink Shared Channel (HS-PDSCH). This channel supports 2 miliseconds Time Transmission Intervals (TTI) where the user can request resources, with adaptive modulation and coding and fast physical layer ARQ. In the HSDPA architecture, the Node B is the responsible of link adaptation and packet scheduling. It has updated online information about the radio environment and the resources requests, and hence it allocates resources and flows according the service priority (by means of packet scheduling) and the best modulation and coding for the TTI due to the radio environment.

For the HSDPA dimensioning the following concepts have to be defined, see [6]:

- **Sub-frame duration:** The Transmission Time Interval-TTI (sub-frame) duration is 2 ms, thus a 10 ms WCDMA frame contains 5 HSDPA sub-frames.
- **Inter-TTI:** refers to the number of TTI between transmissions to the same user equipment (UE). All UEs must support a minimum Inter-TTI of 3 (i.e.: UE must be capable of receiving DL transmission every 3 sub-frames).
- **Modulation and Coding:** each sub-frame may carry data bits modulated with QPSK or 16 QAM levels. The fixed turbo coding rate of 1/3 is always used over both modulation levels.
- **Spreading factor:** is fixed (SF=16). Thus 16 channelization Orthogonal Variable Spreading Factor (OVSF) codes are available under each cell (each cell is identified with a 38400 chips Scrambling Code). Out of 16 codes, one is assigned for HS-SCCH transmission that carries the control data of its corresponding HS-PDSCH. Thus 15 channelization codes are available for data transmission.
- **Multi-code Transmission:** In one TTI, Multiple channel codes can be assigned to one UE in parallel. All UEs must support a maximum transmission of at least 5 parallel channels.

– **Scheduling Mechanism:** There are many of them: Round Robin, Max-Min Fair Throughput, Channel Quality based. In this study we will user Round Robin as the best simple approximation to the problem.

Depending on the type of user terminal, the resources required, the modulation and coding used and the link level parameters, the packet scheduler in the Node B will allocate resources to each user in the HSDPA Node B coverage area, as it is shown in Figure 1.

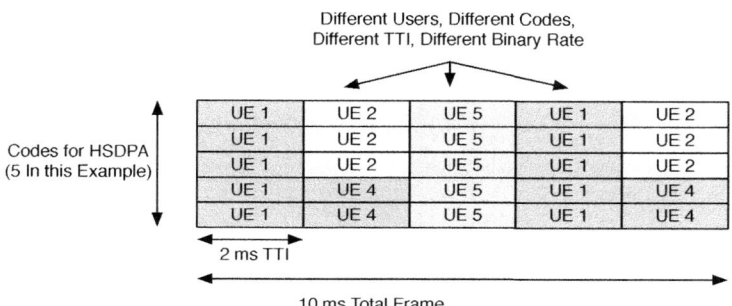

Fig. 1. Allocation of resources in the HSDPA frame to users

Note that depending on the link level parameters, including the Signal to Interference Noise Ratio (SINR) the Node B and the user terminal will agree a specific modulation and coding following Table 1.

Categories 13 and 14 of the table corresponds to HSDPA+. There are some more HSDPA categories but they use multiple input/multiple output (MIMO) techniques that are out of the scope of this work. A relevant parameter in Table 1 is the Channel Quality Indicator (CQI). It is a measure of the mobile channel which is sent regularly from the UE to the Node B. These measurements are used to adapt modulation and coding for the corresponding UE and it can be also used for the scheduling algorithms. It can take one of 30 values, depending on the Transport Block Size and the Modulation and Coding Scheme. Typical values are 15 for QPSK and 30 for 16QAM. This parameter is used for calculating the SINR based on Equation (1).

$$SINR\ (dB) = (CQI \cdot 1.02) - 16.62 \tag{1}$$

The parameter $SINR$ is used to calculate the UE receiver sensibility that is involved in the calculation of the cell range by propagation. This cell range by propagation is used as a starting point to obtain the total throughput required by the HSDPA users in the area under study.

Table 1. HSDPA Modulation and Coding Categories

Category HS-DSCH	Modulation	Number of codes	Inter-TTI	CQI	Speed (Mbps)
Category 1	QPSK	5	3	15	0.553
Category 1	16-QAM	5	3	30	1.194
Category 2	QPSK	5	3	15	0.553
Category 2	16-QAM	5	3	30	1.194
Category 3	QPSK	5	2	15	0.829
Category 3	16-QAM	5	2	30	1.792
Category 4	QPSK	5	2	15	0.892
Category 4	16-QAM	5	2	30	1.792
Category 5	QPSK	5	1	15	1.659
Category 5	16-QAM	5	1	30	3.584
Category 6	QPSK	5	1	15	1.659
Category 6	16-QAM	5	1	30	3.584
Category 7	QPSK	5	1	15	1.659
Category 7	16-QAM	10	1	30	7.205
Category 8	QPSK	5	1	15	1.659
Category 8	16-QAM	10	1	30	7.205
Category 9	QPSK	5	1	15	1.659
Category 9	16-QAM	12	1	30	8.619
Category 10	QPSK	15	1	15	1.659
Category 10	16-QAM	15	1	30	12.779
Category 11	QPSK	5	2	15	0.829
Category 12	QPSK	5	1	15	1.659
Category 13	QPSK	5	1	15	1.664
Category 13	16-QAM	10	1	25	7.212
Category 13	64-QAM	14	1	30	16.132
Category 14	QPSK	5	1	15	1.664
Category 14	16-QAM	10	1	25	7.212
Category 14	64-QAM	15	1	30	19.288

3   Model for HSDPA Cell Range Estimation

This section introduces a novel estimation method for calculating the area covered by a 3G Node B with HSDPA functionalities. This method does not provide exact calculations about the performance of the Node B, but only a simple method to estimate the network investment to be done. For better understanding of the model it is required to introduce before the user service briefcase we consider for the study.

We define a set of services, S resulting from different reports of the UMTS Forum [7]. All these services are based on the set of 34 parameters that we have defined in our simulation tool. The most relevant for our work are described below.

- Service name / Service identifier
- Binary Rate, Rb_i
- Service penetration (Percentage of customers using the service).
- Individual traffic per user in the business hour, a_i
- Percentage of traffic running direcly over HSDPA (only for data services) DR_i.

Last parameter applies only for 3G native data services. In the case of *mobile broadband service*, that is a pure native HSDPA service, this parameter does not apply.

The objective of the algorithm is to find a value of the 3G-HSDPA Node B cell range that makes the Node B resources be enough to fulfil the capacity resources demanded by the users in the area under study.

Calculations are done in terms of bandwidth per user that it is obtained following Equation (3):

$$Throughput = MB_{guaranteed} + \sum_{i=1}^{S} a_i \cdot DR_i \cdot Rb_i \qquad (2)$$

where:

- Throughput: is the demanded binary rate per user
- $MB_{guaranteed}$ is the minimum binary rate of the mobile broadband service that is guaranteed to the customers in the area under study.

From the desired bandwidth calculated in Equation (3) the category that best fits the user's necessities can be obtained from table 1. The SINR can be calculated by knowing the category and the CQI, following Equation (1).

The next step is to obtain the cell radius by propagation, in order to evaluate the number of potential users in the cell area under study, and therefore, the bandwidth demanded by all the potential users of this cell. To do this we consider the Okumura-Hata propagation model described in [8].

Then the maximum bandwidth offered by the cell is calculated, using the maximum number of codes, fifteen for traffic, and the corresponding category inter-TTI. Both bandwidths, demanded and offered, are compared and the calculated radio is iteratively reduced until the required condition is satisfied. Once that the bandwidth offered by the cell is higher than the bandwidth demanded by the users in this area, the final cell radius is obtained. This iterative process is shown in Figure 2.

4   Experiments and Results

In this section we apply the algorithm described in previous section to a nationwide scenario in Spain. This scenario is composed of 5400 entries, named districts, that are composed of one or several cities, towns and villages, representing 100 % of the Spanish population and terrain. Each district is divided into urban, suburban and rural area, depending on the topography and demography of each area.

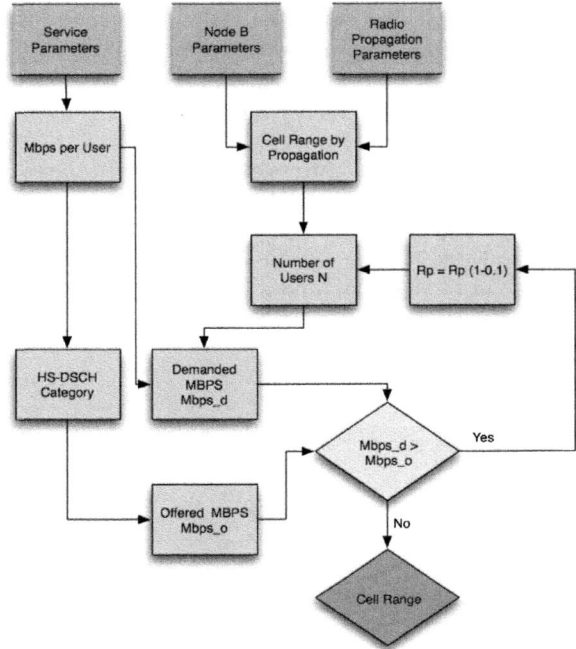

Fig. 2. Scheme of the HSDPA cell range estimation model

The model explained in previous section is implemented in the tool presented in [9]. This tool implements the algorithms for dimensioning 2G and 3G networks. It considers several types of Node B for the dimensioning: macro, micro and pico Node B with 1, 2 and 3 sectors, with different costs. This allows the tool with a large flexibility in order to optimize the estimation of the network investment.

In this study we consider a hypothetical operator with 20% of Martket Share, working with a single 5 MHz spectrum block in the 900 MHz frequency. We have selected this band because, although the native 3G band is in 2.1 GHz, the European countries are involved in a frequency rearranging process. We compare the investment cost of deploying a pure multi-service 3G access network in each area (urban/sub-urban/rural) with the cost related to introduce a mobile broadband service, based on HSDPA, considering different guaranteed binary rates, different market penetrations, and also the different areas of the districts. The service briefcase for the study is shown in Table 2.

Note that we have not fixed the service penetration and the binary rate for the mobile broadband service because they are two parameters that we will vary in our study. For the experiments we have defined the following scenarios, see Table 3.

Note that in *Scen-1*, HSPDA is not provided, and therefore Scen-1 may serve as a reference value to calculate the increment in the investment due to HSPDA

Table 2. User service briefcase for the study

Service	Rb_i Kbps	a_i mErlangs	Service Penetration (SP) %	DR_i %
Voice	12.2	6.22	100	0
RTS	64	0.21	100	50
Streaming	64	0.31	100	80
Guaranteed Data	144	0.62	100	80
Best Effort	144	0.13	100	80
SMS	64	0.25	100	20
MMS	64	0.25	100	20
Mobile Broadband	–	95.86	–	100

Table 3. Scenarios for the Experiments

Scenario	Urban	Suburban	Rural	MB Market %	ME Rb Kbps
Scen-1	UMTS	UMTS	UMTS	0	0
Scen-2	UMTS/HSPDA	UMTS	UMTS	1	400
Scen-3	UMTS/HSPDA	UMTS / HSPDA	UMTS	1	400
Scen-4	UMTS/HSPDA	UMTS	UMTS	10	400
Scen-5	UMTS/HSPDA	UMTS / HSPDA	UMTS	10	400
Scen-6	UMTS/HSPDA	UMTS	UMTS	1	1000
Scen-7	UMTS/HSPDA	UMTS / HSPDA	UMTS	1	1000
Scen-8	UMTS/HSPDA	UMTS	UMTS	10	1000
Scen-9	UMTS/HSPDA	UMTS / HSPDA	UMTS	10	1000

in the rest of scenarios. Table 4 shows the results of the runs of the different scenarios in terms of:

- Absolute investment in the radio access network, Ac.
- Investment increment over a pure 3G access network, Inc.
- Relative increment over a pure 3G access network, $Rinc$.
- Relative cost per Kbps for the HSDPA Service, Rc. This parameter is defined as $(Inc/NC) \cdot TU$, where NC stands for the number of customers and TU stands for the throughput per user.

The experiments offer the following interesting results.

- There are very important investment increments when the target population goes from 1% to 10 % so, the investment depends critically on the traffic, more that in propagation
- If the mobile operator want to offer HSPDA services to a relevant part of the population, 10 %, the additional cost over a pure 3G network will be between 13 to 26 %.
- There are only slight differences (less thant 4 %) in terms of investment between offering a guaranteed service of 400 Kbps or 1000 Kbps.

Table 4. Results of the experiments

Scenario	Ac M€	IncM€	$Rinc\%$	Rc
Scen-1 3669	0	0	0	
Scen-2 3673	46.3	0.13	0.03	
Scen-3 3721	52.1	1.42	0.32	
Scen-4 4149	48.1	13.1	0.3	
Scen-5 4512	84.3	22.9	0.52	
Scen-6 3707	38	1.04	0.09	
Scen-7 3827	15.8	4.31	0.39	
Scen-8 4232	56.3	15.35	0.14	
Scen-9 4653	98.4	26.82	0.24	

– The combination of a high market penetration with a high guaranteed binary rate service will lead to low cost per Kbps, due to the economies of scale
– The decision of having HSDPA on suburban and rural areas is very important, it is required a high market penetration to justify the investment. Otherwise the cost per Kbps will almost double.

5   Conclusions and Future Work

This paper provides a novel method for the estimation of the investment required to provide HSDPA service in a 3G access network. This method, based on a iterative algorithm, provides the network planner with a flexible and powerful tool to make regulatory or economic studies. Note that most of the work related with the calculation of the HSDPA range are based on simulation and therefore they are not valid for the application to a nationwide study.

This method has been applied to a particular study in Spain. We have shown, in different scenarios, the increment of the investment over a pure native 3G network. The conclusion is that to guarantee the HSDPA mobile broadband service for all users nationwide a very important investment in new 3GHSPA Node Bs is required.

The algorithm presented in this paper is a first version of a most depurated one that we are working on. The new version will consider different simultaneous modulation and coding schemes and variable user density. Another improvements are the design of a High Speed Uplink Packet Access (HSUPA) and the extension to more HSPA+ categories.

References

1. Dahlman, E., et al.: 3G evolution: HSPA and LTE for mobile broadband. Elsevier Academic Press (2007)
2. Ergen, M.: Mobile Broadband. Springer, Heidelberg (2007)
3. Kolding, T.E.: Link and system performance aspects of proportional fair packet scheduling in WCDMA/HSDPA. In: IEEE Proc. Vehicular Technology Conference (2003)

4. Pedersen, K.I., Lootsma, T.F., Stottrup, M., Frederiksen, F., Kolding, T.E., Mogensen, P.E.: Network performance of mixed traffic on high speed downlink packet HSDPA/HSUPA for UMTS access and dedicated channels in WCDMA. In: IEEE Proc. Vehicular Technology Conference (2004)
5. 3rd Partnership Project (3GPP), Overview of 3GPP Release 5. 3GPP TSG RAN 20 (June 2003)
6. Holma, H., Toskala, A.: HSDPA/HSUPA for UMTS. John Wiley and Sons (2006)
7. Hewitt, T., et al.: 3G Offered Traffic Characteristics Report. UMTS Forum (November 2003)
8. European Cooperation in the Field of Scientific and Technical Research, EURO-COST 231, Urban Transmissio Loss Models for Mobile Radio in the 900 and 1800 MHz Bands, Revision 2, The Hague (1991)
9. Sánchez-García, J.E., Ahmadzadeh, A.M., Saavedra-Moreno, B., Salcedo-Sanz, S., Portilla-Figueras, A.: Strategic Methods for Radio Access Design in 2G/3G Networks. In: 3 International ICST Conference on Mobile Lightweight Wireless Systems (Mobilight 2011), Spain, May 9-11 (2011)

Challenging Wireless Networks, an Underground Experience

David Palma, Joao Goncalves, and Marilia Curado

Department of Informatics Engineering
Centre for Informatics and Systems University of Coimbra (CISUC)
{palma,marilia}@dei.uc.pt, jccg@student.dei.uc.pt

Abstract. Quite often Wireless Ad-hoc networks are taken for granted as a networking solution in the most challenging environments. Motivated by these networks' self-X properties and the infrastructure-less paradigm, many authors rely on Ad-hoc networks to support a number of applications in remote areas or even in disaster scenarios such as mine collapses, earthquakes, tunnel accidents. However, most of these works are only simulation oriented, disregarding how challenging wireless networks can be in such scenarios. By conducting a set of performance measurements with off-the-shelf netbooks in an underground environment it is possible to see that typical wireless simulation assumptions do not verify, and that covering an area with multi-hop connectivity is not straightforward. These results motivate a tighter interaction between real experiments and future simulation based works in challenging environments which need to be more accurate.

1 Introduction

The dissemination of new portable devices with enhanced communication characteristics has revolutionized the world in many aspects. Not only on the social side, where people are connected by their cell phones, personal digital assistants (PDAs) or laptops, but also on an economical and professional perspective, where these devices have introduced new ways of dealing with different situations. In fact, it is expected that in a near future, users will own several wireless enabled gadgets [9], demanding infrastructures or other connectivity alternatives such as Mobile Ad-hoc NETworks (MANETs).

In order to support the creation of Ad-hoc Networks, several routing protocols have been proposed, using a myriad of approaches to tackle the different aspects of these networks [3]. The existing routing protocols can be grouped into three different main classes: Proactive, Reactive and Hybrid, where the main difference between these classes relies on how routing information is exchanged. An additional class of Position Based routing protocols can also be considered, as no explicit exchange of routing messages is required. However, these protocols may not always be suitable in scenarios where position information is not available or accurate.

J. Del Ser et al. (Eds.): MOBILIGHT 2011, LNICST 81, pp. 312–321, 2012.

Despite all the proposed routing approaches for Multi-hop Networks, since the definition of the concept of Ad-hoc Wireless Networking in the early 70's for battlefields, the actual usage of these networks is very modest even with all the possible advantages. Currently, no particular applications seem to stand out as standard applications [14]. Following the same background on which Ad-hoc Networks were first used, rescue operations have been considered as one of the most suitable scenarios for the implementation of these infrastructure-less wireless networks, supporting monitoring and coordination applications typically required by rescue teams in disaster scenarios [15].

The requirements of wireless networks for rescue operations are very complex and different, depending on the specifics of each rescue operation being considered [22] [20]. In particular, these networks and their behaviour are closely related with the surrounding environment, where the physical scenario characteristics and obstacles are considered a major challenge [4].

Simulations provide an important contribution for the research community but usually depend on a number of assumptions which cannot be verified in the real world [17]. Therefore, the aim of this paper is to provide a new perspective on the behaviour of wireless networks within an underground scenario, presenting real measurements obtained using off-the-shelf "Asus EEE" netbooks, not requiring any additional hardware, representing a configuration likely to be used by any element of a rescuing team. A thorough evaluation of the wireless link performance using this equipment is presented, and an illustrative small scale Ad-hoc network using the Optimized Link State Routing Protocol (OLSR) [10] protocol is also analysed.

In section 2 works related with the creation of wireless networks in challenging environments are presented, followed in section 3 by the definition of challenging wireless networks, and by a real experimental scenario with those characteristics. The registered experiences and obtained results are presented in section 4, leading to the final thoughts on Wireless Communication in Challenged Environments, in section 5.

2   Related Work

Considering the usage of Ad-hoc Networks for dynamic infrastructure-less networks, a number of routing schemes has already been proposed. For topology-based routing protocols, which do not require any additional mechanism for node's position awareness such as the global positioning system (GPS) or other positioning schemes, several proposals have been developed for both proactive [10] [11] and reactive [19] [6] routing protocols.

These protocols aim at providing a reliable self-X networking "infrastructure", allowing the deployment of Ad-hoc networks anywhere at anytime in order to support all the necessary computer communications. Such a flexible and dynamic perspective on wireless networks motivated several authors to investigate on how search/rescue and monitoring applications can be applied to challenging scenarios such as earthquakes, mines, tunnels and even underwater. However, most of

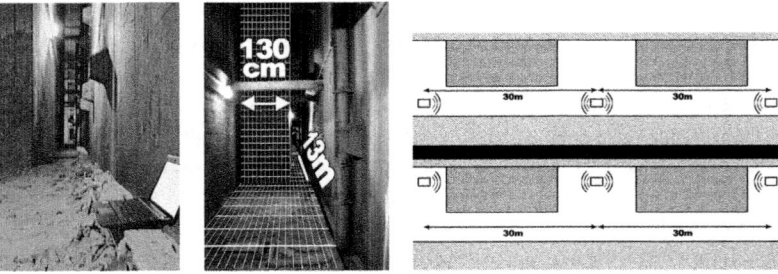

Fig. 1. Underground Gallery Settings

the existing works are based on simulation studies [5] [7], which are important for the development of new revolutionary schemes, specially when dealing with larger networks - as it might be extremely difficult, if not impossible in certain situations, to perform real experiments - but which depend on a considerable number of assumptions, failing many times to provide an accurate perspective on systems' behaviour in particular when these conditions are not predictable in extreme environments.

Other existing works, for instance in underground communication, consist on the proposal of complex architectures which are not representative of the Ad-hoc communication paradigm, creating a wireless infrastructure and relying on hardware specific equipment such as high debit antennas [18] [24]. Through the manipulation of experimental environments, by using obstacles and other means, some authors have presented results in scenarios with similar characteristics of those expected to be found in earthquake or urban search & rescue situations, attempting to accurately portray the possible situations found in a real scenario [8] [21].

The understanding of wireless link quality and how it is influenced represents an important subject of research [12]. This is relevant not only for the existing and upcoming routing protocols (see section 14 of [10]), but also to allow the correct modelling of their behaviour, improving the existing network simulators and their results [13].

3   Challenging Wireless Networks

The creation of wireless networks in challenging scenarios is frequently seen as the most efficient way of allowing networking in such locations. These scenarios are considered challenging due to many different factors, which can be related with the permanent nature of the physical scenario conditions and its vicinities, such as underground chambers, underwater spaces or even remote areas with extremely difficult access. However, the choice of wireless networking in these scenarios is mainly due to destructive natural phenomena such as earthquakes, hurricanes, floods, or even due to human related accidents.

Typically, human built structures, such as mines and tunnels have their own communication infrastructures built "a priori", using many times the most easily maintainable and affordable solution. It is when these solutions fail, due to some of the aforementioned situations, that Ad-hoc wireless approaches are more seriously taken into account in order to support search and rescue operations that require all the possible help in the task of looking for and saving trapped victims.

3.1   Problem Statement

As wireless networks in challenging environments are mainly targeted at providing a good coverage of the possibly affected areas, the presented evaluation is focused on the performance of a wireless link at different distances, with static nodes, using single and multi hop connections. This multi-hop study is particularly important when existing obstacles significantly jam the wireless link between two nodes, requiring at least an additional node in a crossing point, connecting the whole network.

The main focus of this work is to evaluate the wireless link performance, bandwidth and losses. These parameters were measured using several payloads and different physical wireless characteristics. Moreover, the OLSRd protocol [23] was used to evaluate multi-hop performance, even though static routes had to be used in certain situations, as presented in section 3.1. Despite the existence of other routing alternatives exist such as the Babel and the B.A.T.M.A.N protocols [16], the OLSRd protocol was chosen for its popularity.

Scenario Description. With the purpose of providing accurate measurements on a representative scenario of a challenging environment, a set of performance evaluation tests was taken with off-the-shelf netbooks positioned within a subterranean gallery. This gallery was constructed to monitor the underground movements of a main wall supporting the building on against a mountain Moreover, this gallery also constantly drains all the existing groundwater, creating a very humid scenario, with barely any interferences from other wireless equipments.

The gallery where the main tests were performed is presented in figure 1, and consists on a corridor about $2.3m$ wide, interrupted by 1.3 by 13 meter blocks where elevators and stairs were built. These obstacles clearly interrupt the propagation of the wireless signal, being representative of possible debris in disaster scenarios and requiring extra wireless coverage.

In order to assess the quality of the wireless ad-hoc network created in this scenario, several measurements with different characteristics were taken. First, by always having a clear line of sight, a single hop link of 20, 30, 40 and 60 meters was evaluated, varying the link bit rate between $1Mb/s$, $11Mb/s$, $54Mb/s$ and *auto*. The chosen bit rates respectively represent the lowest bit rate possible, the 802.11b standard, the 802.11g and an automatically adapted link rate. This assessment was performed by using the Iperf [2] tool in UDP mode, with 3 different payloads (1, 5 and 10 MByte), in order to determine the overall performance of the link. Secondly, an additional single link measurement was also performed by placing the netbooks such that there would be no line of sight, with each node distanced by $30m$. No other distances were measured with these obstacles as no connectivity was obtained. During this section all the results presented as "30*" represent the values obtained when nodes are placed with no line of sight at a 30m distance.

The multi-hop evaluation was performed between three nodes distanced by 30m from each other, both with obstacles and with clear line of sight. One important experience gathered from this evaluation is related with the usage of the OLSRd protocol for route establishment. Even though the single-hop measurements indicated that virtually no data packets could be transmitted at a distance of 60m, such links were considered as reliable by the OLSRd protocol, leading to single-hop communication and subsequent loss of all the packets. However, this would only happen with a clear line of sight, since with obstacles the 60m link was not detected and multi-hop was correctly established.

Equipment Specification. One of the most important contributions of this work, besides the performed measurements in a real challenging environment, is that no hardware specific assumptions were taken. All the presented results were obtained with an off-the-shelf equipment, an Asus EEE netbook. The particular model used was the 1001 PX with an Intel Atom 450, 1GB Ram and an Atheros Communication Wireless Card (model AR9285, ath9k driver), running the Linux distribution Ubuntu 10.04 Netbook Edition (kernel 2.6.32). No modifications were performed to the hardware, using the default embedded antennas.

Due to the extensive and repetitive amount of performed measurements the nodes were not running on batteries, even though some preliminary tests showed no difference between working with batteries or being connected to an AC Adapter, as long as the card power save option is turned off. In a real rescue scenario these devices would most likely be running on batteries and the expected consumption of the wireless card (with no power saving scheme) would be of 1090mW [1] when not in idle.

4   Underground Measurements and Analysis

All the presented results represent an average sample obtained from 30 measurements taken sequentially by using a script. This allows an accurate understanding of the wireless link characteristics, such that no variations influence the overall interpretation of the results. Moreover, only UDP measurements were performed in order to avoid dynamic adjustment procedures such as TCP congestion control. These results show a 95% confidence interval, obtained from the central limit theorem.

4.1   Link Quality

The measured link quality is characterized by the signal to noise ratio obtained from the network card driver used by the Linux kernel. This value was also collected 30 times with a 5 second interval between each measurement. In figure 2 it is possible to see that there is not much granularity in the link quality value,

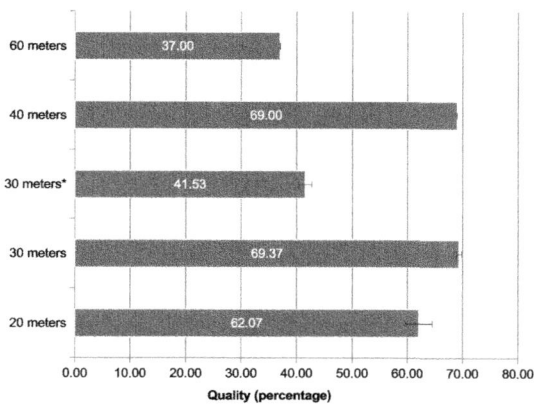

Fig. 2. Average Link Quality Percentage

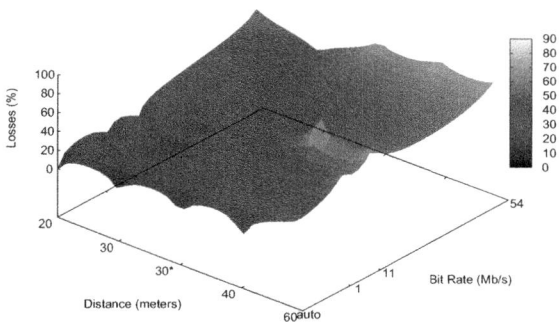

Fig. 3. Average Losses

being hard to see the difference between 30 and 40 meters. However, as shown further in this work, this $10m$ variation in distance has a strong impact in losses, specially for higher data rates. This figure also shows that at 20m the variance is bigger and link quality is worse than for $30m$. Usually the closer the better, but in this scenario, due to the presence of obstacles, the wireless link performance at $20m$ may degrade as a result of being closer to the existing obstacles.

4.2   Losses

The results presented in figure 3 show that the number of losses depends on a combination of the link distance between two wireless nodes and the data bit rate being used. For instance, considering higher distances (e.g. 60m), only lower bit rates such as 1 and $11Mb/s$ are able to send packets with fewer losses, as it would be expected.

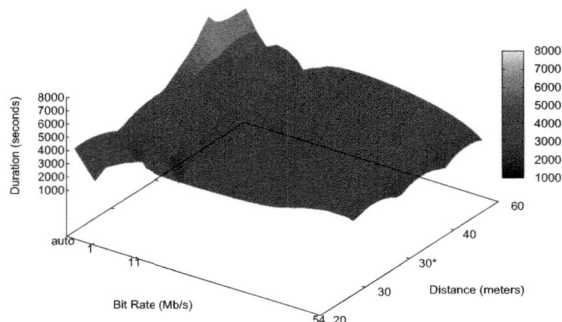

Fig. 4. Average Time Elapsed per Measurement

Focusing on the results obtained for each bit rate, it is clear that the *auto* mode has less losses at any distance. At $1Mb/s$ the obtained results are similar but start to degrade at 60m. For a rate of $11Mb/s$ the number of losses increases at $40m$ while at $54Mb/s$ a high number losses is registered at all distances.

4.3   Elapsed Time

Another aspect that was taken into account while evaluating the perform of wireless Ad-hoc networks was the time required to transmit all the sent data. This parameter depends not only on the amount of data to be transmitted and the used bit rate but also on the wireless medium availability, taking more time to transmit the data when collision avoidance mechanisms are used.

The values presented in figure 4 show the duration, in average, of the total time taken while performing the measurements, averaging the different UDP payloads used. The most interesting result perceived is that the good delivery performance of the *auto* bit rate has a cost since it takes more time to transmit the required data. Moreover, this increase of the total elapsed time is also noticeable even when all the traffic fails to be received. This occurs when the *auto* property is active because the sender keeps trying to adjust the bit rate to send data, even though it is physically impossible.

4.4   Multi-hop Experiments

The obtained quality values for a wireless link are important when considering multi-hop routing such that a routing protocol is able to detect whether a link is reliable or not, allowing it to correctly calculate routing paths. However, from the obtained experience a link's quality does not reflect by any means its reliability. This issue may possibly lead to inaccurate path choices and in fact, this was observed in the performed experiments when using the OLSRd [23] Linux implementation.

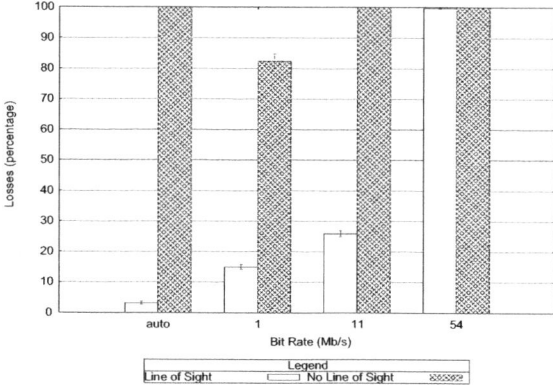

Fig. 5. Average Multi-hop Losses

In figure 5 the multi-hop losses registered between three hops are presented. These nodes, as depicted in figure 1(c) were deployed with and without line-of-sight connectivity. Due to poor link quality assessment, in the scenario with line of sight static routes were used as OLSRd tried to establish single link connection between the end nodes, thus losing all the sent data.

The obtained results show that for lower bit rates, and with line of sight, the amount of losses using multi-hop oscillates from 4 to less than 30%. However, without line of sight the amount of registered losses is greater than expected when comparing with the single-hop case for the same distance, which had a better performance. This increase of losses is explained by incorrect single-hop paths established by the OLSRd protocol when multi-hop had to be used according to the single-hop experiences.

The OLSR protocol specification considers link quality as a "link admission" mechanism such that a node that "hears" a link does not immediately consider it as part of the routing topology. This quality may be measured by analysing signal to noise ratio or by keeping packet reception and loss information, for instance by checking message sequence numbers. However, the obtained results reveal that this scheme is still prone to errors, resulting in incorrect routing paths and consequent losses.

5   Conclusion

Wireless Networking in challenging environments has been long considered as one of the most promising solutions for connecting teams of search and rescue operations where no communication infrastructures may exist. In particular, Ad-hoc networks, for their self-X characteristics, have been analysed by many authors to support different applications in such scenarios. However, the majority of the existing studies are simulation based or are not taken in real challenging scenarios with realistic equipments. In this work a set of thorough wireless performance tests are taken in an underground gallery, using off-the-shelf wireless equipments, which can truly represent a scenario for a challenging rescue operation.

By performing several measurements in a humid underground gallery using Iperf in UDP mode, losses and link quality results for both single and multi-hop cases were gathered. Moreover, different distances separating the nodes with and without line of sight and different bit rates were also considered. The obtained results allow future research works in this area to provide more accurate conclusions on how wireless networks behave in challenging environments and what sort of applications are suitable in these scenarios.

Taking into account the number of registered losses versus the distances and available bit rates, it is clear that providing wireless network coverage in a challenging environment is still an issue. In particular, this issue results from poor multi-hop routing path establishment when using a well known Ad-hoc routing protocol, OLSRd. This protocol failed to successfully manage the existing routes, establishing incorrect single-hop links due to the poor relation between the measured link quality (considering the signal to noise ratio) and actual link performance.

The main registered problems are related with network partitioning, limited bit rates and high delay, suggesting that delay tolerant solutions as well as link quality metrics considerations have to be taken into account for efficient multi-hop routing in real challenging environments.

Acknowledgement. This work was supported by the Portuguese National Foundation for Science and Technology (FCT) through a PhD Scholarship (SFRH / BD / 43790 / 2008) and by the National Project MORFEU (PTDC / EEA-CRO / 108348 / 2008).

References

1. The atheros ath9k power consumption,
   http://wireless.kernel.org/en/users/Drivers/ath9k/power-consumption/
2. NLANR/DAST: Iperf - the TCP/UDP bandwidth measurement tool,
   http://iperf.sourceforge.net/ (accessed 2010)
3. Abolhasan, M., Wysocki, T., Dutkiewicz, E.: A review of routing protocols for mobile ad hoc networks. Ad Hoc Networks 2(1), 1–22 (2004)
4. Ahmed, S., Karmakar, G.C., Kamruzzaman, J.: An environment-aware mobility model for wireless ad hoc network. Computer Networks 54(9), 1470–1489 (2010)
5. Cayirci, E., Tezcan, H., Dogan, Y., Coskun, V.: Wireless sensor networks for underwater survelliance systems. Ad Hoc Networks 4(4), 431–446 (2006)
6. Chakeres, I., Perkins, C.: Dynamic manet on-demand (dymo) routing. Work in progress, internet-draft, Internet Engineering Task Force (July 2010),
   http://www.ietf.org/internet-drafts/draft-ietf-manet-dymo-17.txt
7. Chen, G.Z., Shen, C.F., Zhou, L.J.: Design and performance analysis of wireless sensor network location node system for underground mine. Mining Science and Technology (China) 19(6), 813–818 (2009)
8. Chen, Z., Chen, L., Liu, Y., Piao, Y.: Application research of wireless mesh network on earthquake. In: International Conference on Industrial and Information Systems, IIS 2009, 24-25, pp. 19–22 (2009)

9. Cimmino, A., Donadio, P.: Overall requirements for global information multimedia communication village 10th strategic workshop. Wireless Personal Communications 49(3), 311–319 (2009), http://dx.doi.org/10.1007/s11277-009-9686-3

10. Clausen, T., Jacquet, P.: Optimized link state routing protocol (olsr). RFC 3626, Internet Engineering Task Force (October 2003),
http://www.ietf.org/rfc/rfc3626.txt

11. Eriksson, J., Faloutsos, M., Krishnamurthy, S.V.: Dart: Dynamic address routing for scalable ad hoc and mesh networks. IEEE/ACM Transactions on Networking 15, 119–132 (2007)

12. Gaertner, G., Cahill, V.: Understanding link quality in 802.11 mobile ad hoc networks. IEEE Internet Computing 8(1), 55–60 (2004)

13. Garg, V.K.: Radio propagation and propagation path-loss models. In: Wireless Communications & Networking, pp. 47–84. Morgan Kaufmann, Burlington (2007)

14. Gerla, M.: From battlefields to urban grids: New research challenges in ad hoc wireless networks. Pervasive and Mobile Computing 1(1), 77–93 (2005)

15. Khoukhi, L., Cherkaoui, S., Gaiti, D.: Managing rescue and relief operations using wireless mobile ad hoc technology, the best way? In: IEEE 34th Conference on Local Computer Networks, LCN 2009, 20-23, pp. 708–713 (2009)

16. Murray, D., Dixon, M., Koziniec, T.: An experimental comparison of routing protocols in multi hop ad hoc networks. In: 2010 Australasian Telecommunication Networks and Applications Conference, ATNAC, 31 (2010)

17. Newport, C., Kotz, D., Yougu, Y., Gray, R.S., Jason, L., Elliott, C.: Experimental evaluation of wireless simulation assumptions. Simulation 83(9), 643–661 (2007)

18. Nielsen, Y., Koseoglu, O.: Wireless networking in tunnelling projects. Tunnelling and Underground Space Technology 22(3), 252–261 (2007), http://www.ietf.org/rfc/rfc3561.txt

19. Perkins, C.E., Belding-Royer, E.M., Das, S.R.: Ad hoc on-demand distance vector (aodv) routing. RFC Experimental 3561, Internet Engineering Task Force (July 2003), http://www.ietf.org/rfc/rfc3561.txt

20. Plagemann, T., Munthe-Kaas, E., Skjelsvik, K.S., Puzar, M., Goebel, V., Johansen, U., Gorman, J., Marin, S.P.: A data sharing facility for mobile ad-hoc emergency and rescue applications. In: ICDCSW 2007: Proceedings of the 27th International Conference on Distributed Computing Systems Workshops, p. 85. IEEE Computer Society, Washington, DC (2007)

21. Ribeiro, C., Ferworn, A., Tran, J.: An assessment of a wireless mesh network performance for urban search and rescue task. In: 2009 IEEE Toronto International Conference Science and Technology for Humanity (TIC-STH), 26-27, pp. 369–374 (2009)

22. Sha, K., Shi, W., Watkins, O.: Using wireless sensor networks for fire rescue applications: Requirements and challenges. In: 2006 IEEE International Conference on Electro/information Technology, 7-10, pp. 239–244 (2006)

23. Tnnesen, A., Lopatic, T., Gredler, H., Petrovitsch, B., Kaplan, A., Tcke, S.O., et al.: Olsrd - an ad-hoc wireless mesh routing daemon (July 2010), http://www.olsr.org

24. Wenfeng, L., Jie, G., Peng, B.: Mine multimedia emergency communication system. In: International Conference on Wireless Communications, Networking and Mobile Computing, WiCom 2007, 21-25, pp. 2865–2868 (2007)

Operational Support of Wireless Mesh Networks Deployed for Extending Network Connectivity

Thomas Staub, Benjamin Nyffenegger,
Desislava C. Dimitrova, and Torsten Braun

IAM, University of Bern, Switzerland
{staub,nyffenegger,dimitrova,braun}@iam.unibe.ch

Abstract. Wireless mesh networks (WMNs) have shown high potential to extend the coverage of high bandwidth infrastructure networks. We propose a deployment of a WMN for the needs of higher education institutes. In order to provide extended coverage to campus networks, several open issues such as authentication and authorisation of connected nodes, accounting of network usage and auditing of the network, have to be addressed. This paper outlines our proposal and its relevance in practice and discusses how we intend to solve the mentioned issues.

1 Introduction

Wireless mesh networks (WMNs) have been around for several years as a wireless paradigm between infrastructure networks and ad-hoc networks. Generally, a WMN consists of wireless mesh nodes, which can operate as hosts but also as routers, i.e., working as access points for the mesh network. The mesh nodes are typically fully functional computers with tailored operating systems to match hardware resource constraints. Data is forwarded from node to node towards its destination. The nodes' connectivity is also used for the exchange of management and configuration information. User data forwarding as well as management data dissemination requires that mesh nodes trust each other. Hence, one of the challenges in WMNs is the provisioning of appropriate authentication and authorisation mechanisms. A basic overview of the main research challenges in wireless mesh networks is provided in [2].

WMNs provide efficient, scalable means to offer service coverage to a large number of users with different communication needs [1]. They can complement existing high bandwidth networks and connect devices at remote locations. Since its introduction, WMNs have shown high potential for practical deployment and significant commercial impact. For example, after a successful prototype launch at the Massachusetts Institute of Technology and a pilot deployment in London, Nortel proceeded to deploy commercial networks in, among others, the city of Taipei (Taiwan), the Kennedy Space Center (US) and Edith Cowan University (Australia) [3]. Cisco Systems has also released their own Outdoor Wireless Network Solution[1] as the means to support mission-critical business

---

[1] Consulted in March 2011 at http://www.cisco.com/en/US/products

J. Del Ser et al. (Eds.): MOBILIGHT 2011, LNICST 81, pp. 322–329, 2012.

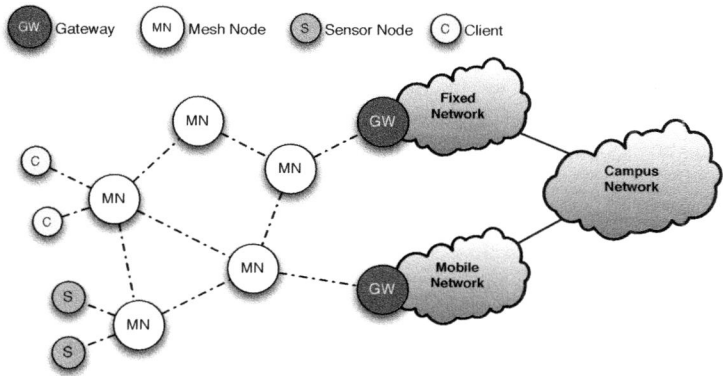

Fig. 1. General composition of a mesh network

applications. Their proposal to ensure outdoor wireless security is described in a white paper released by CISCO on their web-site[2].

Following the experience of others, we propose to use a WMN to supplement an existing university campus fixed network infrastructure with additional wireless coverage. Our goals are to develop, deploy and test a fully functional WMN tailored to the needs of the Swiss institutes of higher education. The proposed network deployment is illustrated in Figure 1. A WMN offers remote users, e.g. notebooks, access to the campus network. The WMN uses dedicated gateways to connect to a network with high bandwidth capacity, e.g., a fixed infrastructure network or a mobile network, which are also connected to the campus network. We propose a fixed infrastructure network as the broadband backbone mainly for two application scenarios, i.e. the extension of a campus network and the connection of remote areas for field experiments.

For an operational network, several requirements have to be met, namely the support of authentication, authorisation, accounting, and auditing (altogether referred to as A4). Authentication and authorisation control the access to the network. Accounting mechanisms enable the correct association of a service to a particular user and perform charging. Auditing should be in place to assist the undisturbed operation of the network by constant monitoring for abnormal behaviour.

The remainder of the paper is organised as follows. First, in Section 2 we address the practical relevance of the research and, in our opinion, its most probable impact areas. Next, Section 3 introduces the A4 functionalities and how we intend to realise their support by the proposed network scenario. Finally, in Section 4 we discuss several challenges that need to be faced and the future steps ahead of the project.

---

[2] Available in March 2011 at http://www.cisco.com/en/US/solutions

2   Practical Relevance

In this section we first describe what are potential applications of the envisioned wireless mesh network. The specifics of the application scenarios partly determine the requirements set towards the WMN. Next, we discuss the expected impact of the proposed research for both practical deployment and scientific contribution.

2.1   Application Scenarios

In our effort to enhance a WMN with the functionalities from Section 3 we are guided by its deployment for two main applications - extending coverage of campus networks and enabling of monitoring in remote areas. We now continue to describe both scenarios in more detail.

WMNs can complement the fixed Internet network in order to extend wireless network coverage in university campus networks. This does not only support individual students and employees in their work but also offers more freedom in organising group activities such as project meetings or educational fairs. To successfully use wireless mesh networks in the area of the Swiss higher education, WMNs have to support authentication, authorisation, accounting, and auditing. Since a campus network belongs to a particular university, access to the network should be provided only to authorised users. Hence, the WMN needs to support authentication and authorisation. As there are usually multiple concurrent users of the network, accounting is necessary in order to enable appropriate billing of the costs to the different users. Last but not least, for a successful operation of a WMN, inconsistent and erroneous states in the network have to be detected and resolved. This requires constant auditing of the network state and configuration such that alarms are triggered or even self-healing can be performed.

In the area of environmental monitoring applications, we distinguish several cases in which wireless mesh networks can provide additional connectivity. In all these monitoring scenarios, the sensor nodes are generally located in remote areas without access to a high bandwidth network infrastructure. In such situations, deploying a wireless mesh network provides a cost-efficient solution to connect these devices to the fixed network infrastructure. Remote environmental monitoring, e.g., related to glaciers, avalanches or landslips, is an intuitive candidate for the deployment of WMNs as cellular networks are not always available or have limited bandwidth. Another excellent candidate is agricultural monitoring, which is often associated with gathering information on soil humidity, precipitation and solar radiation over vast geographical areas. Deployment of a fixed network infrastructure over these areas is not reasonable. However, the collected sensor data still needs to be accessible. Another possibility is the deployment for weather monitoring, in which case the weather/climate monitoring stations need to communicate their collected data to a centralised entity in order to enable, e.g., weather forecasting.

Aside of the two main deployment scenarios described above, WMNs can be used to enable other outdoor monitoring and surveillance applications such as surveillance of restricted areas and service provision for research events of limited duration.

2.2   Expected Impact

We identify two perspectives from which the proposed study is beneficial for the development in wireless mesh networks.

First, the practical point of view is clearly important. The results can be used to extend campus network coverage (in the order of kilometres) by universities. WMNs may further be used for temporarily deployed network services for events and research projects, e.g., excursions and outdoor research experiments. Further, research initiatives in various research areas (climate research, geology and biology) may profit from an easily deployable outdoor wireless network that supports high speed network access as well as authentication, authorisation, accounting, and auditing. The latter two functionalities, i.e., accounting and auditing, are particularly attractive. The ability to charge independent parties enables the concurrent use of the wireless network infrastructure by multiple projects; auditing functions increase awareness on network health and reduce network maintenance costs. In conclusion, we believe that the proposed research has a high potential for providing the basis for commercial WMN services.

Second, there is the scientific value of the research. We intend to provide not only a theoretically sound solution but also a practically feasible one. Hence, the developed mechanisms will consider many practical details and constraints and will be very well suited for deployment.

3   The A4 Concept

This section presents the four basic functionalities - authentication, authorisation, accounting and auditing - required by the WMN in our proposed A4-Mesh network. We also discuss the envisioned architecture, which we have just defined at the present stage of the research, to support these functionalities. However, there are still several options open to explore and as result this initial architecture might undergo redesign. In addition, we expect certain changes may be necessary as a result of future design insights or unforeseen deployment issues.

3.1   Authentication and Authorisation

The goal of the authentication and authorisation functionality is to prevent the unauthorised usage of the network resources. Only authenticated and authorised users and sensors get access to the network. In addition, only authenticated and authorised mesh nodes can join and be part of the WMN. Whereas the network access for end users can be performed by existing approaches such as IEEE 802.11x / RADIUS used in eduroam[3] or a web-based captive portal protected

---

[3] www.eduroam.org

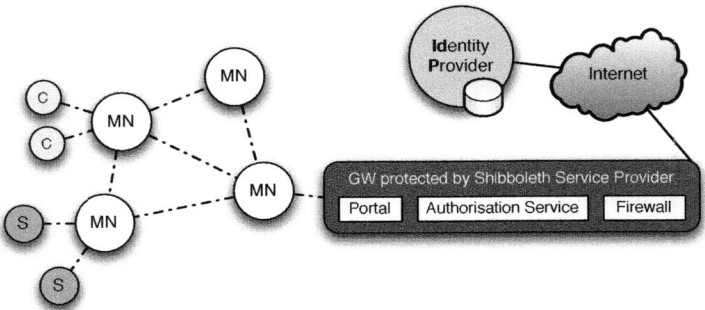

Fig. 2. Schematic representation of the authentication and authorisation mechanism

by SWITCHaai[4], new approaches are necessary for machine authentication and authorisation.

Our concept for machine authentication and authorisation is as follows. The A4-Mesh network implements two (virtual) wireless networks - an unencrypted one for joining the network and an encrypted one for data communication. Per default the A4-Mesh network just allows unencrypted connections for joining the network and encrypted connections for full access to the mesh network and the Internet. The unencrypted mesh network only forwards the necessary traffic for authentication and authorisation towards the gateway and the Shibboleth Identity Provider (IdP). All other traffic is blocked by the firewall of the gateway. The mesh nodes can get authenticated by the IdP through credentials or certificates. After successful authentication, additional attributes stored at the IdP are transmitted to the gateway for authorisation. For the machine authentication we intend to set up an additional IdP within the SWITCHaai Shibboleth federation to have the full flexibility of supporting different attributes for the mesh nodes than for normal SWITCHaai users. For user authentication, the existing IdPs are used.

The machine authentication and authorisation procedure is depicted in Figure 2. A joining mesh node (MN) first requests the key for the encrypted network from the portal through the public network. Upon this request, the portal redirects the node to the corresponding machine authentication IdP, which authenticates the mesh node by a X.509 certificate and replies with a reference to an authentication assertion. Using this reference, the portal on the gateway (GW) requests the attributes from the IdP (e.g. the unique identifier). Using these attributes, the portal can check the authorisation of the mesh node at the authorisation service. After successful authentication and authorisation, the gateway conveys the network key to the mesh node and thus allows traffic from and to the mesh node through its firewall. Finally, the mesh node joins the encrypted network with the received key and is now fully integrated in the WMN.

---

[4] www.switch.ch/aai

In addition, administrators have to be authenticated and authorised before accessing the web-based network management and monitoring of A4-Mesh. The network management therefore uses the Shibboleth service provider for user authentication. Based on the authentication, the network management handles the authorisation.

3.2   Accounting

As accounting, we define the ability to collect and store network-related information for later use. A WMN requires an accounting solution for various reasons such as charging users, network planning or increasing the network performance. To charge individual users for the network usage, the induced traffic of these users must be measured and network statistics may be used for network planning purposes. Finally, a WMN can provide similar characteristics as fixed networks in terms of robustness and reliability. This can be achieved by measuring the traffic on the individual links and exploiting multi-path, multi-channel and multi-interface routing mechanisms, inherent to wireless mesh networks. Path selection procedures are based on Quality of Service (QoS) characteristics such as available bandwidth, experienced bit and packet error rates and interference levels.

An accounting solution has to store the required information with the desired level of detail while simultaneously meeting resource constraints; in particular concerning storage capacities on mesh nodes. We therefore propose a distributed accounting architecture using short-term and long-term data and intend to incorporate parts of the AMAAIS (Accounting and Monitoring of AAI Services [4]) project which acts as an extension to SWITCHaai and provides mechanisms for accounting and monitoring. The architecture of the A4-Mesh accounting mech-

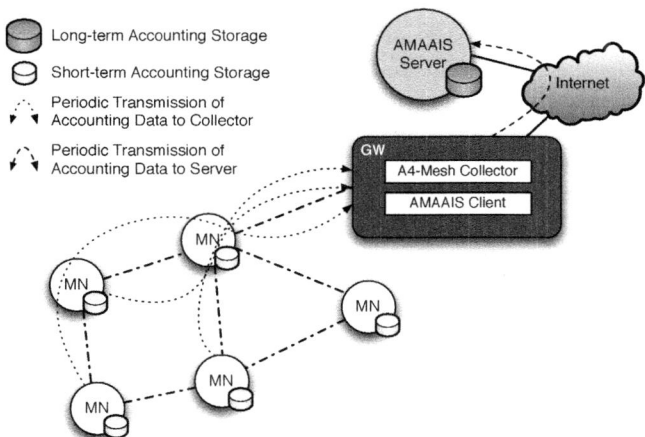

Fig. 3. Schematic representation of the accounting procedure

anism is depicted in Figure 3. The individual mesh nodes will gather short-term accounting data, which includes QoS characteristics of the neighbours and periodically transmit the information to the gateway. This data is also used to select the best communication channel for the transmission. The gateway processes the accounting data with a collector component and passes it on to the AMAAIS client which aggregates the short-term data and periodically transmits it to an AMAAIS server over the Internet. The AMAAIS server stores the long-term accounting data and enables billing of the costs, network dimensioning and planning.

Once available, the accounting data can be used for other purposes as well, such as the automation of path selections. Introducing intelligence locally in the nodes enables them to automatically choose a communication path fitting the needs of the transported data. To assist the selection process we intend to use a recommender system that considers both, the QoS characteristics offered by the network and the QoS characteristics required to transport the data. For example, if a path was able to successfully transport data with specific requirements (low delay, low jitter, low loss) under certain circumstances, it is possible that other paths with similar characteristics can be suitable as well. The benefits of such automated network (re-)dimensioning are reduced administrative overhead and a self-organised network.

3.3   Auditing

Audits are performed to ensure the validity and reliability of information; also to provide an assessment of a system's internal control. Due to (autonomic or manual) network management operations, as well as other external impacts, inconsistent or erroneous node states can occur. Examples for erroneous network states are the selection of frequencies that are not used by any neighbouring nodes or routing tables causing loops. One reason for service disruption can be the inappropriate choice of wireless radio or routing parameters due to, e.g., corrupt self-management mechanisms of the wireless mesh nodes. Another reason might arise from different versions of system software which may lead to compatibility problems and compromise interoperability of mesh nodes. The introduction of auditing functions should allow the detection of wrong and inconsistent configurations and report them to a network operator or perform an appropriate self-healing procedure.

4   Challenges and Further Development

Currently, we are working on building the mesh nodes, which will be used for an outdoor test bed. The outdoor test bed is situated in the region of Crans-Montana, Switzerland with some mesh nodes installed in remote areas. The region has harsh weather conditions with a lot of snow in the winter and requires mesh nodes specifically built for these conditions, i.e. durable cases, solar panels at least 3 meters above ground to prevent them from being covered by snow or batteries large enough in case of poor sunlight. We are also adding a backup mesh node with UMTS access to some of the regular mesh nodes, which are covered by the cellular networks. In the case that an erroneous image gets uploaded to a mesh node, we will be able to connect to the UMTS mesh node and reboot the regular mesh node or upload a new image instead of physically accessing the node.

The next step is completing and testing the implementations of the specific A4-Mesh functionalities. Since one of the goals of A4 is to allow users with a non-technical background to deploy mesh nodes to extend a campus network, special care has to be taken in regard to ease of use and seamless integration in the organisation's own authentication and authorisation infrastructure.

Acknowledgements. The A4-Mesh project is carried out as part of the program "AAA/SWITCH - e-Infrastructure for e-Science" lead by SWITCH, the Swiss National Research and Education Network, and is supported by funds from the Swiss State Secretariat for Education and Research.

References

1. Akyildiz, I.F., Melodia, T., Chowdhury, K.R.: A survey on wireless multimedia sensor networks. Comput. Netw. 51, 921–960 (2007)
2. Moustafa, H., Javaid, U., Meddour, D.E., Senouci, S.M.: A panorama on wireless mesh networks: Architectures, applications and technical challenges. In: Proceedings of International Workshop on Wireless Mesh: Moving towards Applications, Wimeshnets 2006 (2006)
3. Roch, S.: Nortel's Wireless Mesh Network solution: pushing the boundaries of traditional WLAN technology. Notrel Technical Journal (2005)
4. Stiller, B.: Accounting and Monitoring of AAI Services. SWITCH Journal (2), 12–13 (2010)

A Shared Opportunistic Infrastructure for Long-Lived Wireless Sensor Networks

Xiuchao Wu, Cormac J. Sreenan, and Kenneth N. Brown*

Department of Computer Science, University College Cork, Republic of Ireland
{x.wu,cjs,k.brown}@cs.ucc.ie

Abstract. In this paper, a Shared Opportunistic Infrastructure (SOI) is proposed to reduce total cost of ownership for long-lived wireless sensor networks through exploiting human mobility. More specifically, various sensor nodes are opportunistically connected with their corresponding servers through smart phones carried by people in their daily life. In this paper, we will introduce the motivations, present the architecture, discuss the feasibility, and identify several research opportunities of SOI.

1   Background and Motivations

Wireless sensor networks (WSN) are regarded as one of the most promising technologies for the new century. As related technologies mature, we expect that many WSNs will be deployed to provide services (e.g. environmental monitoring, water/gas/electricity meter reading, and structural health monitoring) over long periods. Figure 1 illustrates the architecture of a classical WSN [2], in which a dedicated sink is used to connect sensor nodes with their application server through the Internet. However, in some environments, it could become very expensive to deploy, operate and maintain this kind of wireless sensor networks.

To provide useful services, hundreds or thousands of sensor nodes are normally deployed in a large area (in some cases, sensor nodes may be sparsely deployed and the network is partitioned), and a large amount of data will be continuously generated by these sensors. In order to connect these nodes to the back-end server and/or have enough capacity for timely and energy efficiently forwarding these sensed data, multiple sinks that spread across the covered area must be deployed. These sinks, which are normally much more powerful and expensive than sensor nodes, will increase the hardware cost of the WSN. The monthly subscription fee for their Internet access will also increase the operational cost significantly.

In addition, it is hard to find appropriate sites (with power supply and Internet access point) to install these sinks, especially in outdoor environments. Although large solar panels and cellular networks could solve these issues, these devices

* This work is supported in part by the HEA PRTLI-IV NEMBES Grant (http://www.nembes.com/). Any opinions, findings, and conclusions or recommendations expressed in this material are those of the authors and do not necessarily reflect the views of NEMBES sponsors or the University College Cork.

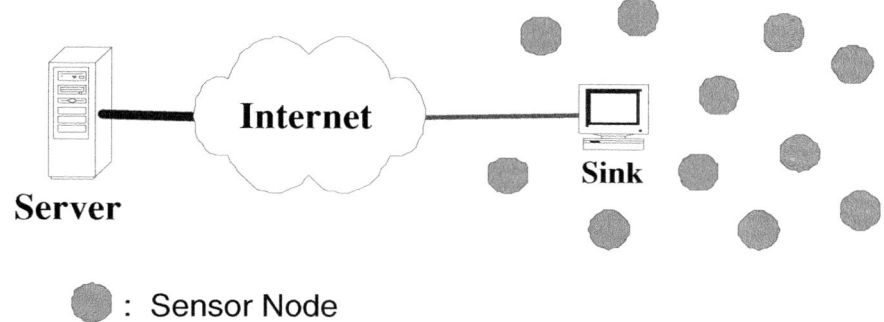

 : Sensor Node

Fig. 1. The Architecture of Classical Wireless Sensor Network

will unavoidably increase the cost of each sink. More importantly, sinks with these devices tend to become attractive targets for theft when they are deployed at unattended sites.

Instead of fixed sinks, mobile robots or employees are also used to collect data directly from the covered area and distribute commands/tasks to sensor nodes [16][17]. However, mobile robots can not be applied in many environments and the salary for employees will significantly increase the operational cost.

Based on the above observations and the fact that many WSN applications are delay tolerant and are normally deployed in the area visited frequently by people, we propose a Shared Opportunistic Infrastructure (SOI) to reduce TCO (total cost of ownership) for these long-lived WSNs through exploiting human mobility. In the following sections, we present the architecture, discuss the feasibility, and identify several research opportunities of SOI.

2   The Architecture

In recent years, human mobility has also been exploited by participatory and opportunistic sensing [3][4]. The sensors on smart phones carried by people are used as mobile sensors to sense the world and many fantastic applications have been proposed [11][14]. However, sensors on smart phones will increase phone price and it may not be inappropriate for many sensors to be installed on smart phones. Hence, the feasible applications of participatory or opportunistic sensing are limited. As for SOI that uses smart phones to opportunistically connect sensor nodes to their corresponding servers, it can complement with participatory/opportunistic sensing and it is worthwhile to be investigated further.

Considering that the sensed data may be totally unrelated with the person that passes through sensor nodes, SOI uses real money as the incentive, and Public Key Infrastructure (PKI) techniques are adopted to satisfy the security requirements. Figure 2 shows the architecture of SOI, in which Data Collection Gateway (DCG), Task Distribution Center (TDC), and smart phones carried

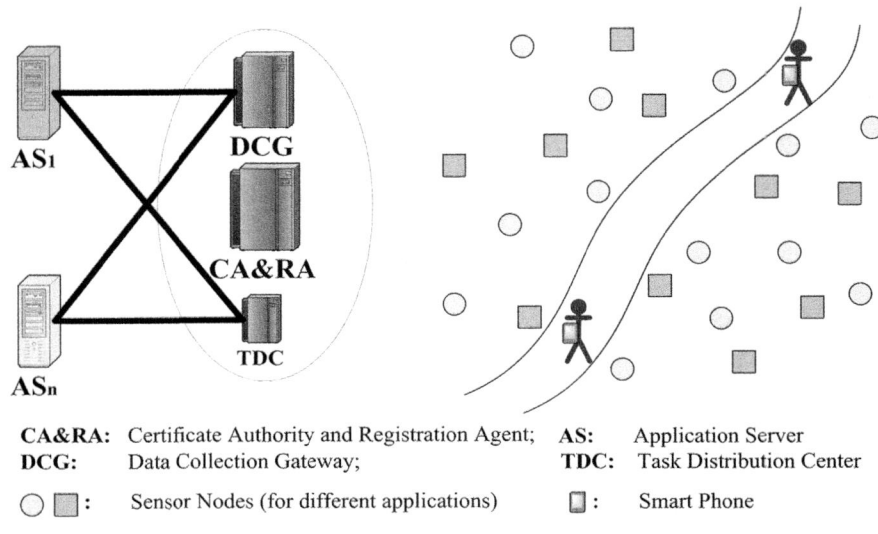

CA&RA: Certificate Authority and Registration Agent; **AS:** Application Server
DCG: Data Collection Gateway;   **TDC:** Task Distribution Center
○ ▢ :   Sensor Nodes (for different applications)   ▢ :   Smart Phone

Fig. 2. The Architecture of SOI

by people connect sensor nodes with their corresponding application servers (AS) opportunistically. CA&RA (Certificate Authority and Registration Agent) is responsible to register these entities and issue certificates to establish trust relationships among them. To alleviate the workload of CA&RA, application provider can use its certificate (issued by CA&RA) to sign one certificate for each sensor node. With the certificate chain installed on a sensor node, it can authenticate itself with smart phones.

The sensed data is collected by smart phones carried by people who pass through the covered areas in their daily life. According to delay requirements and the price paid by application providers, smart phones will then upload these data to DCG immediately through a cellular network, intermittently through open Wi-Fi hot-spots, or through a home broadband network. DCG will then forward these data to their corresponding ASs. In order to control sensor nodes, ASs can post tasks (command, code image, etc.) on TDC so that smart phones can download the appropriate tasks based on their locations and pass to the related sensor nodes. TDC may also broadcast node density in each area and the price offered by application providers. Based on these information, smart phone can decide whether to participate the system and which sensor nodes it will collect data from.

SOI can support many interesting applications. For instance, tourists can opportunistically collect data from sensor nodes deployed in national parks, city residents can be involved with various wireless sensor networks (weather stations, meters in buildings, etc.) deployed for serving the city, etc.

With SOI, application providers need not buy, deploy, or maintain sinks. Communication costs can also be decreased through exploiting human mobility.

Hence, TCO for long-lived WSNs can be reduced significantly. As for people, the intervention from the user can be reduced with carefully designed softwares and it is a strong motivation to earn money automatically during daily life. DCG, TDC, and CA&RA can also survive through claiming their share of the money paid by application providers.

Although SOI could provide many benefits, there are also a lot of issues to be concerned. In the following two sections, we will briefly discuss its feasibility and identify some research opportunities in the context of SOI.

3   Feasibility Analysis

One prerequisite of SOI is that smart phones must be able to talk with sensor nodes that belong to different WSNs. Considering that IEEE 802.15.4 [1] has been well standardized for WSNs, one low-power and cheap IEEE 802.15.4 radio installed on the smart phone is enough. With the maturity of WSN applications (such as health care, home automation, etc.), many devices with IEEE 802.15.4 radio will come to the market. Since the smart phone is a very good candidate as the unified bridge between a user and these devices, we can expect that IEEE 802.15.4 radio will be deployed on smart phones soon. We hope that SOI can also motivate the deployment of IEEE 802.15.4 on smart phones.

SOI should also provide enough incentives for people to participate. Not only the real money paid to people, the burden of user registration should be alleviated, their privacy must be respected, and a payment system is required so that people can get their earned money conveniently. Fortunately, since smart phones are used in SOI, mobile telecoms are the best candidates for operating CA&RA and these issues can be solved easily in this case.

The current registration process for phone services can be reused for SOI user registration. During this process, the certificates can be issued and installed on the smart phone, and the necessary softwares can also be installed. These softwares can also be pre-installed by the smart phone manufacturers.

As for privacy, compared to phone service registration, SOI does not need any further personal information and all personal information is held only by CA&RA. Each smart phone will get multiple certificates without any personal information. Hence, DCG and TDC can not identify the person based on the certificates. DCG and TDC will also conceal the certificates from ASs. With this scheme, to infer the movement of a specific person, application providers must let sensor nodes transmit user certificates inside the data forwarded by smart phones. Since multiple certificates without any personal information are issued to each user, application providers must pay for transmitting these certificates and they still cannot easily violate user privacy (locations, tracks, movement patterns, etc.). To avoid being tracked through wireless interface card, MAC address pseudonyms [8] can also be adopted by smart phones.

As for payment, the phone billing system can be utilized to pay the money to users. Through the cooperation among DCG, TDC, and CA&RA, application providers will be charged and the money can be deposited to phone bill accounts of the users correspondingly.

Since mobile telecom acts as CA&RA, misbehavior entities, such as smart phone users who discard data and application providers who refuse to pay, can be kicked out from the system through revoking their certificates. In serious situations, further punishments are possible since these entities had registered with mobile telecom.

In addition, data delivery latency in SOI could be huge since smart phone users may not visit some sensor nodes for a long period. However, many wireless sensor network applications are delay-tolerant and sensor nodes could be equipped with large-capacity flash cards for buffering the sensed data. For example, analysis of environmental monitoring data is rarely urgent, and meter readings for billing purposes can be delayed by weeks. In the case that data delivery latency must be low, some alternative methods must be provided. SOI still could be helpful in the periods with abundant user mobility.

4   Research Opportunities

In SOI, there are some unique challenges to be solved. Many problems, that have been studied in delay-tolerant network [10][18], wireless sensor network [13][15], participatory or opportunistic sensing [5][6], etc., should also be re-scrutinized in the context of SOI. In the following sub-sections, we will discuss several topics that should be worthwhile to be investigated further.

4.1   Architecture and Protocols

The architecture of SOI should be refined. The functions of each entity and their trust relationship must be clarified. The protocols used among these entities should also be designed carefully. The main point is to ensure user privacy, detect abnormal behaviors of each kind of entity, and provide security schemes required by an E-Payment system, such as non-repudiation and non-replicability.

4.2   Computation Cost of Sensor Nodes and Smart Phones

PKI is used by SOI to satisfy the security requirements. Although smart phones are now very powerful and some PKI chips have already been developed for sensor nodes, it is still worthwhile to study how to reduce the number of slow and expensive private key operations [9]. Due to the regularity of human mobility [7], some sensor nodes and smart phones will encounter frequently. Hash chains [12] could be used to reduce the number of private key operations carried out by these familiar strangers for authentication.

It should also be desirable to decouple private key operations from the process of data exchange so that protocols will not be interfered and sensor nodes could schedule these operations in a better way. For example, data can be signed by sensor nodes when there is no smart phone nearby and/or when energy supply from the environment is abundant in the case that energy harvesting (solar panel, etc.) is exploited.

4.3   Contact Probing

Contact probing had been well studied in delay tolerant network [18]. With the assumption that mobile nodes are always active, mobile nodes learn contact arrival process and adjust the frequency of probing accordingly with the aim to save energy and reduce contact miss probability. However, sensor nodes in SOI are normally duty-cycled and contact probing scheme should be re-scrutinized. It is worthwhile to study how smart phones and sensor nodes can find each other timely for collecting data as much as possible. Contact probing scheme should also be energy efficient for both smart phones and sensor nodes.

In the context of SOI, smart phones are not suitable for initiating contact probing. If so, smart phones must probe with a high frequency as sensor nodes are duty-cycled. Consequently, smart phones will waste their energy and wireless channel will be congested by these probing packets. Hence, it should be better to let smart phones keep listening and let sensor nodes probe when they wake up according to their duty-cycle. This issue has been studied further in [20][19].

When sensor nodes could form a connected network, these nodes may also cooperate to improve the performance further. For example, sensor nodes may predict the mobility of the smart phone that they are talking with. They can then activate the corresponding nodes at appropriate time so that the smart phone can be probed timely.

Within SOI, CA&RA can also be used to deduce the density of sensor nodes and the number of SOI users in the current cell. This information can be broadcasted through base station and smart phones can adjust their behaviors correspondingly. For example, according to user density, a smart phone can adjust its back-off window, that is used before responding to a probing packet broadcasted by a sensor node, with the aim to reduce packet collision.

4.4   Data Replication and Distribution

As sensor nodes in SOI may not be visited by smart phones for a long period, the sensed data should be replicated and distributed for data reliability under node failure [10].

In the context of SOI, not only the cost of sensor node's energy and storage, application providers must also pay for data replications collected by smart phones. Hence, replication overhead should be deterministic and should be as small as possible. Instead of multiple copies, source-based erasure coding approaches, such as Reed-Solomon Codes, could be adopted. These data slices are then distributed among sensor nodes for reliability and each data slice should be signed separately for non-replicability.

When distributing these data slices, not only data reliability, many other issues should also be considered. For example, sensor nodes in SOI may have different probability to be visited by smart phones. If data slices can be distributed according to contact and storage capacity of these nodes, network throughput can be increased and the latency of data delivery can be decreased. In addition, due to network topology, some sensor nodes need forward data for other nodes. For load balancing, they may ask other nodes to sign their data slices.

A two-phase data distribution should be suitable for SOI. In the first phase, for both data reliability and load balancing, data slices are forwarding to several nodes with less contact capacity, at which data slices are signed. In the second phase, the signed data slices are forwarded to nodes with more contact capacity for higher throughput and lower latency.

Data replication and distribution will unavoidably consume storage and energy resources. Hence, there are many tradeoffs to be considered when designing these schemes. When solar panels are used by sensor nodes for energy harvesting, it is also very valuable to study how to synchronize these operations with energy supply of the deployment environment.

5 Future Work

This paper presents the concept and the architecture of SOI. In future work, we will refine the architecture and design the protocols and algorithms used by the entities of SOI. We also plan to implement a prototype of SOI for investigating this topic further.

References

1. IEEE 802.15.4, Part 15.4: Wireless Medium Access Control (MAC) and Physical Layer (PHY) Specifications for Low-Rate Wireless Personal Area Networks (LR-WPANs). IEEE (2006)
2. Akyildiz, I.F., Su, W., Sankarasubramaniam, Y., Cayirci, E.: Wireless sensor networks: a survey. Computer Networks 38, 393–422 (2002)
3. Burke, J., Estrin, D., Hansen, M., Parker, A., Ramanathan, N., Reddy, S., Srivastava, M.B.: Participatory sensing. In: World Sensor Web Workshop (in conjunction with Sensys) (2006)
4. Campbell, A.T., Eisenman, S.B., Lane, N.D., Miluzzo, E., Peterson, R.A.: People-centric urban sensing. In: ACM/IEEE WICON (2006)
5. Cornelius, C., Kapadia, A., Kotz, D., Peebles, D., Shin, M., Triandopoulos, N.: Anonysense: Privacy-aware people-centric sensing. In: Mobisys (2008)
6. Edalat, N., Xiao, W., Tham, C.-K., Keikha, E., Ong, L.-L.: A price-based adaptive task allocation for wireless sensor network. In: MASS (2009)
7. Gonzlez, M.C., Hidalgo, C.A., Barabsi, A.-L.: Understanding individual human mobility patterns. Nature 453, 779–782 (2008)
8. Gruteser, M., Grunwald, D.: Enhancing location privacy in wireless lan through disposable interface identifiers: a quantitative analysis. Mobile Networks and Applications 10(3), 315–325 (2005)
9. Hu, W., Corke, P., Shih, W.C., Overs, L.: secFleck: A Public Key Technology Platform for Wireless Sensor Networks. In: Roedig, U., Sreenan, C.J. (eds.) EWSN 2009. LNCS, vol. 5432, pp. 296–311. Springer, Heidelberg (2009)
10. Jain, S., Demmer, M., Patra, R., Fall, K.: Using redundancy to cope with failures in a delay tolerant network. In: SIGCOMM (2005)
11. Kang, S., Lee, J., Jang, H., Lee, H., Lee, Y., Park, S., Park, T., Song, J.: Seemon: Scalable and energy-efficient context monitoring framework for sensor-rich mobile environments. In: Mobisys (2008)

12. Lamport, L.: Password authentication with insecure communication. Communications of the ACM 24, 770–772 (1981)
13. Liu, C., Wu, J., Cardei, I.: Message forwarding in cyclic mobispace: the multi-copy case. In: MASS (2009)
14. Mun, M., Reddy, S., Shilton, K., Yau, N., Burke, J., Estrin, D., Hansen, M., Howard, E., West, R., Boda, P.: Peir, the personal environmental impact report, as a platform for participatory sensing systems research. In: MOBISYS (2009)
15. Pasztor, B., Musolesi, M., Mascolo, C.: Opportunistic mobile sensor data collection with scar. In: MASS, pp. 1–12 (2007)
16. Shah, R.C., Roy, S., Jain, S., Brunette, W.: Data mules: Modeling a three-tier architecture for sparse sensor networks. In: IEEE SNPA Workshop, pp. 30–41 (2003)
17. Somasundara, A.A., Kansal, A., Jea, D.D., Estrin, D., Srivastava, M.B.: Controllably mobile infrastructure for low energy embedded networks. IEEE Transactions on Mobile Computing 5(8), 958–973 (2006)
18. Wang, W., Srinivasan, V., Motani, M.: Adaptive contact probing mechanisms for delay tolerant applications. In: Mobicom, pp. 230–241 (2007)
19. Wu, X., Brown, K.N., Sreenan, C.J.: Exploiting rush hours for energy-efficient contact probing in opportunistic data collection. In: The 4th International Workshop on Sensor Networks (in conjunction with IEEE ICDCS) (2011)
20. Wu, X., Brown, K.N., Sreenani, C.J.: SNIP: A sensor node-initiated probing mechanism for opportunistic data collection in sparse wireless sensor networks. In: The First International Workshop on Cyber-Physical Networking Systems (in conjunction with IEEE INFOCOM) (2011)

Wireless Sensor Networks for a Zero-Energy Home

Rangarao Venkatesha Prasad, Vijay S. Rao, Ignas Niemegeers,
and Sonia Heemstra de Groot

Faculty of EEMCS, Delft University of Technology,
Mekelweg 4, 2628 CD Delft, The Netherlands
{R.R.VenkateshaPrasad,V.SathyanarayanaRao,I.G.M.M.Niemegeers,
S.M.HeemstradeGroot}@tudelft.nl

Abstract. Energy, especially *Electricity* has become a critical concern of society. The generation and distribution are, to a large extent, not matched with each other. There are many initiatives to balance the load on the Grid vis-à-vis production and consumption. Recently, technology has made it possible to generate small amounts of energy in each household with solar, thermal, wind and other sources. With storage devices, it is possible to use the generated energy at the sources i.e., localized generation and consumption of energy. The system can be made efficient and possibly design a zero or positive--energy home with the use ICT infrastructure, wireless sensor networks, in particular. This paper proposes a three-layered architecture for this system and also lists several associated issues and challenges.

1 Introduction

Electrical energy has become a critical concern of society. The generation and distribution are, to a large extent, not matched with each other. In fact the generation and consumption locations are also geographically isolated and to a large extent these two are always seen in isolation. There are many initiatives to balance the load on the Grid vis-à-vis production and consumption. This is not sufficient in the current age since the centralized production of power implies that we are heavily dependent on fossil fuels, nuclear fuel, windmills and large hydro projects, followed by a complex distribution system that needs to be managed. Thus it is imperative to look beyond this existing system for better solutions [1].

Recently, technology has made it possible to generate small amounts of energy in each household. There are ample opportunities to harvest energy, e.g., solar, thermal, wind, using micro-turbines, photo-voltaic cells, fuel cells etc. There are possibilities of storing the generated small amount of energy and possibly feeding it back to the Grid [2, 6, and 11]. Moreover, if the network is designed well, the generated (even small amounts) could be spent in the neighborhood, thereby avoiding distribution losses. The capability to generate and leverage the small amounts of energy in neighborhoods autonomously provides a subsystem with potential to reduce or avoid the centralized grid and its associated complexity and losses. Having sources closer to loads contributes to enhancement of the voltage profile, reduction of distribution and

J. Del Ser et al. (Eds.): MOBILIGHT 2011, LNICST 81, pp. 338–346, 2012.
© Institute for Computer Sciences, Social Informatics and Telecommunications Engineering 2012

transmission bottlenecks, lower losses, exploits wasted heat,. It also allows to avoid or postpones investments in new transmission and large scale generation systems. Therefore, smartgrids and microgrids are being actively researched [2].

Major companies have started developing systems for Smart Metering [8]. As an ex ample, Cisco and IBM [9, 10] started a joint project in which 500 households will have smart metering implemented.

There are few implementations of microgrids and they are complex; however, there is a need for a non-complex model and control methods that take care of the generation, distributed storage and consumption in a localized manner. Such a system should, ideally, realize zero-energy (or even positive-energy) homes and buildings. Information and Communication Technologies (ICT), in particular, Wireless Sensor (and Actuator) Networks (WSNs) have been widely used to increase energy efficiency of systems. We believe this technology could be employed in these microgrids to achieve cooperation and collaboration between the sources and the loads, and facilitate building the zero-energy homes. Moreover, with sensors it is possible to get context information at the loads. Along with the sensors at the sources, it is possible to enhance energy management functionality of the system. In this paper, we present an overview of microgrids and their benefits; we also propose an ICT architecture for realizing zero-energy homes using Wireless Sensor Networks and discuss important issues involving its implementation. Finally we brief about the projects on microgrids.

1.1   Microgrids

There is ample opportunity for small scale energy harvesting in homes, buildings, and public places. The very fact that the living environment is controlled opens many doors to harvest small amount of energy. For example, the thermal difference occuring during severe winters enables us to harvest some amount of the energy spent on heating the living space. Small wind turbines are also part of this in many countries. Energy can also be harvested using solar and thermal gradient systems [11]. With these micro-generators, the reduced energy cost as well as having a clean environment is making people conscious of the opportunity of harvesting energy. At present these initiatives are isolated.

Further to the above description, we observe that energy is not required at a uniform rate during a day. The energy consumption also varies depending on the number of people, their habits and requirements in any particular building or home. Domotics and many recent initiatives are bringing newer ICT technologies to households to save energy. For example, the sleeping rooms are not heated till sometime before the persons go to bed to save energy. The lighting is switched off and intelligent lighting is also used in many buildings and long corridors where it is turned on only when required, or lighting is adapted to the context, e.g., the activities or the mood of people. Thus, the energy load in buildings or houses varies depending on the available technology as well as the habits and context. For example, households use the least amount of energy in the day time but offices and factories use mire energy at that time (see Fig. 1).

The second aspect is that the micro-generators that people use to harvest small amounts of energy are always stored in small storage elements such as batteries, flywheels, and super-capacitors [11]. Since the energy harvesting will not usually yield usable power continuously there must be some devices to store the generated energy and to refine it for later use. Thus wherever there is a generator there would be usually a small battery to store the energy generated. Thus we see that there is a huge potential for storing the generated energy locally and in a distributed way.

The third aspect is that the cost of connecting to the Grid and maintaining the power flow from - and to - the Grid. Energy transportation also introduces heavy losses. When small amounts of energy are generated there is no reason to transport and maintain that energy connecting the generator with the Grid. Thus local storage and consumption of the locally generated energy is the best possible way to minimize the losses and reduce the maintenance overhead.

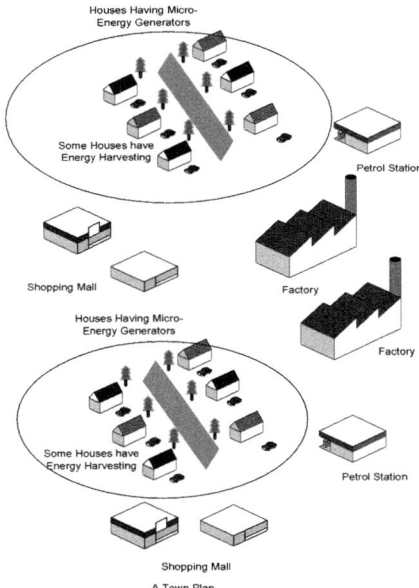

Fig. 1. A view of various energy generation and consumption zones

1.2   Sensors

We envisage that sensors are going to be an important aspect of any ICT enabled architecture or service or applications in the future. As we have seen the applications of sensors in many ICT enabled application, we need to integrate the sensors seamlessly into the design of systems which takes care of balancing generation and consumption. We identify the sensors under three different classes. Please note that sensors here mean an integrated system of sensors that could also accomplish some computation. Three classes are: (i) source sensing; (ii) load/sink sensing and (iii) context sensing. The distinguishing aspect between many micro grids and the architecture we propose is that the complete system is seen holistically. We explain here each of the sensors we envisage here in this architecture.

(i) *The Source* sensor gives the information on amount of energy generated. The sensors which monitor micro energy generation also give an indication as to the amount of energy that could be generated. We envisage that these sensors are also equipped with prediction algorithms and also they have the capability to learn. For example, a sensor that is attached to photovoltaic panels would also learn and then predict when there would be some energy available (based on day night periodicity).

(2) *The Load* sensor has the information of energy requirement if the load is turned on. In fact the sensing is done in realtime based the amount of load that each equipment has. As in the source sensing this can also predict to an extent how much load would be required. For example, based on the temperature difference between the outside temperature and usually required temperature inside, required amount of load can easily be predicted in case of room heaters.

(3) *The Context* sensor gives or predicts when to turn on or off the load. Basically we group all the sensors that generate the context of the environment including the users' locations, day/night temperature, requirements and usual settings of users, etc. Indeed for the sake of convenience, we also group together some of the max-min settings of users.

The energy manager can consider these parameters, to budget the energy. If any of these three parameters are missing, then accurate budgeting is not possible.

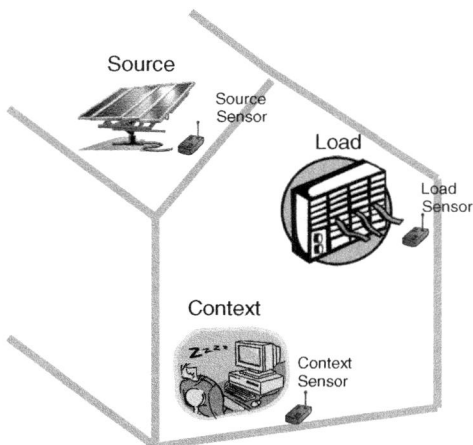

Fig. 2. Schematic of types of sensing

2   Proposed Architecture

An aspect that enhances the efficiency of the whole system is ICT - especially wireless communication has become pervasive and is being used in various domains for different purposes. The purpose of microgrids and ICT is twofold: (1) enabling the generation and storage of (small amount of) energy harvested in homes and buildings; and (2) using the communication and networking technologies to achieve the control and balancing of the micro generation of energy which would ultimately support the large Grids along with efficient usage of the energy.

Wireless sensor and actuator networking (WSAN) has been steadily growing and permeating into many spheres of life. Many techniques such as localization of events, clustering, self organization of the nodes and various other facets of WSANs have been studied in depth. Now the very same methods and the devices could be coupled to the energy generation and consumption modules to derive maximum benefits. Here we list some commonalities and the motivation to bring seemingly different domains together:

1. WSN and WSAN have been studied from the point of view of information generation and consumption (source and sink). In case of energy harvesting too, we can draw analogies for source and the sink i.e., the generation unit to the nearest place of consumption.
2. The generation and consumption areas and points are to be grouped so as to easily manage the energy requirements. This is similar to the clustering in WSNs.
3. There is always a provision of distributed computation and storage in WSNs. Here too the batteries are storage devices which are distributed and one can use them for temporary storage of harvested energy before converting it to usable form.
4. Dynamic self-organization on the fly, based on the requirements and available resources, is a hallmark of ad hoc networks and WSNs. Here too, since energy consumption and generation is varying with time (at micro layers), self-organization plays an important role.
5. Further, WSNs have been used to gather context i.e., movements or activities of people. The context information can be used to increase the efficiency of the system [12].

Therefore, it is but natural to use the techniques that have been addressed in depth in this different domain in the case of balancing and utilizing the harvested energy. Thus we propose the following conceptual 3-layer architecture (see Fig. 3). At the lowest layer, there are instruments, modules and storage equipments in which some are energy generators and others are energy consumers. These equipments are connected to switches, actuators, transducers (current transformers, etc) and sensors. They are grouped in the second layer. These sensors and actuators act as the glue connecting the lowest layer and the highest layer that contains the brain of the system. The top layer has communication and computing devices. The sensed data is actually sent to this layer. The context of the situation such as, sleep time, office time, someone is present in the building or moving, etc., is given to these computing nodes. They interpret the sensed data collectively and individually to make decisions. A decision is sent across the devices in the neighborhood. The data also contains the energy consumption and generation rates at various places. Since these computing devices, which are equipped with communication capability, can quickly reorganize themselves and collectively decide to take actions. The actions are in the form of triggers to be passed to the actuators/switches. Thus the energy generation, consumption, storage and requirements (using the context) are all addressed holistically in this concept. The localization of the generation and consumption as well as the balancing is to be done in real-time.

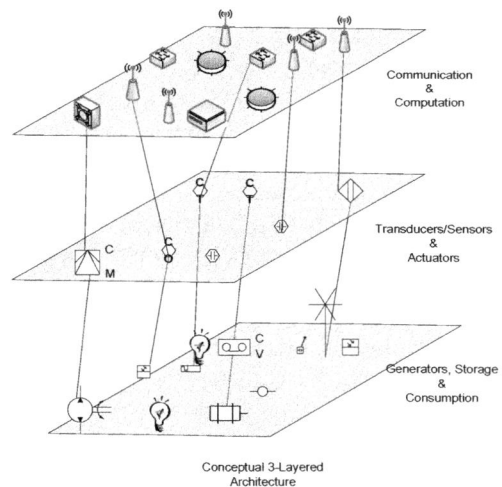

Conceptual 3-Layered
Architecture

Fig. 3. Three layered architecture

In Layer 2 of the architecture, we can distinguish between three types of sensing operations (see Fig. 2) which have described earlier.

To realize a zero-energy system, the following aspects need to be addressed:

1. Investigating the possible avenues of harvesting energy in homes and other buildings.
2. Localization of energy harvesting and consumption i.e., getting an accurate view of where energy can be generated and where it can be used.
3. Development of a model of energy storage in small quantities for a large group of harvesters (say in each building).
4. Development of simple, reliable, efficient and cost-effective energy balancing methods so that efficient systems could be designed that take into account local energy needs (microgrids).

5. The ICT infrastructure to manage and control the energy system i.e., the communication and computing backbone to support the above activities.
6. Efficiently using the small scale energy storage devices such as batteries stationed in various locations in houses that act as distributed storage devices.

The above issues can further be delved in to gather further set of technical issues such as ensuring stable operation during faults, handling large mismatches between generation and loads (a frequency and voltage control problem), system protection to line faults, [9, 11] etc.

We emphasize that the effective management of the system is the key to efficient system. This effective management can be achieved through ICT. Issues such as coordination of the distributed sources and loads (with possibly conflicting requirements), dispatch control, load-shedding control, dynamic management etc are some issues to be handled by the energy management subsystem. In addition to the above tasks, the system should enable:

1. Analyzing the energy generation, storage and consumption in homes and buildings and in turn in localities and cities.
2. Modeling generation and consumption in small localities and studying its effect on the grids including the prediction of energy generation and usage.
3. Modeling the energy storage in small amounts in a distributed way.
4. Distributed control and management of the complete system. We envisage that the sensors, actuators, source and the load may decide about the actions to be taken to micromanage the complete system. Since this is an ambitious goal, in the initial stages, we can bring in a centralized simple *manager* who does take appropriate actions in real-time.

3   Related Work

The EU funded two major projects "Microgrids: Large Scale Integration of Micro-Generation to Low Voltage Grids"and "More Microgrids: Advanced Architectures and Control Concepts for More Microgrids" [13] for researching, designing and developing microgrids. Several deployment sites were also identified and tested.

In the U.S., the Consortium for Electric Reliability Technology Solutions (CERTS) funded a project CERTS Microgrid. A Microgrid testbed was also developed [14]. An implementation of energy manager design done under CERTS project is given in [5].

In the industry, GE Energy funded a project to develop and demonstrate a microgrid energy management (MEM) [4] framework for a broad set of microgrid applications that provides a unified controls, protection, and energy management platform.

The IEEE Standards Coordinating Committee 21 is supporting the development of IEEE P1547.4, Draft Guide for Design, Operation, and Integration of Distributed Resource Island Systems with Electric Power Systems [15]. This covers microgrids connected to both local and area island electric power systems (EPS). This document provides alternative approaches and good practices for the design, operation, and integration of the microgrid, including the ability to separate and reconnect, while providing power to the island local EPSs.

4   Ongoing Work

There is an ongoing work on which this article is based. It is called as L-BEES -- *Load Balancing and Efficient Energy Sourcing*. This is under the exploration and initial stage at the moment. However, under another Dutch national project IOPGencom-GoGreen, we are investigating the use of energy harvesting sensors that could provide the context as well as sensing to help in reducing the total energy consumption in indoor environment. This work enhances the aspects in GoGreen by looking at the generation, consumption along with intelligent sensing together. We also mention here that the batteries (used for storing energy from micro-generators before conversion) in different houses and buildings act as small energy storage devices. This is another novelty of this proposed architecture. Context of the users and also the environment parameters are another aspects that is being considered in the above projects.

5   Conclusions

Microgrids have been actively pursued and researched for local energy generation and consumption. These systems enable enhancement of the voltage profile, reduction of distribution and transmission bottlenecks, lower losses, exploiting waste heat, and they allow postponing investments in new transmission and large scale generation systems. Three main aspects of microgrids are energy generators, storage of the generated energy and distribution. We propose the use of ICT, in particular, wireless sensor networks for increased efficiency of the system. One recognizes similarities in the WSN applications and microgrids. Further context information can be another useful parameter for the energy management which is made possible by the usage of sensor networks. Issues such as coordination of the distributed sources and loads (with possibly conflicting requirements), dispatch control, load-shedding control, dynamic management etc, can be handled better with wireless sensor network, which increases the overall efficiency of the system. A three-layer architecture comprising of these is also proposed. Finally, recent projects on microgrids have been discussed.

Acknowledgements. The authors would like to deeply thank Hermes Partnership (http://www.hermes-europe.net/home/) and Trans research academy (http://www.trans-research.nl/) for supporting this work.

References

1. Electric Power Research Institute, Grid 2030: a national vision for electricity's second 100 years,
   http://www.oe.energy.gov/DocumentsandMedia/Electric_Vision_D ocument.pdf
2. Hatziargyriou, N., et al.: Microgrids. An overview of ongoing research, development and demonstration projects. IEEE Power & Energy Magazine, 78 (2007)
3. Marnay, C., Siddiqui, A.S., Rubio, F.J.: Shape of the microgrid. In: Proc. IEEE Power Eng. Soc. Winter Meet, p. 150 (2001)
4. Pogaku, N., Prodanovic, M., Green, T.C.: Modeling, analysis and testing of autonomous operation of an inverter-based microgrid. IEEE Trans. Power Electron. 22, 613 (2007)
5. Firestone, R., Marnay, C.: Energy manager design for microgrids (2005)
6. Lasseter, R.H.: Microgrids. In: Proc. Power Eng. Soc. Winter Meeting, vol. 1, pp. 305–308 (January 2002)
7. Hatziargyriou, N., Jenkins, N., Strbac, G., et al.: Microgrids – Large Scale Integration of Microgeneration to Low Voltage Grids. In: Proc. CIGRE 2006, Paris, paper C6-309 (August 2006)
8. EURELECTRIC's Position Paper, Building a European Smart Metering Framework suitable for all Retail Electricity Customers (2008)
9. http://www.ibm.com/smarterplanet/us/en/smart_grid/nextsteps/solution/K626033G66635H91.html
10. http://www.cisco.com/web/strategy/docs/energy/aag_c45_539956.pdf
11. Lasseter, R., Akhil, A., Marnay, C., Stevens, J., Dagle, J., Guttromson, R., Meliopoulos, A.S., Yinger, R., Eto, J.: White Paper on Integration of Distributed Energy Resources: The CERTS MicroGrid Concept. Lawrence Berkeley National Laboratory Report LBNL-50829, Berkeley (2002)
12. Taylor, K., Ward, J., Gerasimov, V., James, G.: Sensor/actuator networks supporting agents for distributed energy management. In: 29th Annual IEEE International Conference on Local Computer Networks, November 16-18, pp. 463–470 (2004)
13. http://www.microgrids.eu/
14. http://certs.lbl.gov/certs-der-micro.html
15. http://grouper.ieee.org/groups/scc21/

Analysis Methodology for Flow-Level Evaluation of a Hybrid Mobile-Sensor Network

Desislava C. Dimitrova[1], Geert Heijenk[2], and Torsten Braun[1]

[1] Univeristy of Bern, Bern, Switzerland
[2] University of Twente, Enschede, The Netherlands

Abstract. Our society uses a large diversity of co-existing wired and wireless networks in order to satisfy its communication needs. A cooperation between these networks can benefit performance, service availability and deployment ease, and leads to the emergence of hybrid networks. This position paper focuses on a hybrid mobile-sensor network identifying potential advantages and challenges of its use and defining feasible applications. The main value of the paper, however, is in the proposed analysis approach to evaluate the performance at the mobile network side given the mixed mobile-sensor traffic. The approach combines packet-level analysis with modelling of flow-level behaviour and can be applied for the study of various application scenarios. In this paper we consider two applications with distinct traffic models namely multimedia traffic and best-effort traffic.

1 Introduction

In the last few years a fast development in the area of wireless applications is observed. On the one hand, traditional infrastructure-based cellular networks have undergone major changes with the shift toward UMTS and its successor LTE. On the other hand, wireless networks with local geographical coverage, such as wireless sensor networks, attracted the interest of both industry and academia. Indeed, a survey of the ongoing research efforts by both industry and academia shows that the topic of network interoperability is very attractive, see [7,11]. We recognise that both types of networks have merits and propose to combine them in one hybrid mobile-sensor network.

In this paper we are specifically interested in the impact of sensor traffic on the performance of mobile traffic and how the phenomenon can be best studied. To this end we propose an approach that can capture both mobile traffic specifics and Quality of Service (QoS) requirements of sensor traffic. Special attention is paid to the impact of flow dynamics, i.e., the generation of flows at arbitrary moments and arbitrary locations, which leads to a changing number of service requests. In previous studies, see [3,4], it was shown that flow dynamics are crucial for the performance of data traffic. We expect this to hold even stronger when several traffic classes are considered.

A hybrid mobile-sensor network arises many research challenges among which also resource management and QoS provisioning. These two topics have received

J. Del Ser et al. (Eds.): MOBILIGHT 2011, LNICST 81, pp. 347–355, 2012.
© Institute for Computer Sciences, Social Informatics and Telecommunications Engineering 2012

much attention in the literature as [2,5,8] and [12] attest. Although providing much insight the authors, e.g., [1,6,10,11], often consider only one type of network - mobile or sensor. However, in a hybrid network there are more factors that contribute to the complexity of the problem. There are several proposals, e.g., [7,9], which describe a complex new architecture to support the cooperation between wireless networks. Unfortunately, in the general case such studies provide only a highly abstract view without discussing in detail specific network functionality. To the best of our knowledge there are no studies specifically dedicated to the topic of resource management in a hybrid mobile-sensor network, independently of the presence of flow-level analysis.

An interesting study related to our proposal is [7]. The authors evaluate the cooperation between UMTS and WLAN networks with an architecture similar to the one we propose. However, [7] focuses on protocol support issues and not on resource management. Another interesting reference is [10] which addresses QoS provisioning in a LTE network with mixed traffic, e.g. voice, video and data. The authors provide a theoretical as well as a simulation approach to evaluate performance but do not account for the dynamic behaviour of the users and their main focus is on the voice traffic.

We propose to use a combined analysis approach which includes modelling of flow-level behaviour (left out by many studies) and can therefore offer new perspectives on performance. In contrast to other authors we are not interested in developing a new platform for QoS provisioning but consider QoS parameters as factors that set constraints on resource management and hence have an effect on performance. To achieve the latter we intend to use a modified version of the combined analytical/simulation approach proposed in [3]. Advantages of the approach are its ability to capture in sufficient detail both packet and flow-level behaviour while still supporting quick evaluation.

The paper continues as follows. The next Section 2 presents the concept of a hybrid mobile-sensor network and discusses potential applications. In Section 3 we describe the specific research scenario chosen to investigate in this paper. The proposed analysis methodology is explained in Section 4. Finally, the planned future development is given in Section 5.

2   Hybrid Mobile-Sensor Network

The proposed hybrid network is shown in Figure 1 and consists of a sensor and a mobile domain. The mobile domain is typically represented by a single cellular network with its entities - base stations and a core infrastructure. The specific choice of mobile technology in the current study falls on the Long Term Evolution (LTE). The sensor domain can be formed by several wireless sensor networks, which in turn can have sub-networks. In Figure 1 we indicate two connectivity modes between the domains. In the first case each sensor node can communicate directly to a base station. Such deployment is challenging in terms of implementation, i.e., each sensor node needs to be equipped with two wireless interfaces. Besides, not every sensor node might be able to 'see' the

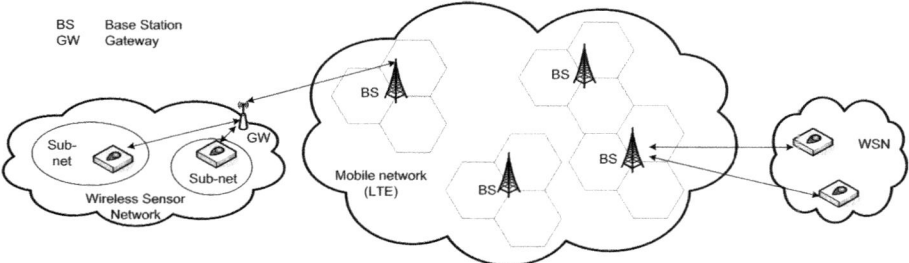

Fig. 1. Hybrid network consisting of sensor and mobile domain

mobile network, e.g., due to shadowing effects. In the second case the sensor nodes communicate to a dedicated gateway which provides connection to the mobile network. In this approach only a single entity, i.e., the gateway, needs to have two interfaces. The gateway can be positioned either at the sensor network side or at the mobile network side.

2.1   Deployment Potential

We consider a hybrid mobile-sensor network interesting for practical deployment due to the individual merits of each participant. A mobile network is characterised by ubiquitous coverage and a reasonably high throughput. Furthermore, base stations generally do not experience energy constraints or buffer shortage contrary to sensor nodes. In terms of scalability, however, sensor networks are cheaper and easier to deploy. They are also versatile in terms of application purposes. Hence, the benefits that a hybrid mobile-sensor network may offer and the challenges that it may pose depend on the taken perspective.

From a sensor network perspective there are multiple expected gains. For example, reporting via a mobile network can positively affect the lifetime of the sensor network. A phenomenon typically observed in sensor networks is the waste of energy by sensor nodes while waiting for the sink to wake up. If a dedicated gateway, constantly powered electrically and always awake, is used as a sink a sensor node can send to it directly, thus decreasing energy consumption. Further, the much larger buffer capacity of the mobile network can offer an excellent temporal storage solution to the sensor network. However, the reverse situation - the mobile network wants to send (large) data to the sensor network this may lead to problems such as buffer overflow and increased collisions in the wireless 'sensor' channel.

From a mobile network perspective there are arguably fewer benefits from the cooperation with a sensor network, partly explained by the technical superiority of the mobile network. The potential opportunities however are worth exploring. For example, a mobile network can use the sensor network to perform measurements, which can then be offered to interested third parties. Sensor nodes can be deployed easily and quickly, which can be used by mobile operators for the monitoring and maintenance of mobile network sites. There are few challenges as well, e.g., support of traffic differentiation and priority service. These required changes, however, can benefit the service of mobile traffic as well, e.g., to guarantee the QoS requirements of voice calls or to offer preferential service to emergency traffic. Note that LTE is developed for packet-switched traffic only which drives researchers to investigate possibilities for QoS support, see [1,10]

2.2   Application Scenarios

We identify the following big groups of possible applications for a hybrid mobile-sensor network - monitoring, streaming, event detection and remote activation.

Monitoring. Wireless sensor networks (WSNs) can be used, e.g., for environmental monitoring, health monitoring, construction monitoring or weather forecasting. Such applications are associated with periodic transmission of the collected sensor data. Still, the frequency of reporting may differ from every few minutes to every few hours or even days and depends on the specific application. For example, a health monitoring system needs to report every few minutes in order to ensure the safety of the patient. A sensor network that collects data for weather forecasting, however, may aggregate the data and report only once a day. Independently of the scenario a mobile network can be used to temporally store and deliver the collected sensor data to a central collection point.

Streaming. Monitoring for security or safety purposes generates continuous stream of data over long periods of time and we give it in an individual category. Streaming applications, in contrast to monitoring, is characterised by long lifetimes and larger data traffic. Some examples of streaming are video surveillance of restricted areas or areas with high crime risk, video surveillance for traffic control purposes or remote audio surveillance. Such applications generate multimedia type of traffic, e.g., voice, audio, video, with strict (and divers) QoS requirements and can therefore benefit significantly from the high bandwidth capacity of a mobile network.

Event detection. WSNs can be also used to detect the occurrence of an event. In this more passive mode a senor node is typically asleep and awakes only in the occasion of a specific event. Upon event detection multiple sensors transmit data at the same time. The total amount of generated sensor data depends on the number of deployed sensor nodes and the type of measurements and can be quite bursty as well. If only an alarm needs to be activated, e.g., in case of fire detection, the data is much less compared to data gathered during a parameter reading, e.g., in case of landscape profiling by echo sounding.

Remote activation. It is possible that the mobile network wants to communicate to all nodes in a particular sensor network, e.g., to remotely (re-)configure the sensor network, to disseminate software updates or to request measurements by the sensors. The latter can be used to acquire information on channel fading at a specific frequency by measuring the received signal of each sensor and comparing it to the transmitter metrics that the sensor reported.

In order to illustrate the potential of the proposed hybrid network we describe few feasible applications but a hybrid network can be deployed with success in many more situations. Our first example is a health monitoring application. A health monitoring unit (consistent of sensors) allows high-risk patients to have normal life while still providing an appropriate medical supervision. A health monitoring system should be at all times connected to a central server that can alert medical personnel if necessary. If the patient is on the move a mobile network can provide uninterrupted connectivity.

In the second example we consider a disaster situation involving a building collapse. A wireless sensor network is easily deployed and can provide local emergency services with valuable information. By connecting to a mobile network the collected sensor data can be further distributed towards experts at remote locations. If the local mobile network infrastructure was fully or partially destroyed other intermediate solutions, e.g. a mesh network, can be deployed to bridge communication.

Finally, an interesting example is a habitat monitoring application. Despite being used for monitoring purposes it is in fact an event detection - the sensors are activated by an approaching animal. If a mobile network is in range the data can be directly relayed to a central database enabling the more buffer-challenging video monitoring. Note however that energy consumption is still an issue. If mobile coverage is scares, e.g., in large isolated areas, observation is kept to the basic data collection.

3 Research Topic

Before a hybrid mobile-sensor network can be deployed there are several challenges that need to be addressed; some were already mentioned, e.g., buffer management, load balancing and QoS provisioning, others include network discovery, radio resource control, signalling and sensor data protection. In this study we are only interested in the challenges related to radio resource management (and scheduling in particular) in the mobile network. Note that resource management in the mobile and in the sensor network domains are two separate issues with network-specific challenges. Mobile network scheduling is affected by mainly two aspects of the sensor network namely the traffic size and the QoS requirements both determined by the application type and the deployment scenario. QoS requirements, e.g. delays, throughput, packet loss, of a sensor network traffic can vary significantly; for example health monitoring data is critical and should be served as first at all times while weather forecasting data only needs guaranteed delivery.

In the current study we focus on a streaming scenario and an event detection scenario with large data traffic. The two scenarios differ significantly in terms of traffic characteristics, including QoS parameters, which is why we chose them. In the streaming scenario the sensor nodes are distributed over a large geographical area, e.g., an industrial area. Each sensor connects via a gateway to the closest base station from where the sensor data is propagated to a central database.

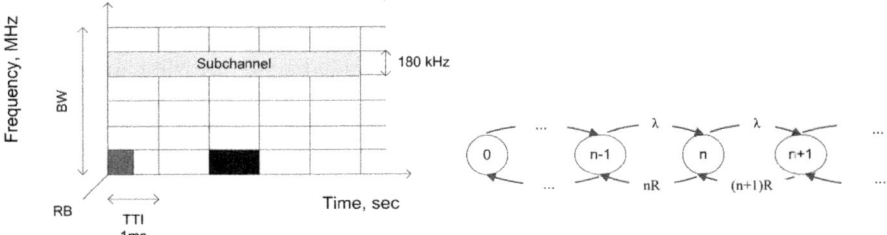

Fig. 2. LTE radio resource structure

Fig. 3. Markov model of a single cell with mobile traffic only

It is assumed that the sensors are connected to the power grid and hence energy consumption is not an issue. Streaming applications generate a continuous traffic stream which is sensitive to transmission delays. Therefore, they would require to be served with highest priority.

For the event detection scenario we are interested in a situation where a large number of sensors is spread over an area of limited size and each sensor collects large amounts of data. An example of such an application is landscape profiling by using echo sounding. Both the number of sensors and the amount of sensor data cause a significant increase in traffic for the mobile network. The data is bursty and of best effort type, i.e., it can tolerate delays as long as the data is correctly transmitted.

In both scenarios the mobile network perspective is considered, i.e., what changes are observed in the mobile network performance after the introduction of sensor network traffic. The choice of LTE as the mobile technology has the advantage of possibility to schedule in the time and frequency domain. This offers flexibility to meeting the various traffic demands and QoS requirements, associated with the streaming and event detection scenario. The possibility to schedule in the frequency domain may also provide easy means to separate mobile from sensor traffic handling.

In order to investigate how the performance of mobile users is affected by the introduction of sensor network traffic we need an approach that can capture both the specifics of the data traffic in the mobile networks as well as the specifics of the sensor network traffic. Further, we argue that it is important to take into account the change in number of ongoing flows in both networks; this is especially important for the streaming application. The proposed methodology is further described in the following Section 4.

4   Analysis Methodology

In order to enable performance evaluation in the mobile network for both scenarios, streaming and event detection, we propose a modified version of the analysis approach proposed in [3]. The basic form of the approach, which combines system analysis at packet level with modelling of user behaviour at flow level, is presented in Section 4.1. Later, in Section 4.2 we discuss the modifications that each scenario needs, due to specific QoS requirements.

4.1   Basic Approach

The system analysis at **packet level** captures the impact of several factors among which scheduler specifics (e.g., resource size), environmental characteristics (e.g., interference) and user specifics (e.g., power headroom). The performance of a mobile user is characterised by an *instantaneous rate* r and a *state-dependent throughput* R. In order to explain the differences we need to refer to the radio resource management in LTE. Recall that in LTE scheduling in two dimensions is possible, see Figure 2. In the time dimension access in organised in Transmission Time Intervals (TTIs) of 1 ms while in the frequency dimension the system bandwidth is divided over 'sub-channels' of 180 kHz. The intersection of a TTI and a sub-channel defines the smallest radio resource unit that can be allocated to a user.

The instantaneous rate r is the data rate realised by a user within a TTI and it is calculated over all resource units allocated to that user. It is determined by the signal-to-interference-plus-noise ratio(SINR), the possible modulation and coding schemes (MCS) and the receiver characteristics related to that MCS. If the user was not scheduled for service in the TTI intuitively $r = 0$. The state-dependent throughput R captures the impact of the number of ongoing flows n on the service, i.e., the instantaneous rate, of a particular user. Referring back to Figure 2, depending on the allocation strategy and the number of users n it may happen that several TTIs are needed to serve all requests. This effect is captured by R, which is derived as $R = r/c$ where c is the *transmission cycle*, i.e., the total number of TTIs necessary to serve all users at least once.

The dynamic behaviour of the system at **flow level** is modelled by continuous time Markov chains. They are very appropriate for the task since a state in the chain can be mapped to a state in the system, i.e., the number of ongoing flows n. In Figure 3 we give an example for a single cell. Each state in the Markov chain corresponds to a number of currently active users in the cell. The jumps in the Markov chain represent the initiation and completion of flow transfers. The corresponding transition rates (for a particular state) are determined from the (a-priori) given user arrival rate λ and the state-dependent throughput R (in that state).

Eventually, the evolution of the number of mobile data users corresponds to a sequence of state transitions in the Markov chain. From the steady-state distribution of this Markov chain we can derive performance parameters, e.g., (long-term) flow throughput or fairness index. The steady-state distribution of the Markov chain can be found, in special cases, by analytical approaches leading to explicit closed-form expressions. When the resulting form of the Markov chain is very complex and not trivial to solve, simulation of the state transitions can provide the means to derive the steady state distribution.

Working with Markov models has many advantages among which time-efficient performance evaluation, scalability in the system size and easy parameter modification due to the division of packet and flow level analysis.

4.2 Modified Approach

The introduction of sensor traffic to the analysis requires different changes to be made depending on whether the streaming or the event detection scenario is addressed. As previously discussed streaming applications need to be served with highest priority meaning that part of the radio resource is exclusively reserved for the sensor traffic. The size of the reservation depends on the number of active sensor nodes and on the multimedia class, i.e., audio, video, and determines how much resource is left for the service of the mobile users. In the event detection scenario however no priority service is required and sensor traffic is treated equally with mobile data traffic. All sensor data generated at a particular moment can be managed as a single data flow.

For both scenarios we do not foresee explicit changes of the expressions for the instantaneous rate and the state-dependent throughput since the effects are hidden in the calculation of the available bandwidth. The form of the Markov chain, however, needs to be adapted in order to capture the behaviour of sensor traffic. For the streaming scenario we need to add a new dimension, which represents the changes in the number of multimedia flows, while in the event detection scenario a new dimension with only two states - no sensor data and data - suffice.

The proposed methodology can be also easily applied to study a mobile network with monitoring traffic. In our opinion a monitoring scenario is less interesting since: (i) monitoring generates traffic at a predefined time interval, generally bigger than a minute; and (ii) monitoring data is of relatively small size. As result we did not expect a big impact on the performance of an LTE mobile network, given its large bandwidth capacity and scheduling periods of 1ms.

5   Future Steps

As a next step in the research we will systematically work out the analysis approach for the streaming and event detection scenarios. This will provide us with the corresponding Markov chains and will allow us to observe performance in the mobile network for various evaluation scenarios, e.g., video only traffic, combined video and audio traffic, mobile data traffic only (reference case).

Acknowledgements. We want to thank Nirvana Meratnia (University of Twente) for her feedback on the application scenarios and Markus Anwander and Philipp Hurni (University of Bern) for the insights on deployment issues in wireless sensor networks.

References

1. Anas, M., Rosa, C., Calabrese, F.D., Pedersen, K.I., Mogensen, P.E.: Combined admission control and scheduling for QoS differentiation in LTE uplink. In: VTC-Fall 2008, pp. 1–5. IEEE Computer Society Press (2008)
2. Chen, D., Varshney, P.: QoS support in wireless sensor networks: a survey. In: International Conference on Wireless Networks (2004)
3. Dimitrova, D.C.: Analysing uplink scheduling in mobile networks. A flow-level perspective. PhD thesis, Wöhrmann Print Service (2010)
4. Dimitrova, D.C., van den Berg, J.L., Heijenk, G.: Performance of relay-enabled uplink in cellular networks - a flow level analysis. In: International Conference on Ultra Modern Telecommunications, pp. 1–8. IEEE Computer Society Press (2009)
5. Ekstrom, H.: QoS control in the 3GPP evolved packet system. IEEE Communications Magazine 47(2), 76–83 (2009)
6. Harrold, T.J., Nix, A.R.: Capacity enhancement using intelligent relaying for future personal communication systems. In: VTC-Fall 2000, pp. 2115–2120. IEEE Computer Society Press (2000)
7. Jacobsson, M., Hoebeke, J., Heemstra-de Groot, S., Lo, A., Moerman, I., Niemegeers, I., Munoz, L., Alutoin, M., Louati, W., Zeglache, D.: A network architecture for personal networks. In: 14th IST Mobile and Wireless Communications Summit (2005)
8. Li, Y., Chen, C.S., Song, Y.-Q., Wang, Z.: Real-time QoS support in wireless sensor networks: a survey. In: 7th IFAC International Conference on Fieldbuses & Networks in Industrial & Embedded Systems (2007)
9. de Poorter, E., Latr, B., Moerman, I., Demeester, P.: Symbiotic networks: Towards a new level of cooperation between wireless networks. Wireless Personal Communicaitons, 479–495 (2008)
10. Siomina, I., Wanstedt, S.: The impact of qos support on the end user satisfaction in lte networks with mixed traffic. In: PIMRC 2008, pp. 1–5. IEEE Computer Science Press (2008)
11. Wei, H.Y., Ganguly, S., Izmailov, R.: Ad hoc relay network planning for improving cellular data coverage. In: PIMRC 2004, pp. 769–773. IEEE Computer Science Press (2004)
12. Xia, F.: Qos challenges and opportunities in wireless sensor/actuator networks. Sensors 8(2), 1099–1110 (2008)

Author Index

GPSR Compliance

The European Union's (EU) General Product Safety Regulation (GPSR) is a set of rules that requires consumer products to be safe and our obligations to ensure this.

If you have any concerns about our products, you can contact us on ProductSafety@springernature.com

In case Publisher is established outside the EU, the EU authorized representative is:

Springer Nature Customer Service Center GmbH
Europaplatz 3
69115 Heidelberg, Germany

Batch number: 09478804

Printed by Printforce, the Netherlands